This newly updated book shows how the various approaches to ocean governance can be integrated to improve social, economic, and environmental outcomes. It should be essential reading for all those of us who strive to better manage and protect our home-planet ocean.

Dan Laffoley, Marine Vice Chair, International Union for Conservation of Nature (ICUN) World Commission on Protected Areas

Citing examples from around the world, this book provides a timely and useful perspective on the challenges facing effective marine policy, including the complexities of climate change, along with new and emerging issues such as ocean energy and deep-sea mining.

Jon C. Day, James Cook University, Australia and formerly Director with the Great Barrier Reef Marine Park Authority

Marine Policy

This book provides readers with a foundation in policy development and analysis, describing how policy, including legal mechanisms, are applied to the marine environment. It presents a systematic treatment of all aspects of marine policy, including climate change, energy, environmental protection, fisheries, mining and transportation.

The health of marine environments worldwide is steadily declining, and these trends have been widely reported. *Marine Policy* summarizes the importance of the ocean governance nexus, discussing current and anticipated challenges facing marine ecosystems, human activities, and efforts to address these threats. This new, fully revised edition has been updated throughout, including content to reflect the recent advances in ocean management and international law. Chapters on shipping, energy/mining and integrated approaches to ocean management have been significantly reworked, plus completely new chapters on the United Nations Convention on the Law of the Sea, and the impacts of climate change have been added. Pedagogical features for students are included throughout.

Aligned with current course offerings, this book is an ideal introduction for undergraduates and graduate students taking marine affairs, science and policy courses.

Mark Zacharias is the Deputy Minister of the Ministry of Environment and Climate Change Strategy with the province of British Columbia, Canada. He is also Adjunct Professor at the University of British Columbia Fisheries Centre and Associate Adjunct Professor in the Department of Geography at the University of Victoria. He is the co-author of the book *Marine Conservation Ecology* (Earthscan, 2011).

Jeff Ardron is an Adviser on Marine Governance for the Commonwealth Secretariat, based in London, UK. He leads the Commonwealth Blue Charter, which was adopted by all 53 Commonwealth countries in 2018. With more than 25 years' experience in marine planning and conservation, he has worked in governmental and non-governmental sectors, as well as in research and academia. He has more than 60 peer-reviewed publications and serves on various boards and committees of marine-focused organizations.

Earthscan Oceans

Marine Biodiversity, Climatic Variability and Global Change
Grégory Beaugrand

Transboundary Marine Spatial Planning and International Law
Edited by S.M. Daud Hassan, Tuomas Kuokkanen, Niko Soininen

Marine Transboundary Conservation and Protected Areas
Edited by Peter Mackelworth

Marine and Coastal Resource Management
Principles and Practice
Edited by David R. Green and Jefferey Payne

Citizen Science for Coastal and Marine Conservation
Edited by John A. Cigliano and Heidi L. Ballard

Ocean Energy
Governance Challenges for Wave and Tidal Stream Technologies
Edited by Glen Wright, Sandy Kerr and Kate Johnson

Offshore Energy and Marine Spatial Planning
Edited by Katherine Yates and Corey Bradshaw

Coral Reefs: Tourism, Conservation and Management
Edited by Bruce Prideaux and Anja Pabel

Marine Extremes
Ocean Safety, Marine Health and the Blue Economy
Edited by Erika Techera and Gundula Winter

Marine Policy
An Introduction to Governance and International Law of the Oceans
2nd Edition
Mark Zacharias and Jeff Ardron

For further details please visit the series page on the Routledge website: www.routledge.com/books/series/ECOCE

Marine Policy

An Introduction to Governance and International Law of the Oceans

Second edition

Mark Zacharias and Jeff Ardron

Second edition published 2020
by Routledge
2 Park Square, Milton Park, Abingdon, Oxon, OX14 4RN

and by Routledge
52 Vanderbilt Avenue, New York, NY 10017

Routledge is an imprint of the Taylor & Francis Group, an informa business

© 2020 Mark Zacharias and Jeff Ardron

The right of Mark Zacharias and Jeff Ardron to be identified as authors of this work has been asserted by them in accordance with sections 77 and 78 of the Copyright, Designs and Patents Act 1988.

All rights reserved. No part of this book may be reprinted or reproduced or utilised in any form or by any electronic, mechanical, or other means, now known or hereafter invented, including photocopying and recording, or in any information storage or retrieval system, without permission in writing from the publishers.

Trademark notice: Product or corporate names may be trademarks or registered trademarks, and are used only for identification and explanation without intent to infringe.

First edition published by Routledge 2014

British Library Cataloguing-in-Publication Data
A catalogue record for this book is available from the British Library

Library of Congress Cataloging-in-Publication Data
Names: Zacharias, Mark, 1967– | Ardron, Jeff, author.
Title: Marine policy : an introduction to governance and international law of the oceans / Mark Zacharias and Jeff Ardron.
Description: Second edition. | New York, NY : Routledge, 2019. | Includes bibliographical references and index.
Identifiers: LCCN 2019011651 (print) | LCCN 2019012926 (ebook) | ISBN 9781351216227 (eBook) | ISBN 9780815379263 (hardback : alk. paper) | ISBN 9780815379270 (pbk. : alk. paper)
Subjects: LCSH: Law of the sea.
Classification: LCC KZA1145 (ebook) | LCC KZA1145 .Z33 2019 (print) | DDC 341.4/5—dc23
LC record available at https://lccn.loc.gov/2019011651

ISBN: 978-0-8153-7926-3 (hbk)
ISBN: 978-0-8153-7927-0 (pbk)
ISBN: 978-1-351-21622-7 (ebk)

Typeset in Goudy
by Apex CoVantage, LLC

Contents

List of illustrations	ix
Preface to the second edition	xii
Acknowledgements	xvii
Acronyms and abbreviations	xviii

1	An introduction to the world ocean: human impacts and trends	1
2	A brief introduction to the marine environment: the structure and function of the world ocean	13
3	An introduction to international law and international marine law: how nations cooperate to govern the oceans	37
4	The United Nations Convention on the Law of the Sea and related agreements: moving from 'free seas' to codifying jurisdiction	65
5	An introduction to policy and policy development: instruments for marine law and policy	93
6	Marine environmental protection policy: international efforts to address marine pollution and protect marine biodiversity	146
7	Addressing climate change and its impacts on the world ocean: international efforts to mitigate and adapt to a changing planet	167
8	International fisheries policy: sustaining global fisheries for the long term	177
9	Marine transportation and safety policy: the operation and regulation of international shipping	220

10 International law and policy of the Polar oceans: governing the
    Southern and Arctic Oceans   256

11 International law and policy related to offshore energy and mining:
    renewable and non-renewable resource development on continental
    shelves and in the Area   277

12 Integrated approaches to ocean management: putting policy into
    practice and managing across sectors   288

    Conclusions: a report card on ocean management   310
    Index   312

Illustrations

Figures

2.1	Diagram of the pelagic and benthic realms of the marine environment	17
2.2	Global distribution of diversity of reef building coral genera	19
2.3	Global species richness of bivalve molluscs	20
2.4	Global map of pelagic provinces	25
2.5	Global map of Large Marine Ecosystems (LMEs)	26
2.6	Surface ocean circulation and the major surface ocean currents	27
4.1	Maritime jurisdiction under the United Nations Convention on Law of the Sea	69
4.2	Exclusive Economic Zones of the world	70
4.3	Globally important shipping straits and volume of crude oil and petroleum products shipped in 2010 for comparison purposes	75
4.4	Landlocked developed states and landlocked developing countries	77
4.5	Least developed countries	77
5.1	The policy cycle and associated actors and institutions	99
5.2	Flow chart of counterfactual analysis process	116
6.1	United Nations Environment Programme (UNEP) Regional Seas Programmes	159
8.1	International initiatives that contributed to the ecosystem approach to fisheries management	190
8.2	The Food and Agriculture Organization of the United Nations Major Fishing Areas	197
8.3	The Food and Agriculture Organization of the United Nations Major Fishing Subareas	197
8.4	Regional Fisheries Management Organizations	198
9.1	Compensation levels established by international conventions to fund oil spill clean up	248
10.1	Overview of the Antarctic maritime area	258
10.2	Overview of Arctic circumpolar area (north of the Arctic Circle (66° 33′N))	266
10.3	Northern Sea Route (A) and Northwest Passage (B)	272
12.1	Various definitions and extents of the coastal realm and coastal zone	292
12.2	Global map of current marine protected areas	297

Tables

2.1	Summary of some major marine communities	31
3.1	Distribution of legal systems among United Nations member states	39
3.2	List of international marine conventions and non-binding agreements	48
4.1	List of small island developing states (SIDS)	79
5.1	Examples of different interpretations/definitions of the stages of policy development (the 'policy cycle')	98
5.2	Application of selection criteria for different categories of policy instruments	104
5.3	Main types of performance criteria for policies and marine examples	105
5.4	Key considerations for the selection of performance criteria	105
5.5	Criteria and considerations in the selection and design of policy instruments	109
5.6	Common models/methods/approaches/theories used to describe how public policy decisions are made	110
5.7	Types of policy instruments	121
5.8	Taxonomy of policy instruments	121
5.9	Sample applications of policy instruments for marine management	122
5.10	Summary of advantages and disadvantages of market-based instruments' application in the marine environment	129
6.1	State parties to the Convention for the Protection of Natural Resources and Environment of the South Pacific Region (Noumea Convention), including overseas parties	163
8.1	Traditional fisheries management compared with EAFM	194
8.2	List of current marine regional fisheries management organizations (RFMOs) grouped by purpose and listed by year of establishment	200
9.1	World fleet size in 2016 by principal types of vessel	223
9.2	Global ship classifications and sizes	224
9.3	Largest marine oil spills from ships, pipelines and oil production platforms	246
10.1	Articles of the Antarctic Treaty	260
10.2	Summary of national Arctic interests	268
10.3	Declarations of the Arctic Council and key provisions	270
12.1	Selected list of Marine Spatial Planning initiatives	305

Boxes

0.1	Governance challenges resulting from the unique characteristics of the marine environment	xiv
1.1	Importance of the oceans	2
1.2	Marine goods and services that support humanity	2
1.3	Human perceptions on ocean health	3
2.1	Divisions in the marine environment	15
3.1	Key terms as defined by the Vienna Convention on the Law of Treaties	43
3.2	Differences between marine law and other legal systems	63
4.1	International Court of Justice marine-related cases	82
4.2	International Tribunal for the Law of the Sea rulings	86
5.1	The Weitzman Theory	106
5.2	Constructing a green GDP	117

5.3	Regulating fisheries discards	123
5.4	Liability and marine oil spills in the United States	125
5.5	Considerations in establishing permit trading systems	132
5.6	World Wildlife Fund's (WWF) International Smart Gear Competition	137
5.7	Biodiversity audits	138
5.8	Product certification example: Marine Stewardship Council	139
8.1	Types of fisheries based on gear types	180
8.2	Catch limits	181
8.3	The case of deep-sea fisheries	199
8.4	Whales species covered by the International Convention for the Regulation of Whaling	203
8.5	Types of marine property rights	212
8.6	Successful examples of catch share programs	214
9.1	Methods of categorizing ship size	222
9.2	An overview of commercial shipping	227
9.3	Der Blaue Engel ('the blue angel') eco-label and shipping	251
10.1	List of parties to the Antarctic Treaty and their status	260
10.2	Organizations formed to advise the Antarctic Treaty System	262
10.3	China and the Polar oceans	273
12.1	An example of an integrated policy approach: the European Commission Blue Growth Strategy	289
12.2	An example of a troubled integrated policy approach: Australia's Ocean Policy	290
12.3	Large Scale Marine Protected Areas	298
12.4	Four stages to be considered in the development of representative networks of MPAs	303

Preface to the second edition

In his classic text *Unpopular Essays*, Bertrand Russell contended that humanity confronts two different problems: The first is mastering nature to provide for human needs, which is firmly in the domain of science and technology (Russell, 1950). The second is wisely utilizing the fruits of science and technology to improve the human condition. Prudence, however, has not always prevailed. Russell provided examples of technological advances throughout history that led to disastrous outcomes. In particular, he evoked the domestication of the horse (conquest), mechanization (slavery) and modern physics (nuclear weapons) as specific advances that were accompanied by unfortunate consequences. Russell stated that while science and technological achievements and skills are important, 'something more than skill is required, something which may perhaps be called "wisdom". This is something that must be learnt, if it can be learnt, by means of other studies than those required for scientific technique'.

Russell's observations are especially apropos to our relationship with the ocean. Evidence confirms that humans consumed seafood 164,000 years ago, and early pre-Western cultures developed sufficient technical expertise to overharvest species they could access. The advent of industrialization created the opportunity to fish in remote oceans, extract oil and gas from continental shelves, transport goods over great distances and use the ocean as a repository for waste. Yet, even as late as the early 20th Century, many still thought the ocean immutable and immune to human activities.

Paralleling Russell's examples, scientific and technological advancements have outpaced humanity's ability to understand the consequences of our activities on the marine environment. Nor have we been able to envision how to regulate these activities as a whole. Even after evidence emerged on the effects of overfishing and pollution, piecemeal responses took decades. This unfortunate delay is exacerbated by the fact that, even to this day, most of the ocean is owned by no one and thus must be managed as a commons resource where the negligence of a few can come at the expense of many: a tragedy of the commons in general, but specifically a tragedy of international cooperation. In every international forum that we have visited or participated in, we have witnessed a great many states that wish to improve the lot of the ocean, only to be thwarted by an intransigent minority who have held other interests.

This book explores what Russell termed the 'wisdom' necessary to ensure that humanity's scientific and technical achievements – and the consequences of these achievements – lead to what he termed 'contentment'. In the context of the nine years of difficult negotiations that led to the signing of the Law of the Sea Convention (LOSC) in 1982, the word 'contentment' might have been perhaps described then as 'peaceful equity' – an ocean free of conflict, with resources available for all of humanity. More than 30 years later, the peace has held – a testament to the international diplomacy of that time. However, the question of equity is not so easily answered. For some states, the rich fisheries in the 1970s are but a distant memory. Some continue to subsidize their commercial fleet to fish deeper and further away, whilst other states place new hope in an old idea, wagering that deep-seabed mining's time has finally arrived. If Ambassador Arvid Pardo, who urged the United Nations General Assembly in 1967 to develop a fair governance regime for the ocean, were alive today, would he say, 'job well done'?

To be fair, the management of the global ocean is incredibly complex. Even those employed full-time in the field of ocean governance find it difficult to be well-informed in all its aspects. At present, there are over 150 multilateral international agreements that govern the use of the marine environment. In turn, many of these agreements establish dozens of multinational agencies, commissions, organizations and secretariats to oversee their implementation. These administrative bodies then create policies, guidelines and reporting requirements. Increasing the complexity is that many of these agreements nest hierarchically within other agreements or require other agreements to be fully implemented. The blue Byzantine maze wavers and shifts over time, such that its international conventions are frequently amended, or buttressed by key decisions that are not always easy to find. As if this was not enough, most geographic regions of the ocean are governed by multiple, separate regional agreements, which may or may not have the same states as parties.

Unsurprisingly, no single person can have a comprehensive understanding of what might make this international ocean governance more effective. Staring down the double barrels of climate change and ocean acidification, we will need to better listen to one another, trust one another and learn together.

Purpose of this book

This intent of this book is to provide the reader with an understanding of how nation-states currently work together to address socio-economic, environmental and cultural issues with respect to the management of the world ocean. Marine governance is an amalgam of historical practice and custom, economic and trade considerations, domestic law and policy, and regional and international agreements. How oceans are governed is considered here through the lens of 'marine policy'. Related but distinct from maritime law or marine science, marine policy is the stated intent and subsequent actions, usually by governments, to influence ocean-related development, activities and outcomes. In this way, marine policy often includes the use of science and laws to achieve its stated goals.

While this text introduces the legal context and international law that underpins ocean governance, it is not a law textbook nor is it an authoritative guide to any particular sector or convention. Dozens of books have been written on each of the international conventions discussed here and our intent is not to duplicate these scholarly works. Instead, this textbook explores the how various approaches to ocean governance have emerged in various maritime sectors (e.g. fishing or transportation) and the opportunities to learn from, and integrate, these approaches to improve social, economic and environmental outcomes.

This text primarily focuses on the international aspects of marine management. While each nation-state (even landlocked states) has legal standing in international maritime law and may have its own set of marine laws and policies, this book will cast only occasional glances at individual nations and their domestic marine agendas by way of example. There are many excellent 'coastal management' texts, to which the curious reader is directed. However, this book attempts to provoke the reader, who may be a specialist in one particular aspect of marine research, law or policy, towards thinking about the opportunities and benefits of taking a more holistic view.

Why this book now and why a second edition

The current environmental condition of marine environments worldwide is steadily declining, and these trends have been widely chronicled. Traditional, sector-based approaches to ocean governance (such as managing fisheries or regulating transportation) have stumbled when attempting to resolve the bigger social and environmental predicaments facing humanity (see Box 0.1). Climate change, sovereignty claims and regional economic disparity are not addressed by drafting an improved single-sector regulation. Complex, interrelated problems require sophisticated interconnected decision-making, aimed towards common goals.

Dramatic changes in the world of ocean governance have occurred since the publication of the first edition in 2014. For the first time ever, the ocean was recognized as an international priority, and provided with its own development goal and targets, in the form of United Nations Sustainable Development Goal 14. In 2017, the United Nations followed up with its first Ocean Conference since the days of the Law of the Sea negotiations. Meanwhile, climate change has been found to be more rapid and damaging than previously thought, warranting a completely new (nearly) global agreement under the UNFCC. In addition, significant progress has occurred on regional bi- and multilateral agreements to address fisheries management and marine health. The first edition provided only a cursory treatment of marine pollutants, which is rectified in this edition. Finally, the rapid evolution of marine planning approaches continues and is expanded upon in this new edition.

> **Box 0.1 Governance challenges resulting from the unique characteristics of the marine environment**
>
> - Sector-based single-focus solutions are often poor fits for the biophysical complexity of marine environments and interconnections with the economy and human communities.
> - Social, economic and ecological interactions at multiple geographical scales are not well understood, but one size (scale) of governance is unlikely to be the solution. Therefore, a cooperative multi-scalar approach is required.
> - Management based on 'command and control' regulatory approaches on land is less effective in marine systems because of their unpredictable dynamics and complexity.
> - Short-term (human scale) attitudes and timeframes can hobble planning for long-term sustainable solutions compatible with marine ecological timeframes.
> - A 'frontier mentality' of a limitless ocean that assumes few ecological, thermodynamic and economic constraints to resource development is centuries-old thinking that does not acknowledge the explosion of human technological power over the past 50 years.
> - Marine systems transcend national and regional boundaries, leaving management systems developed for fixed-boundary property rights (ownership) inadequate.
>
> Source: Significantly modified from Glavovic (2008)

Book outline

Until the 20th Century, the regulation of maritime activities was widely perceived as unnecessary and unachievable given the widespread belief that the vast ocean was impervious to human efforts. **Chapter 1** summarizes the importance of the world ocean to humanity and the modern human footprint on the world ocean. The chapter then discusses threats to biodiversity, which can be broadly categorized as a result of overharvesting, pollution, habitat loss, introduced species, ocean acidification and global climate change.

The discrete inner workings of the oceans are largely invisible to us and often counter-intuitive to our terrestrial experience. As such, a basic knowledge of the biological and physical characteristics of marine systems is required to understand how the application of marine law and policies may impact marine environments. An 'Oceans 101' is provided in **Chapter 2** to familiarize the reader with the structure, function and processes that operate in marine environments. It begins with an exploration of the differences between terrestrial and marine environments. The intent of this section is to evince how our 'land referenced' perspective instils a bias that must be overcome if we are to fully fathom how marine environments could be governed.

Chapter 3 introduces the history and operation of international law. Within this chapter is a discussion of the types of ocean ownership regimes, an introduction to the world's different legal systems and a brief explanation of the workings of international law. The chapter then introduces international marine law and the differences between marine law and other legal systems. It concludes with a summary of international marine conventions and non-binding agreements.

Chapter 4 introduces the United Nations Convention on the Law of the Sea (LOSC). The LOSC is one of the most comprehensive international agreements created, with an extensive literature all its own. This chapter provides an overview of the history, content and operation of the LOSC while subsequent chapters will delve further into the specific functions of the Convention.

An introduction to the field of policy encompasses **Chapter 5**. It explores the purpose of public policies, types of policies and the relationship between law and public policy. The unique characteristics of marine law and policy are then discussed with particular attention to areas beyond national jurisdiction (the high seas and deep seabed). The fifth chapter then reviews how policies are formed, the characteristics of policy development, considerations in selecting policies and methods for selecting and analysing policies. Central to this textbook, this chapter is its longest.

The primary international mechanisms that govern the protection and preservation of the oceans – along with the important regional conventions and examples of bi-lateral and multilateral agreements – are presented in **Chapter 6**. The chapter outlines the primary international and regional agreements governing the protection of marine environments.

Addressing climate change and its impacts on the world ocean is the subject of **Chapter 7**. The chapter first reviews the impacts of climate change on the oceans and then discusses the international efforts to mitigate and adapt to a changing planet.

The operation and regulation of marine fisheries is discussed in **Chapter 8**. The chapter provides an overview of marine fisheries management, focussing on divergent approaches (single-species, multi-species, ecosystem-based) that have been applied – sometimes successfully and sometimes not. A history of international fisheries law and policy is then provided along with a summary of the organizations that oversee fisheries management. The primary international agreements that govern fishing are then discussed.

The **ninth chapter** explores the operation and regulation of international marine transportation. First, it begins with an overview of contemporary shipping and introduces marine transportation law and policy. Organizations that regulate marine transportation are then discussed along with the key international conventions and agreements dealing with the ship, shipping company and seafarer. The chapter then addresses international laws and policies related to maritime security (including piracy) and the environmental effects of shipping.

The Polar Regions (Arctic and Antarctic) warrant a separate treatment in **Chapter 10** since they have unique governance arrangements due to their remoteness, inhospitable climate and unique biological systems. While the Polar Regions are similar in climate, they are distinctive in almost all other respects, be they biological structure, governance or human use. Chapter 10 introduces the Antarctic and Southern Ocean and the Antarctic Treaty System that governs these regions. Governance of the Arctic Regions is then discussed along with the key institutional arrangements that assist with Arctic management.

Chapter 11 introduces the international law and policy instruments that apply to ocean energy and mining activities. It first differentiates energy and mining opportunities between the continental shelves and the deep seabed; it then discusses the role of the LOSC in energy and mineral development. Organizations that oversee energy and mining activities are then reviewed. A brief summary of marine energy and mineral resources and their management in the continental shelf and deep-sea areas ensues.

Integrated approaches to ocean management are elicited in **Chapter 12.** These approaches are loosely defined as tools to assist marine decision-making and the analysis of policies. Coastal management, ecosystem approaches to management, marine protected areas, systematic conservation planning, marine spatial planning and cumulative effects are analysed.

Terminology used in this book

Lawyers, policy analysts, environmental planners, engineers, diplomats and bureaucrats have their own lexicons related to ocean governance. This text borrows from all of these disciplines and uses the following terms based on recent practice:

> **States** are equivalent to countries, nations, nation-states and economies, including states that are still in the process of achieving full recognition by the international community.
> **Marine** and **ocean** are used interchangeably. However, 'maritime' has specific connotations with human use, particularly the marine transportation sector and is used only in this context.
> **Law** as an umbrella term may either connote domestic or international law and includes treaties, legislation, rules and regulations.
> **Marine/ocean policy** is the stated intent and subsequent actions, usually by governments, to influence ocean-related development, activities and outcomes.
> **Governance** is the design, creation and implementation of laws or policies.
> **Convention** is synonymous with an international agreement or treaty. When a particular convention is discussed, it is capitalized.
> **Law of the Sea Convention (LOSC)** is used to abbreviate the United Nations Convention on the Law of the Sea as per common practice in the international law literature. The term 'UNCLOS', often used in marine policy literature, is synonymous here with 'LOSC'.

Further reading

Glavovic, B. (2008) 'Ocean and coastal governance for sustainability: Imperatives for integrating ecology and economics', in M. Patterson and B. Glavovic (eds.), *Ecological Economics of the Oceans and Coasts*, Edward Elgar, Cheltenham, UK.
Russell, B. (1950) *Unpopular Essays*, Routledge, New York.

Acknowledgements

I cannot hope to thank everyone that shaped the development of this book. However, a number of key individuals were critical in its execution. Laura Feyrer, Kevin Knox and Meherzad Romer were instrumental in assisting with research, editing and the development of figures. I am similarly grateful to Rod Dobell, Lorne Kriwoken, Martha McConnell, Ted McDorman, Andrew Stainer and Rashid Sumaila for donating their time to review the various chapters. Above all, I would like to thank John Roff for his continued support over the years and for his permission to incorporate modified sections of our previous text, *Marine Conservation Ecology*, into this book.

Many colleagues and friends contributed indirectly to this endeavour and include: Jackie Alder, Hussein Alidina, Jeff Ardron, John Baxter, Rosaline Canessa, Villy Christensen, Chris Cogan, Phil Dearden, Jon Day, Dave Duffus, Zach Ferdana, Leah Gerber, Ed Gregr, Ben Halpern, Ellen Hines, Don Howes, David Hyrenbach, Sabine Jessen, David Kushner, Nancy Liesch, Josh Laughren, Olaf Niemann, Carol Ogborne, Charlie Short, Rashid Sumaila and Mark Taylor. I apologize to anyone I have forgotten.

Lastly, I would like to acknowledge the past contributions of the authors, editors and staff of *The International Journal of Marine and Coastal Law*, *Marine Policy*, *Ocean & Coastal Management*, and *Ocean Development & International Law*, without which this book would not have been possible.

Acronyms and abbreviations

AAC	Arctic Athabaskan Council
ABNJ	areas beyond national jurisdiction
ACAP	Agreement on the Conservation of Albatrosses and Petrels
ACAP	Arctic Contaminants Action Program
ACIA	Arctic Climate Impact Assessment
AEPS	Arctic Environmental Protection Strategy
AHP	analytic hierarchy process
AIA	Aleut International Association
AMAP	Arctic Monitoring and Assessment Programme
AOP	Australia's Ocean Policy
APFIC	Asia-Pacific Fisheries Commission
ARREST	International Convention on Arrest of Ships
ASBAO	Regional Convention on Fisheries Cooperation among African States Bordering the Atlantic Ocean
ASOC	Antarctic and Southern Ocean Coalition
AT	Antarctic Treaty
ATCM	Antarctic Treaty Consultative Meeting
ATS	Antarctic Treaty System
BAS	British Antarctic Survey
BP	British Petroleum
BUNKERS	International Convention on Civil Liability for Bunker Oil Pollution Damage
BWM	International Convention for the Control and Management of Ships' Ballast Water and Sediments
CAFF	Conservation of Arctic Flora and Fauna
CAM	coastal area management
CBD	Convention on Biological Diversity
CCAMLR	Convention on the Conservation of Antarctic Marine Living Resources
CCAS	Convention for the Conservation of Antarctic Seals
CCSBT	Commission for the Conservation of Southern Bluefin Tuna
CDM	clean development mechanism
CDQ	community development quotas
CE	cumulative effects
CECAF	Committee for the Eastern Central Atlantic Fisheries
CEP	Caribbean Environment Program
CEP	Committee for Environmental Protection

CER	corporate environmental responsibility
cif	cost, insurance and freight
CITES	Convention on International Trade in Endangered Species of Wild Fauna and Flora
CLC	International Convention on Civil Liability for Oil Pollution Damage
CLCS	Commission on the Limits of the Continental Shelf
CLEE	Convention on Civil Liability for Oil Pollution Damage Resulting from Exploration for and Exploitation of Seabed Mineral Resources
CLL	International Convention on Load Lines
CMI	Comité Maritime International
CMS	Convention on the Protection of Migratory Species of Wild Animals
COFI	Committee on Fisheries
COLREG	Convention on the International Regulations for Preventing Collisions at Sea
COLTO	Coalition of Legal Toothfish Operators
COMNAP	Council of Managers of National Antarctic Programs
COP	conference of the parties
CPPS	Permanent Commission on the South Pacific
CPs	consultative parties
CPUE	catch per unit effort
CRM	coastal resource management
CSI	Container Security Initiative
CZM	coastal zone management
DSM	deep-seabed mining
DFO	Fisheries and Oceans Canada
DMC	dangerous maritime cargoes
dwt	dead weight tonnes
EA	enterprise allocations
EAF	ecosystem approach to fisheries
EAFM	ecosystem approach to fisheries management
EAGGF	European Agricultural Guidance and Guarantee Fund
EAM	ecosystem approaches to management
EAP	Action Plan for the Protection, Management and Development of the Marine and Coastal Environment of the Eastern African Region
EbA	ecosystem-based adaptation
EBSA	ecologically and biologically significant area
EEC	European Economic Community
EEDI	Energy Efficiency Design Index
EEZ	exclusive economic zone
EFR	ecological fiscal reform
ELECRE	elimination and choice expressing reality
EPPR	Emergency Prevention, Preparedness and Response
ERDF	European Regional Development Funds
ERFEN	Protocol on the Regional Program for the Study of the El Niño phenomenon in the Southeast Pacific
ESF	European Social Fund
ESG	environmental, social and governance
ETR	ecological tax reform
EU	European Union

EUR	Euro
EVI	economic vulnerability index
FAL	Convention on Facilitation of International Maritime Traffic
FAO	Food and Agriculture Organization
FCWC	Fishery Committee of the West Central Gulf of Guinea
FFA	South Pacific Forum Fisheries Agency
fob	free on board
FSC	flag state control
GBRMPA	Great Barrier Reef Marine Park Authority
GDP	gross domestic product
GDS	geographically disadvantaged states
GFCM	General Fisheries Council for the Mediterranean
GGI	Gwich'in Council International
GHG	greenhouse gasses
GIS	geographic information system
GPA	Global Programme of Action
GPS	global positioning system
GSPs	green shipping practices
GT/G.T./gt	gross tonnage
GW	gigawatts
HAFS	International Convention on the Control of Harmful Anti-Fouling Systems on Ships
HAI	human assets index
IAATO	International Association of Antarctica Tour Operators
I-ATTC	Inter-American Tropical Tuna Commission
IBSFC	International Baltic Sea Fisheries Commission
ICC	Inuit Circumpolar Council
ICCAT	International Commission for the Conservation of Atlantic Tunas
ICES	International Council for the Exploration of the Sea
ICJ	International Court of Justice
ICM	integrated coastal management
ICNAF	International Commission for the Northwest Atlantic Fisheries
ICRW	International Convention for the Regulation of Whaling
ICZM	integrated coastal zone management
IFQ	individual fishing quotas
ILA	International Law Association
ILC	International Law Commission
ILO	International Labour Organization
IM	integrated management
IMO	International Maritime Organization
IOPC	Funds International Oil Pollution Compensation Fund
IOS	Indian Ocean Sanctuary
IOTC	Indian Ocean Tuna Commission
IPHC	International Pacific Halibut Commission
IPOA	international plans of action
ISA	International Seabed Authority
ISMC	International Safety Management Code
ISO	International Organization for Standardization
ISPS	International Ship and Port Facility Security Code

ITLOS	International Tribunal for the Law of the Sea
IUCN	International Union for the Conservation of Nature
IUU	illegal, unreported, and unregulated
IVQ	individual vessel quotas
IWC	International Whaling Commission
kg	kilogram
lb	pound
LDC	Convention on the Prevention of Marine Pollution by Dumping of Wastes and Other Matter
LDS	landlocked developing states
LLDCs	landlocked developing countries
LLMC	Convention on Limitation of Liability for Maritime Claims
LOSC	Law of the Sea Convention
LSMPA	large scale marine protected area
M/V	or MV motor vessel
MAP	Mediterranean Action Plan
MARPOL	International Convention for the Prevention of Pollution from Ships
MAUT	multi-attribute utility theory
MBI	market-based instruments
MBTA	Migratory Bird Treaty Act
MCDA	multi-criteria decision analysis
MEPC	Marine Environment Protection Committee
MLC	Maritime Labour Convention
MPA	marine protected area
MSC	Marine Stewardship Council
MSC	Maritime Safety Committee
MSP	marine spatial planning
MSY	maximum sustainable yield
MW	megawatts
NAFO	North Atlantic Fisheries Organization
NAMMCO	North Atlantic Marine Mammal Commission
NASCO	North Atlantic Salmon Conservation Organization
NEAFC	North East Atlantic Fisheries Commission
nmi	nautical mile
NOAA	National Oceanic and Atmospheric Administration
NDC	nationally determined contributions
NPA	National Programmes of Action
NPAFC	North Pacific Anadromous Fish Commission
NT/N.T./nt	net tonnage
OCA/PAC	Oceans and Coastal Areas Programme Activity Centre
OLDEPESCA	Latin American Organization for the Development of Fisheries
OPA	Oil Pollution Act
OPC	Oil Pollution Convention
OPRC	International Convention on Oil Pollution Preparedness, Response, and Co-operation
OSPAR	Convention for the Protection of the Marine Environment of the Northeast Atlantic
OSY	optimum sustained yield
P&I	Clubs protection and indemnity clubs

PAME	Protection of the Arctic Marine Environment
PCA	Permanent Court of Arbitration
PCBs	poly-chlorinated biphenols
PERSGA	Programme for the Environment of the Red Sea and Gulf of Aden
PICES	North Pacific Marine Science Organization
PM	particular matter
PROMETHEE	preference ranking organizational method for enrichment evaluation
PSC	Pacific Salmon Commission
PSC	port state control
PSI	Proliferation Security Initiative
PSSA	particularly sensitive sea area
PSU	practical salinity units
RAIPON	Russian Arctic Indigenous Peoples of the North
ReCAAP	Regional Cooperation Agreement on Combating Piracy and Armed Robbery against Ships in Asia
RFMOs	regional fisheries management organizations
RMP	Revised Management Procedure
RMS	Revised Management Scheme
ROPME	Regional Organization for the Protection of the Marine Environment
ro-ro	roll on/roll off
SAP	BIO Strategic Action Plan for the Conservation of Marine and Coastal Biodiversity in the Mediterranean
SAR	Convention International Convention on Maritime Search and Rescue
SC	Saami Council
SCAR	Scientific Committee on Antarctic Research
SCP	systematic Conservation Planning
SDC	Seabed Disputes Chamber
SDR	Special Drawing Rights
SDWG	Sustainable Development Working Group
SEAFO	South East Atlantic Fisheries Organization
SECAs	sulphur emission control areas
SEEMP	Ship Energy Efficiency Management Plan
SIDS	small island developing states
SIOFA	South Indian Ocean Fisheries Agreement
SOLAS	International Convention for the Safety of Life at Sea
SOS	Southern Ocean Sanctuary
SPA	Protocol Concerning Specially Protected Areas and Biological Diversity in the Mediterranean
SPLOS	State Parties to the LOS Convention
SPR	spawning potential ratio
SPREP	South Pacific Regional Environmental Programme
SPRFMO	South Pacific Regional Fisheries Management Organization
SRCF	Sub-Regional Commission on Fisheries
SSB	spawning stock biomass
SSBR	spawning stock biomass per recruit
STCW	International Convention on Standards of Training, Certification and Watchkeeping for Seafarers
TAC	total allowable catch
TBT	tributyltin

TCF	trillion standard cubic feet
TEU	twenty-foot equivalent units
TURFs	territorial user rights in fisheries
UBC	University of British Columbia
UK	United Kingdom
ULCC	ultra large crude carrier
UN	United Nations
UNCED	United Nations Conference on Environment and Development
UNCHE	United Nations Conference on the Human Environment
UNCITRAL	United Nations Commission on International Trade Law
UNCLOS	I 1958 United Nations Conference on the Law of the Sea
UNCLOS	II 1960 United Nations Conference on the Law of the Sea
UNCLOS	III 1982 United Nations Conference on the Law of the Sea
UNCLOS	United Nations Convention on the Law of the Sea
UNCTAD	United Nations Conference on Trade and Development
UNEP	United Nations Environment Program
UNESCO	United Nations Educational, Scientific and Cultural Organization
UNFCC	United Nations Framework Convention on Climate Change
UNFSA	UN Fish Stocks Agreement
UNGA	United Nations General Assembly
UNGAR	United Nations General Assembly Resolution
UNHCR	United Nations High Commission for Refugees
UNSCR	United Nations Security Council
USA	United States of America
USSR	Union of Soviet Socialist Republics
UV	ultra-violet
VLCC	very large crude carrier
VOCs	volatile organic compounds
WACAF	Abidjan Convention for Co-operation in the Protection and Development of the Marine and Coastal Environment of the West and Central African Region
WCED	World Commission on Environment and Development
WCPFC	Western and Central Pacific Fisheries Commission
WECAFC	Western Central Atlantic Fisheries Commission
WiP	with policy
WMD	weapons of mass destruction
WRC	Nairobi International Convention on the Removal of Wrecks
WSSD	World Summit on Sustainable Development
WTO	World Trade Organization
WWF	World Wildlife Fund

Chapter 1

An introduction to the world ocean

Human impacts and trends

Key topics

- The ocean is biogeochemically downstream from terrestrial environments, as such, most human activities eventually impact the world ocean.
- Threats to biodiversity can be broadly categorized as a result of overharvesting, pollution, habitat loss, introduced species, ocean acidification and global climate change. Threats can cumulatively interact with negative consequences.
- Until recently, different human impacts on the world ocean were managed in isolation, resulting in seemingly rational management actions that sometimes have unintended consequences.

Introduction

Humanity requires the services of the world ocean in a multitude of ways. For centuries, the oceans were envisaged as immutable and immune to human activities (see Box 1.1 and Box 1.2), and, until the 20th Century, the regulation of human activities affecting the world ocean was widely perceived as unnecessary and unachievable, given the absence of state ownership or hard boundaries in the marine environment. In addition, unlike the human footprint on the terrestrial realm, which is often easy to observe, most marine environments are hidden from us and therefore out of sight and out of mind (see Box 1.3).

The scale of the human footprint on the world ocean is enormous. Many stocks of once globally abundant fishes such as cod, herring and tuna have since become 'commercially extinct;' i.e. no longer commercially fishable due to their greatly reduced numbers. Over one million whales were harvested during the 19th and 20th Centuries; to date, only the eastern Pacific grey whale has recovered to near pre-exploitation levels. The global seabird population has declined by 69.7 per cent since 1950; over the past 40 years, the abundance of marine vertebrates (fish, seabirds, sea turtles and marine mammals) has declined by an average of 22 per cent (Paleczny et al., 2015; Sumalia et al., 2016). Elevated levels of pollutants are found in most marine species, even those living in the Polar Regions. Ocean temperatures are rising – aggravated by the addition of greenhouse gasses from the combustion of fossil fuels – with many deleterious effects. Tens of thousands of square kilometres of coral reefs have bleached (lost their photosynthetic symbionts), with large associated die-offs in recent years. Important breeding, feeding, mating and resting areas for migratory species have been affected by human activities. The continuing human-induced degradation of marine ecosystems is as wide and deep as the ocean itself (Roberts, 2013).

Box 1.1 Importance of the oceans

Globally, the oceans are the:

- Main reservoir of water: 71 per cent of the earth's surface is covered by oceans; less than 0.5 per cent is freshwater.
- Main place for organisms to live; they comprise over 99 per cent of the inhabitable volume of the 'earth'.
- Main planetary reservoir of O_2.
- Possible main planetary producer of O_2 from phytoplankton.
- Planetary thermal reservoir and regulator.
- Medium for longitudinal heat transfer and circulation.
- Major reservoir of CO_2 especially in HCO_3^-, $CO_3^=$ forms.
- Habitat for enormous diversity of living organisms, from bacteria to whales.
- Reservoirs of enormous resource potential, both renewable and non-renewable, oil, minerals, etc.; also, the location of half the planet's carbon fixation.

Source: After Roff and Zacharias (2011)

Box 1.2 Marine goods and services that support humanity

Production services

- Food provision: Marine organisms for human consumption.
- Raw materials: Marine organisms for non-consumptive purposes, such as pharmaceuticals and genetic materials; non-living resources including minerals, sand and petroleum.

Cultural

- Identity/cultural heritage: Value associated with the marine environment, e.g. for cultural and spiritual traditions.
- Psychological health, leisure and recreation: Revitalizing the human body and mind through leisure activities ranging from visiting the beach to SCUBA diving with marine organisms in their natural environment.

Value

- Research value: Studying the many aspects of the ocean to gain a greater understanding of the world around us.
- Future unknown and speculative benefits: Currently unknown potential future uses of the marine environment and associated biodiversity.

Regulation services

- Gas and climate regulation: Balance and maintenance of the chemical composition of the atmosphere and oceans by marine living organisms and chemical processes.

- Flood and storm protection: Dampening of environmental disturbances by biogenic structures such as coral reefs, kelp beds and mangroves.
- Bioremediation of waste: Removal of pollutants through storage, dilution, transformation and burial.

Supporting services

- Nutrient cycling, including carbon sequestration: Storage, cycling and maintenance of nutrients by living marine organisms and physical processes.
- Biologically mediated habitat: Habitat that is provided by marine organisms.
- Resilience/resistance: Extent to which ecosystems can absorb recurrent natural and human perturbations and continue to offer the above-listed services without degrading or unexpectedly switching to alternate states.

Source: Beaumont et al. (2007) and Ruiz-Frau et al. (2011)

Box 1.3 Human perceptions on ocean health

A recent review of research into the public attitudes and perceptions on threats to the world ocean found that regardless of which state the survey was conducted in, there was agreement (70 per cent) that oceans are threatened by human activities, and 45 per cent of respondents ranked these threats as high or very high. The review collated over 32,000 respondents across 21 nations and found that while the public ranked the threats of pollution and overfishing respectively as the greatest threats, there were regional disparities based on past events (e.g. oil spills in Scotland), media campaigns (e.g. ocean plastics) and differences in opinion between the public and marine scientists (e.g. the threats of climate change). In addition, most respondents wanted to see more of the world ocean protected but were unsure how much is already protected or what the socio-economic costs of additional protection would be.

Public opinion surveys are important to policy-makers as they identify electorate priorities, where to focus education and awareness initiatives and how management and conservation programs may be improved with limited available resources.

Source: Lotze et al. (2018)

Although human capacity for massive disturbance in marine environments has been known at least since the extinction of the Steller's Sea Cow in 1868, the plight of the oceans did not become a public concern until the appeals in the 1950s and 1960s by authors such as Rachel Carson (1962) and Jacques Cousteau. Since the 1970s, the steadily increasing output of books, films and television series along with the establishment of organizations such as Greenpeace have resulted in heightened public awareness and concern. Inter-governmental efforts to better manage the marine environment began in earnest with international conventions such as the London Dumping Convention (1972), the International Convention for the Prevention of Pollution from Ships (1978), the United Nations Convention on the Law of the Sea (1982), and the Convention on Biological Diversity (1992). More recently, under the Paris Agreement (2015) of the United Nations Framework Convention on Climate Change (UNFCCC, 1992), states have gradually begun to recognize the importance of the global ocean in meeting targets and 'nationally determined contributions' (NDCs).

Notwithstanding these efforts, human uses and impacts on the ocean continue unabated and the diversity of life in our oceans is now being dramatically altered by rapidly increasing and potentially irreversible human activities. Approximately 40 per cent of the world's population and 60 per cent of the world's economic production are concentrated in a 100km swath along the world's coasts. Twenty-one of the world's 33 mega cities are coastal, and it is estimated that by 2020 up to 75 per cent of the world's 7.5 billion people may be living within 60km of the coastal zone (Martínez et al., 2007; UN, 2017).

Human impacts on the oceans can be broadly categorized as a result of overharvesting, pollution, habitat loss, introduced species and global climate change/ocean acidification. The following sections provide a brief discussion of these impacts on the marine environment. It should be noted, however, that these impacts do interact, often in ways detrimental to ocean health. These negative interactions are termed cumulative or synergistic effects. The seemingly rational regulation and management of individual sectors generally fails to consider the collective impacts of other sectoral activities on either the environment, public health or human use of the resource, sometimes resulting in unintended negative consequences to ecological systems and human well-being. Cumulative effects may be localized (e.g. pollution discharge) or global (e.g. climate change) and may be acute (e.g. overharvesting a fishery) or enduring (e.g. disposed radionuclides).

Overharvesting

Unsustainable harvesting is perhaps the most serious threat to marine environments worldwide. Overharvesting is not a new phenomenon in the oceans. Historically, human cultures either removed particular species and moved on to harvesting others, or other areas, or had to develop methods of regulating the timing and amount of harvest in order to avoid over-exploitation. With the advent of the industrial revolution in the mid to late 1800s came the increasing power to fish in ways hitherto unheard of: Steam-trawling the seafloor, fishing further offshore, and mechanically winching on board large nets and animals. Species such as whales and offshore pelagic fish (tuna, swordfish) became accessible to those willing to take the risks. Most whale species have been reduced to levels where they are considered endangered or threatened. Most populations of palatable fish stocks, particularly the larger fish, have been seriously depleted (e.g. Pauly et al., 1998; Worm et al., 2006).

From a western perspective, the importance of marine fisheries to global food security is often overlooked. In North America and many parts of Europe, seafood is simply another source of protein for consumers to choose. Elsewhere, over one billion people depend on seafood as their primary source of protein, and 3.1 billion people rely on seafood for at least 20 per cent of their protein. Fishing has recently been estimated to employ 34 million people in full- or part- time jobs, producing 81.5 million tonnes of seafood in 2014. The first-sale value of the world's fisheries is estimated at $US 100 billion (FAO, 2017).

Be that as it may, the sustainability of marine capture fisheries has recently been shown to be at risk. The Food and Agriculture Organization of the United Nations (FAO) currently reports that 31.4 per cent of the world's fisheries are now overexploited, depleted or recovering, 58.1 per cent are fully exploited and only 10.5 per cent under exploited or moderately exploited (FAO, 2017). For context, in 1974, the corresponding percentages were 10, 50 and 40 per cent, respectively. Global catches peaked in 1996 and, for the past

15 years, have been relatively stable at a level approximately 10 per cent less than 1996 (FAO, 2017). The European Commission has determined that for the European fish stocks where stock status is known, only a third are currently managed for sustained future production (CEC, 2008).

The status of high seas stocks (i.e. those beyond national jurisdictions) is even less well known. Of the 17 Atlantic fish stocks managed under the North Atlantic Fisheries Organization (NAFO), six are collapsed and only two are considered to be sustainable (NAFO, 2016). Atlantic cod biomass is estimated at 6 per cent of historical levels and North Sea cod stocks are depressed to such a degree that recently fishers have been unable to harvest enough to meet their allowable catches, and each year further reductions are recommended. In 2018, the International Council for the Exploration of the Sea (ICES) recommended a 47 per cent reduction to catches of cod in the North Sea, eastern English Channel and Skagerrak Region (ICES, 2018).

A closer look at certain aspects of marine capture fisheries shows that 50–70 per cent of pelagic predators (i.e. tunas, swordfish) have been removed by fishing. Fishing pressure continues to shift towards lower trophic levels as apex predators decline; this has been termed 'fishing down marine food webs' (Pauly et al., 1998). Furthermore, the global fishing fleet is far larger than what necessity dictates, especially since technological innovations increase the ability to catch fish. This surplus fishing capacity is underwritten by $US 30–40 billion in annual subsidies by most fishing nations, thus providing no incentive to reduce fishing effort (Sumaila et al., 2016). In addition, evidence suggests that fishing efforts from port-based fisheries have increased over the past three decades at upwards of 3 per cent per decade. As such, more fishers are chasing fewer fish; if one were to imagine the global fishing fleet as a country, it would be the 18th largest oil-consuming nation on earth.

Another serious problem is the unintended harvest of species. An estimated 20 per cent of the global fisheries' catch constitutes unwanted by-catch that is discarded. Shrimp trawl fisheries produce the largest by-catch and small pelagic fisheries the least. By-catch also consists of the incidental take of endangered marine mammals, turtles and seabirds although recent advances in gear technology are reducing these impacts.

There are signs that some fisheries are on a path to sustainability. In the United States and New Zealand, the numbers of overfished stocks have been steadily improving, from over 20 per cent a decade ago to about 15 per cent currently (Fisheries New Zealand, 2017; NOAA, 2017). Canadian northern cod stocks off of Newfoundland and Labrador – fished to collapse between the 1960s and 1990s – have seen recent growth rates of up to 30 per cent, indicating that even stocks thought to be fished past the point of no return can recover if protected (Rose and Rowe, 2015).

Pollution

Pollution – from a myriad of sources – has impacted every marine system on earth. As a result of global transport mechanisms, contamination has spread to all continents and depths. Indigenous human populations in Arctic areas have the highest contaminant levels of any people on earth, a result of ingesting marine fish and mammals which bioaccumulate toxins due to their high trophic levels. Up to 90 per cent of pollution entering the ocean is derived from land-based sources; the remainder is a result of deliberate or accidental dumping. It has been assumed that land-based sources of pollutants reaching the oceans is primarily via rivers; however, transport of pollutants through the atmosphere to the oceans is also significant, particularly for metals such as mercury. Sources of pollution in the ocean may be point source (from a single source, such as an industrial facility or sewage discharge) or non-point source (where the exact source is unknown, e.g. agricultural runoff).

Most researchers recognize the primary types of pollution as biological (pathogens), toxic substances (contaminants), nutrients, marine debris, hydrocarbon spills and underwater noise. Certain scientists also suggest that ocean acidification, as a result of climate change, is a type of pollution.

Biological pollution (pathogens) includes increased levels of bacteria, viruses and parasites, occurring naturally or introduced as a result of human activities. They may reside in seawater, the seabed or live on or within marine organisms. Anthropogenic (human) sources of pathogens in the ocean include sewage and agricultural runoff. Pathogens affect human health through direct ingestion (e.g. shellfish, seawater when swimming), skin contact or inhalation. Shellfish are known to transmit gastroenteritis and hepatitis, as well as cholera and typhoid fever. Marine mammals are known to have been infected by terrestrial animal diseases, likely transmitted from freshwater into the ocean. Fish hatcheries and aquaculture operations may also inadvertently introduce pathogens into the environment. Climate change may exacerbate the effects of certain marine pathogens mainly through increased temperatures providing optimal growth conditions (Burge et al., 2014).

Over the past 100 years of industrial chemistry, hundreds of *synthetic compounds* have found their way into the world ocean. Thousands of new compounds are created each year with sometimes unknown implications should the compound be released in the environment and without a regulatory regime to apply precaution to their use. Synthetic compounds have been categorized in many ways, but generally they can be separated into persistent organic pollutants (POPs), pesticides, pharmaceuticals, flame retardants and other substances captured under the general heading of 'contaminants of emerging concern' (or CECs). Unlike pathogens, the consequence of synthetic compounds on the environment and human health are subtler, including long-term bio accumulative effects leading to potential chronic illness and cumulative/delayed effects.

Persistent organic pollutants (POPs) capture a large group of chemicals that are similar in that they are halogenated (containing fluorine [F], chlorine [C], bromine [Br], iodine [I] and astatine [At]) organic compounds that degrade very slowly in the environment, bioaccumulate in fatty tissues and are highly toxic. Persistent organic pollutants have a wide range of effects including changes in behaviour, development, growth and reproductive capacity.

Uses of POPs have varied since their introduction, but common uses have included pesticides (e.g. DDT), industrial chemicals (e.g. PCBs), solvents and pharmaceuticals. Over time, the use of the more persistent POPs has slowly been phased out through international agreements or individual nations passing bans, resulting in a gradual decrease of these substances in the marine food chain. However, while the compounds replacing the 20+ banned POPs are certainly less persistent and toxic, they continue to impact marine food webs, often in new and unknown ways (Desforges et al., 2015).

'First generation' pesticides, such as DDT, were banned by many countries in the 1970s and internationally in 2001 due to persistence and toxicity. 'Second generation' pesticides consist primarily of organophosphates and carbamates, which are cholinesterase inhibitors, meaning they inactivate a key enzyme necessary for nervous function. These compounds break down much quicker in the environment; however, they can still be toxic to marine life. 'Third generation' pesticides either work via mimicking juvenile hormones so the insect cannot reach adulthood or through preventing insect larvae from moulting and are the safest pesticide for marine environments.

The effects of pharmaceuticals on the marine environment is a relatively new branch of study. Humans use over 6,000 different prescription and non-prescription drugs that enter the marine environment through either direct disposal (e.g. flushing unused medicines) or human wastes. Approximately 90 per cent of oral drugs pass through the body that in turn pass unaffected through sewage treatment systems and ultimately enter the oceans. While there have been demonstrated effects on marine life from certain pharmaceuticals, such as birth control pills reducing reproductive success in invertebrates due to endocrine disruption or development of antibiotic resistance in marine microbes, the overall impacts of drug inputs into the marine environment is poorly understood. However, early evidence suggests that many drugs are biologically active in marine environments at very low concentrations and have the capacity to affect behaviour and reproductive success of marine biota that reside near human settlements (Mezzelani et al., 2018). In addition to pharmaceuticals, personal care products such as cosmetics, antibacterial soaps, lotions and insect repellents are not removed in the sewage treatment process and exhibit many of the same properties as prescription drugs.

Lastly, CECs include a range of compounds that include: Flame retardants used in furniture, carpets, clothing and toys; plasticizers that increase the flexibility or fluidity of materials; solvents capable of dissolving other substances; surfactants that break down the interface between water and oils and therefore aid cleaning substances; and nanoparticles, which are substances smaller than one hundred nanometres used in a range of industrial processes and consumer products.

Metals are naturally occurring elements that, when increased above background levels by human activities, may have significant impacts on environmental and human health. Certain metals such as copper (Cu) and zinc (Zn) are necessary to sustain life; however, they can be toxic at higher concentrations. Other metals, including cadmium (Cd), chromium (Cr), lead (Pb) and mercury (Hg) have no role in biological processes but are highly toxic. Metals are introduced into the ocean via land-based processes (mining, industrial processes), the atmosphere (e.g. coal combustion and cement production releasing mercury) and anti-fouling compounds used to keep vessels and other marine structures free of marine plants and invertebrates.

Metals affect the marine environment and humans in a number of ways. First, metals – like certain other pollutants – bioaccumulate up the food chain and concentrate in top predators (e.g. tuna, swordfish, marine mammals, seabirds). Metals may bind to small clay and silt particles that are ingested by benthic (bottom) animals that are in turn consumed by other animals or stored in nearshore plant and marine algae tissues and in turn consumed. At certain concentrations, metals may affect respiration, digestion, nervous and reproductive system functions, leading to changes in behaviour, slower growth and reproductive failure (Burger et al., 2014).

While copper, cadmium, chromium and lead are generally found in localized areas (e.g. marinas, industrial areas) with localized impacts on environmental and human health, mercury is a global problem. Mercury is a potent neurotoxin that easily bioaccumulates, which is why most countries now recommend limits on tuna and swordfish consumption for children. In 2017, the global Minamata Convention on Mercury came into force given the rapid rise in mercury in marine biota and corresponding human health issues.

Nutrients include human inputs of nitrogen (N) and phosphorus (P) that exceed natural background levels resulting in algal blooms. Aside from upsetting the natural state of the food web, these algal blooms may also be toxic ('harmful algal blooms' or HABs) to marine animals (fish and shellfish) or humans or consume available dissolved oxygen creating 'dead zones' (hypoxia or anoxia) due to a lack of oxygen for other species, a process termed 'eutrophication'. Harmful algal blooms have expanded in recent decades, primarily due to increasing coastal human populations and the resultant creation of N from agriculture (animal wastes and fertilizers), sewage and the deposit of airborne N from fossil fuel combustion into rivers and oceans. Climate change may also be exacerbating HABs due to increases in water temperatures, fostering the increased growth of primary producers such as phytoplankton.

Marine debris is any intentionally or unintentionally discarded solid manufactured item in the marine environment. NOAA estimates that approximately 600 million kilograms of trash enters the ocean each year, 60–80 per cent of which are plastics (NOAA, 2016). Polystyrene foam or 'styrofoam' is another major source of marine debris that tends to float while the majority of plastics sink. The majority of marine debris (about 70 percent) is found on the seabed while the remainder is split evenly between floating on the ocean surface or found on beaches.

The primary types of marine debris are domestic materials (e.g. shopping bags, plastic bottles), industrial materials and fishing gear. Plastics are durable. Though many will break apart into ever smaller pieces, becoming largely invisible 'micro-plastics', they will not break down chemically for hundreds of years. Additionally, plastics absorb and concentrate hydrophobic organic pollutants, thus acting as reservoirs and transport mechanisms for these substances. It is estimated the economic impact of ocean plastics is $US 13 billion dollars annually (UNEP, 2014). Approximately 80 per cent of marine debris comes from land-based sources.

Marine debris impacts the marine environment in a number of ways. Ingestion and entanglement are the primary effects on marine mammals, seabirds, sea turtles and fishes. In highly populated coastal areas, plastics may affect plants and animals living on or in the seabed through accumulation resulting in degradation of habitat through smothering and toxicity. It is estimated that marine debris is responsible for up to 100,000 marine mammal deaths annually and 90 per cent of seabirds have plastics in their bodies. Ingested plastics inhibit growth and reproduction through lacerating or displacing food in animals' stomachs or transferring toxic substances to the host. Lost or abandoned fishing gear can continue to 'ghost fish' where fish persist to be caught in nets that may have been adrift for many years (Gall and Thompson, 2015).

Hydrocarbon (oil) spills are a unique type of pollution. Hydrocarbon spills occur as a result of the transport of different forms of oil or the exploration and development of offshore oil resources. Approximately half of the world's oil production (approximately 1.6 billion tonnes) is transported by sea. Much of this transport is over long distances in large tankers that pass through straits and transit along coastlines. This means of transport has resulted in many serious accidents and spills over the past several decades; however, due to improvements in safety, currently about 12 per cent of oil entering the ocean is a result of transportation accidents with the remainder from primarily land-based sources. While the average number of major oil spills per year has dropped from 25 in the 1970s to 3 today, smaller spills continue to impact fisheries, tourism and coastal economic activities.

Marine environments also currently produce one third of the world's hydrocarbons and have the potential for significant expansion. The shallow marine (< 400 metres) and deep marine (> 400 metres) environments are estimated to contain 36 per cent and 11 per cent of the global distribution of oil-bearing reservoir rocks respectively, though much of the deep-sea has not yet been explored. Offshore reserves currently account for about 30 per cent of oil production globally. It is estimated that 300 billion barrels may yet be discovered in offshore areas.

In addition to shipping, a significant amount of oil is transported short distances by seabed pipelines from production wells to offshore or onshore facilities.

Habitat loss and degradation

Approximately 40 per cent of the world's population and 60 per cent of the world's economic production are concentrated in a 100km swath along the world's coasts. The confluence of such a significant proportion of the world's population on the narrow fringe separating the marine and terrestrial realms has significant ecological and socio-economic implications.

Habitat loss due to the destruction of larger vascular vegetation upon which many species depend for food and shelter is probably the most serious threat to biodiversity in terrestrial environments. Loss of marine habitat is primarily a concern in coastal nearshore and intertidal marine environments. Increasing pressure on coastal systems results from a combination of shipping and its attendant infrastructure, modification of natural coastlines, bottom contact fishing (e.g. trawling), recreational activities and increased land runoff – including nutrients and suspended solids. The types of habitats in these areas that can be 'lost' include marine macrophytes (kelp), mangroves, seagrasses, corals and other biotic communities (e.g. sponges, seapens, sea fans, aphotic corals). Abiotic habitats can be impacted as well, such as intertidal and estuarine mud flats and other areas that are dredged or subject to dumping. Habitat loss in deeper marine environments and the pelagic ocean is more difficult to determine as these habitats are primarily composed of either oceanographic (e.g. currents, gyres, fronts) or physiographic (e.g. seabed composition) structures and processes that are either not immediately impacted or more resistant to human activities (however, deep-seabed mining, should it progress to commercial operations, could radically alter deep-sea habitats). Loss of marine habitat is significant not only from an ecological perspective but increasingly from a socio-economic perspective as well. The interaction of human effects and natural marine processes is most evident in coastal waters where strategies to prevent habitat loss or restore habitats are integrated within coastal zone management initiatives.

Habitat loss can be rapid and its consequences significant. An estimated kilometre of Europe's coastline was developed each day between 1960 and 1995, resulting in a 50 per cent loss of coastal wetlands and seagrasses (Gibson et al., 2007). Much of the world's continental shelves have experienced habitat degradation due to bottom contact fishing (bottom trawling) where nets are dragged over the ocean floor, causing damage to sessile organisms such as corals and seapens, as well as flattening benthic (bottom) biogenic structures (e.g. oyster reefs). Furthermore, bottom contact fishing can result in the fine sands, silts and clays being resuspended in the water column for extended periods of time.

As natural resources become depleted or more difficult to access on land, the pressure to extract material from the oceans increases. The nearshore marine environment has long been utilized as a source for materials. For centuries, beaches and nearshore areas have been mined for gold (Au), tin (Sn), diamonds, sand and gravel. As mining and shipping technologies improve, exploration and development has radiated seaward onto the continental shelves. Offshore oil drilling was pioneered in the 1930s in the Gulf of Mexico and the first offshore windfarm established in Denmark in the 1990s. Currently, almost one third of the world's petroleum is produced from continental shelves. This output is likely to continue to increase as sea ice retreats in the Polar Regions, opening up new areas for exploration and development.

It is estimated that deep-sea areas (outside of continental shelves) contain two thirds of the world's accessible mineral wealth. For the past 60 years, anticipated mineral shortages combined with increased commodity prices have created periods of interest in mining the mineral-rich deep-sea. However, these prognosticated shortages have failed to transpire, due to a combination of increased metal recycling, refinements in industrial processes and new terrestrial discoveries. Additionally, the technological challenges of mining the deep seabed have proven more difficult than some anticipated. To date, lower than expected commodity prices combined with higher than expected technological costs have made it cost prohibitive for deep-sea mining to become economically attractive to conventional mining firms, though interest remains with some states and newcomer companies. Long-term demand – based on forecasted development of land-based deposits – suggests that marine deposits high in cobalt (for batteries) could become commercially viable at some point in the future.

A final potential future impact on continental shelf environments is methane hydrates. Produced by the decomposition of organic materials, methane hydrates consist of molecules of natural gas (predominately methane) trapped in ocean sediments by a combination of low temperature and high pressure. They are generally found in water depths greater than 400–500 metres and up to 1,100 metres below the seafloor. One cubic metre of hydrate is equivalent to 164 cubic metres of methane at atmospheric pressure and temperature and the energy potential of the world's reserves is estimated between 10^5 and 10^8 trillion standard cubic feet (TCF). These astounding numbers make the energy potential of methane hydrates greater than all the world's known coal, oil and natural gas reserves.

Introduced species

Species introductions (also termed invasive, exotic and non-native species) likely have occurred since humans first used the oceans for exploration and trade. There is evidence that many species believed to be native were introduced through marine transportation prior to the industrialized era. More recently, species introductions are facilitated through trade, travel and transport because of globalization and population growth. The rate of species introductions has grown by an order of magnitude over the past three decades. Currently, 1,369 non-native species are reported in European seas, whereas a decade earlier there were only 737. California, with a much smaller coastline, is host to 235 known invasive species (Williams et al., 2013).

The ballast water of ships appears to be the main mode of travel (vector) for introduced species, and impacts are generally observed mainly in coastal waters and estuaries. Once established, introduced species can have a number of detrimental impacts on their new environments including out-competing native species for food and habitat, introducing new pathogens, hybridizing with native species or disruption of entire food webs. Some introduced species, such as the Japanese oyster (*Crassostrea gigas*), have been intentionally introduced throughout the world for commercial aquaculture purposes and, although they displace native species, are generally regarded as a net benefit. Others, such as the American comb jellyfish (*Mnemiopsis leidyi*), have had disastrous consequences. The jellyfish arrived in the Black Sea in 1982 in the ballast water of ships and – without predators or competitors – quickly consumed most of the available zooplankton, leading to the collapse of fisheries and significant economic impacts. By the 1990s, the comb jellyfish accounted for 90 per cent of the biomass in the Black Sea.

Some of these invasive species can have dramatic local socio-economic effects. Different species of jellyfish have had substantial impacts on fisheries and even coastal human recreation. Outbreaks of jellyfish have now been reported worldwide due to species invasions, overharvesting (removing jellyfish predators), nutrient enrichment (increasing primary production resulting in increases in available food) and increases in local water temperature. Most species introductions are less obvious, found in the phytoplankton and zooplankton, with ecosystem impacts throughout the food web.

Global climate change

There is no doubt that the earth's climate has changed throughout history and that cyclical changes transpired long before humans dominated the planet. Global climate changes have been responsible for mass extinctions in the past and will likely result in future extinction events. During the Quaternary period, sea levels deviated as much as 85m, which inhibited the evolution of established marine communities in coastal and shelf environments. Unquestionably, humans have impacted global temperatures; since the 1980s, there has been considerable debate on differentiating the natural and anthropogenic contributions to climate change. Recent evidence, however, concludes that climate change since the 1900s is human-induced, and it is estimated that the oceans have lost 2 per cent of their oxygen since 1960. (Gilbert, 2017)

Human activities that influence change include the release of carbon dioxide through the burning of fossil fuels and large-scale deforestation, which decreases the removal rate of CO_2 from the atmosphere. Changes in ocean temperatures impact marine biological communities in various ways, including changes in geographical distribution, behaviour and life history (e.g. reproduction, growth and dispersal). It is estimated that 892 commercially important fish species are shifting into new territories at a rate of 70km per decade and that the world's exclusive economic zones (EEZs) are anticipated to receive up to five new fish stocks by the end of the century (Poloczanska et al., 2013). Coral reefs, for example, are particularly vulnerable to ocean warming, and even minor temperature increases may cause many coral species to expel their algal symbiotic zooxantheallae (termed 'bleaching') that provide energy through photosynthesis, ultimately causing coral death if temperatures stay elevated and the zooxantheallae do not return. There is also evidence that the ranges of some species have shifted towards the north and south poles as temperatures increase in equatorial and temperate waters (Pinsky et al., 2018).

The global shipping industry emits significant amounts of greenhouse gasses. The International Maritime Organization (IMO) estimates that shipping was responsible for 796 million tonnes of CO_2 emissions, or about 2.2 per cent of global emissions in 2012 (IMO, 2014). If the global shipping industry were a country, it would be the sixth largest emitter of CO_2 after the US, China, Russia, India and Japan. Shipping moves approximately 70 per cent of global freight while producing around 15 per cent of global CO_2 emissions. Without additional regulations on CO_2 emissions or the adoption of more efficient shipping technologies, shipping emissions are forecasted to increase by a factor of 2 to 3 by 2050. On the other hand, if technical and operational improvements are implemented, they could reduce CO_2 emissions by 25 per cent to 75 per cent below current levels.

Discussion questions

- Of the five main threats to the world ocean, which is the most serious to the continued ecological function of the ocean and the goods and services it provides to humanity?
- Of the five main threats to the world ocean, which do you believe individual states (countries) can address independently, and which will require states to work together to address?
- If status quo management of the world ocean continues, what might be the impacts to humanity in 50 years?

Further reading

Beaumont, N. J., Austen, M. C., Atkins, J. P., Burdon, D., Degraer, S., Dentinho, T. P., Derous, S., Holm, P., Horton, T., van Ierland, E., Marboe, A. H., Starkey, D. J., Townsend, M. and Zarzycki, T. (2007) 'Identification, definition and quantification of goods and services provided by marine biodiversity: Implications for the ecosystem approach', *Marine Pollution Bulletin*, vol 54, pp. 253–265.

Burge, C. A., Eakin, C. M., Friedman, C. S., Froelich, B., Hershberger, P. K., Hofmann, E. E., Petes, L. E., Prager, K. C., Weil, E., Willis, B. L., Ford, S. E. and Harvey, C. D. (2014) 'Climate change influences on marine infectious diseases: Implications for management and society', *Annual Review Marine Science*, vol 6, pp. 249–277.

Burger, J., Gochfeld, M., Jeitner, C., Pittfield, T. and Donio, M. (2014) 'Heavy metals in fish from the Aleutians: Interspecific and locational differences', *Environmental Research*, vol 131, pp. 119–130.

Carson, R. (1962) *Silent Spring*, Houghton Mifflin, New York.

CEC (2008) *Council Facts and Figures on the CFP: Basic data on the Common Fisheries Policy: Edition 2008*, Office for Official Publications of the European Communities, Luxemburg.

Desforges, J. P., Sonne, C., Levin, M., Siebert, U., De Guise, S. and Dietz, R. (2015) 'Immunotoxic effects of environmental pollutants in marine mammals', *Environment International*, vol 86, pp. 126–139.

FAO (2017) *The State of World Fisheries and Aquaculture 2018*, FAO, Rome, www.fao.org/documents/card/en/c/I9540EN, accessed 24 July 2018.

Fisheries New Zealand (2017) 'Fish stock status', www.mpi.govt.nz/growing-and-harvesting/fisheries/fisheries-management/fish-stock-status/, accessed 04 January 2019.

Gall, S. C. and Thompson, R. C. (2015) 'The impact of debris on marine life', *Marine Pollution Bulletin*, vol 92, pp. 170–179.

Gibson, R. N., Atkinson, R. J. A., Gordon, J. D. M. and Beck, M. W. (2007) 'Loss, status and trends for coastal marine habitats of Europe', *Oceanography and Marine Biology*, vol 45, pp. 345–405.

Gilbert, D. (2017) 'Environmental science: Oceans lose oxygen', *Nature*, vol 542, pp. 303–304.

ICES (2018) 'Mixed-fisheries advice for Subarea 4, Division 7.d, and Subdivision 3.a.20 (North Sea, Eastern English Channel, Skagerrak)', www.ices.dk/sites/pub/Publication%20Reports/Advice/2018/2018/mix-ns.pdf, accessed 4 February 2019.

IMO (2014) *Third IMO Greenhouse Gas Study 2014*, IMO, London.

Lotze, H. K., Guest, H., O'Leary, J., Tuda, A. and Wallace, D. (2018) 'Public perceptions of marine threats and protection from around the world', *Ocean and Coastal Management*, vol 152, pp. 14–22.

Martínez, M. L., Intralawan, A., Vázquez, G., Pérez-Maqueo, O., Sutton, P. and Landgrave, R. (2007) 'The coasts of our world: Ecological, economic and social importance', *Ecological Economics*, vol 63, pp. 254–272.

NAFO (2016) 'Stock advice', www.nafo.int/Science/Stocks-Advice, accessed 29 July 2018.

NOAA (2016) 'Modelling Oceanic transport of Marine Debris', marinedebris.noaa.gov/sites/default/files/publications-files/Modeling_Oceanic_Transport_of_Floating_Marine_Debris.pdf, accessed 04 January 2019.

NOAA (2017) '2017 Report to congress on the status of U.S. fisheries', www.fisheries.noaa.gov/national/2017-report-congress-status-us-fisheries, accessed 04 January 2019.

Paleczny, M., Hammill, E., Karpouzi, V. and Pauly, D. (2015) 'Population trend of the world's monitored seabirds, 1950–2010', *PLOS One*, vol 10, no 6, pp. 1–11.

Pauly, D., Christensen, V., Dalsgaard, J., Froese, R. and Torres, F. (1998) 'Fishing down marine food webs', *Science*, vol 279, no 5352, pp. 860–863.

Pinsky, M. L., Reygondeau, G., Caddell, R., Palacios-Abrantes, J., Spijkers, J. and Cheung, W. W. L. (2018) 'Preparing ocean goverance for species on the move', *Science*, vol 360, no 6394, pp. 1189–1191.

Poloczanska, E. S., Brown, C. J., Sydeman, W. J., Kiessling, W., Schoeman, D. S., Moore, P. J., Brander, K., Bruno, J. F., Buckley, L. B., Burrows, M. T., Duarte, C. M., Halpern, B. S., Holding, J., Kappel, C. V., O'Connor, M. I., Pandolfi, J. M., Parmesan, C., Schwing, F., Thompson, S. A. and Richardson, A. J. (2013) 'Global imprint of climate change on marine life', *Nature Climate Change*, vol 10, no 1038.

Roberts, C. (2013) *The Ocean of Life: The Fate of Man and the Sea*, Penguin, New York.

Roff, J. C. and Zacharias, M. A. (2011) *Marine Conservation Ecology*, Earthscan, London.

Rose, G. A. and Rowe, S. (2015) 'Northern cod comeback', *Canadian Journal of Fisheries and Aquatic Sciences*, vol 72, pp. 1789–1798.

Ruiz-Frau, A., Edward-Jones, G. and Kaiser, M. J. (2011) 'Mapping stakeholder values for coastal zone management', *Marine Ecology Progress Series*, vol 434, pp. 239–249.

Sumaila, R. U., Lam, V., Le Manach, F. and Pauly, D. (2016) 'Global fisheries subsidies: An updated estimate', *Marine Policy*, vol 69, pp. 189–193.

UN (2017) 'Ocean fact sheet', www.un.org/sustainabledevelopment/wp-content/uploads/2017/05/Ocean-fact-sheet-package.pdf, accessed 04 January 2019.

UNFCC (1992) *United Nations Framework Convention on Climate Change*, Bonn, Germany, treaties.un.org/doc/Treaties/1994/03/19940321%2004-56%20AM/Ch_XXVII_07p.pdf, accessed 04 January 2019.

UNEP (2014) 'Valuing plastics Nairobi', www.trucost.com/publication/valuing-plastic, accessed 13 September 2018.

Williams, S. L., Davidson, I. C., Pasari, J. R., Ashton, G. V., Cartlon, J. T., Crafton, R. E., Fontana, R. E., Grosholz, E. D., Miller, A. W., Ruiz, G. M. and Zabin, C. J. (2013) 'Managing multiple vectors for marine invasions in an increasingly connected world', *BioScience*, vol 63, no 12, pp. 952–966.

Worm, B., Barbier, E. B., Beaumont, N., Duffy, E., Folke, C., Halpern, B. S., Jackson, J. B. C., Lotze1, H. K., Micheli, F., Palumbi, S. R., Sala, E., Selkoe, K. A., Stachowicz, J. J. and Watson., R. (2006) 'Impacts of biodiversity loss on Ocean ecosystem services', *Science*, vol 314, no 5800, pp. 787–790.

Chapter 2

A brief introduction to the marine environment
The structure and function of the world ocean

Key topics
- Temperature governs the distribution of marine biological communities more than any other factor.
- Water masses are defined by homogenous temperature and salinity regimes and are analogous to the major terrestrial regions.
- High biological productivity is found in areas where water masses converge or diverge.
- The great majority of phyla of multicellular animals are found only in the marine environment.
- Phytoplankton are responsible for > 95 per cent of the annual primary production in the ocean
- While the ocean's habitable volume is hundreds of times greater than that of the land, only the upper 50–200 metres of the oceans – the photic zone – will support the growth of primary producers.
- In terrestrial ecosystems, primary producers (mainly vascular plants such as trees and grass) constitute the great majority of biomass, and individual organisms are often large. In contrast, the dominant primary producers in marine ecosystems – the phytoplankton – are generally microscopic and can reproduce rapidly and are rapidly processed, either by consumers or reducers (e.g. bacteria).
- In the oceans, species and communities can be passively transported long distances, whereas on land, colonization of distant areas generally requires active migration and passive dispersal.
- Species diversity decreases from west to east in both the Atlantic and Pacific Oceans. Furthermore, faunas in the Pacific Ocean (e.g. coral reefs) are on the whole more diverse than in the Atlantic Ocean.
- Water motion in the surface waters of the oceans (down to a depth of about 600m) is driven primarily by winds and tides. Below approximately 600m, water motion and horizontal currents are much slower and driven primarily by density differences.

Introduction

Understanding of the structure and function of the oceans is foundational to their effective management. For example, an understanding of ocean temperature and ocean chemistry is fundamental to comprehending the impacts of greenhouse gas emissions on marine environments (e.g. sea level rise, ocean acidification). This knowledge also contributes to evaluating opportunities to either mitigate (reduce) greenhouse gas emissions or adapt (respond to) to climate change. Another example is the need to understand how deep-seabed mining may affect marine environments. This requires knowledge of the oceanographic, physiographic and biological structures and processes that contribute to the ecological health of the deep-sea and associated communities.

The unique structure and function of the ocean suggest that approaches to terrestrial management may not be successful when applied in the marine environment. For example, management approaches to emissions (e.g. pollution, noise) from terrestrial transportation activities (e.g. aircraft, vehicles) have no marine analogue. Noise can have far reaching implications to marine mammal communities, and pollution from ballast water discharges can have significant economic and environmental impacts, should species transported across long distance become established. Another example is the difference in complexity between marine and terrestrial food webs. Terrestrial approaches to wildlife management utilizing models with few trophic (food web) levels may have limited utility in marine environments that are comprised of five to six trophic levels, which shift according to annual and decadal oceanographic regimes.

This chapter provides a short overview of the structure and function of marine environments and, in doing so, informs discussions on marine management approaches in the following chapters. Knowledge of the workings of marine environments is necessary to understand how marine management can overcome humanity's biased 'land referenced' perspective. The chapter is not a comprehensive treatise on oceanography or marine biology, and readers interested in a more detailed treatment should refer to, among others, Valiela (2015), Barnes and Hughes (1999), Bertnes et al. (2001), Knox (2001), Sverdrup et al. (2003), and Mann and Lazier (2005).

Similarities and differences between marine and terrestrial environments

Marine systems differ significantly from terrestrial systems. Consequently, attempts to manage marine environments simply by applying terrestrial theories, paradigms and concepts are likely to fail. For this reason, this section provides a brief overview of the similarities and differences between terrestrial and marine ecosystems (adapted from Day and Roff, 2000).

Similarities between terrestrial and marine ecosystems

At a very broad conceptual level, marine and terrestrial systems do have some similarities:

- Both are composed of interacting physical and biological components with energy from the sun driving almost all ecosystems.
- Both are complex patchworks of differing environments and habitats that are occupied by different communities and species.
- Both marine and terrestrial species show gradients in diversity (number of species) with latitude – generally species diversity, at least in shallower marine environments, increases with decreasing latitude.
- In both types of ecosystems, the primary zones of biological activity tend to be concentrated nearer the surface (i.e. the sea – air or land – air interface).

Differences between terrestrial and marine ecosystems

Difference in size

The oceans are far larger in area than all the land masses combined, covering 71 per cent of the earth's surface. The Pacific Ocean alone could easily contain all the continents. Even more marked is the difference in vertical extent and volumes between marine and terrestrial habitats. Life on land generally extends from a few metres underground to the tops of trees – a vertical extent of perhaps no more than 30–40 metres. Although birds, bats, insects and bacteria may periodically rise above these heights, the atmosphere above the trees is a medium of temporary dispersal only. Since species must return to the terrestrial environment for resources, reproduction and shelter, it is not considered a habitat. In contrast, the average depth of the oceans is 3,700 metres and contains life at all depths. The sea's habitable volume is therefore hundreds of times greater than that of the land. Though difficult to measure exactly, some estimates put the global ocean as occupying about 95% of all habitat on the earth.

Difference in physical properties

Terrestrial and marine ecosystems differ markedly in their physical properties. For example, the water overlying the seabed is 60 times more viscous than air and has greater surface tension. Water is also about 850 times denser than air; this provides buoyancy and allows organisms to survive without the need for powerful supporting structures. Buoyancy enables organisms to exist in the sea that are morphologically and anatomically very different from those on land. Seawater's buoyancy and viscosity keep food particles suspended and result in an environment – the pelagic realm (Box 2.1) – without an analogue in the terrestrial environment.

Box 2.1   Divisions in the marine environment

The ocean is composed of two fundamentally distinct habitats: the pelagic (water column) and the benthic (seafloor). A secondary division is the intertidal or estuarine realm which exhibits characteristics of both the pelagic and benthic realms along with contributions (inputs) from terrestrial environments.

Pelagic realm

The pelagic realm (pelagos meaning 'open sea' in ancient Greek) is the water column itself and all the organisms that inhabit it. It is massive, with a volume of $1,330 \times 10^6 km^3$, an average depth of 3.68km and a maximum depth of 11km (Charette and Smith, 2010). A fully three-dimensional world, the pelagic realm is capable of functioning independently from either terrestrial or benthic environments. It generates its own primary production from planktonic organisms that, in turn, often engender complex food webs with birds, sharks and marine mammals as top predators.

Benthic realm

The benthic realm (benthos meaning 'depths of the sea' in Greek) is loosely defined as the seabed (or lakebed), sediments on the seabed and adjacent waters and their associated biological community. Benthic environments are primarily structured by depth, substratum

> (bottom type) and nutrient inputs from the water column. These inputs are driven by both abiotic (e.g. water motion) and biotic mechanisms (e.g. carbon flux raining down from above) at broader scales and biological interactions (e.g. competition, predation) at finer scales. Nearshore benthic environments generally exhibit higher energy, and hence higher spatial and temporal variability, whereas offshore deep environments generally experience lower energy and variability.
>
> **Intertidal and estuarine environments**
>
> Intertidal environments, while fundamentally a type of benthic environment, are those coastal marine areas that are exposed during low tides. As such, they are structured by atmospheric, marine and terrestrial (e.g. runoff) processes and exhibit characteristics of both marine and terrestrial environments. Estuarine environments may be intertidal environments, but they are defined as areas at the confluence of freshwater and marine systems. Estuarine environments are characterized by low salinities, high biological productivity and low species diversity (i.e. few species adapted for this brackish water environment).

Differences in temperatures

The strong seasonal and inter-annual fluctuations in the terrestrial climate contrast with the much more moderate fluctuations in the marine environment. Seawater has much greater heat capacity than air; therefore, temperatures change more slowly than they do on land. The higher viscosity of seawater also causes it to circulate more slowly than air. As such, 'seasons' for the ocean are more often determined by solar energy that triggers primary production, allowing for growth of all organisms throughout the food web, than by temperature, though it also plays a role, particularly at high latitudes.

Light and vertical gradients

Compared to the terrestrial environment, a far greater proportion of the ocean is light-limited. Due to changes in availability of light and nutrients, only the upper ~50m and ~200m of the oceans – the photic and mesophotic zones, respectively – will support the growth of primary producers and light-dependent symbiots (Figure 2.1). Light is necessary for primary production through photosynthesis, both within the water column and in the benthos. The exceptions are a few unique communities such as hydrothermal vents that support chemosynthetic producers that gather energy from the sulphur-rich soup of chemicals pouring out of the vents. Notwithstanding these exceptions, essentially no organic matter is produced within the vast unlit depths of the ocean (the dysphotic and aphotic zones). In these dark areas, the entire biological economy of the sea is dependent on the low-density flux of detrital organic matter from the productive surface layer.

Mobile and fluid nature of water

The fluid nature of the marine environment means that most marine species are widely dispersed, and individuals can be far ranging. In addition to enhancing cross fertilization and dispersal of larvae and other propagules, water motion also enhances the migration and aggregation of marine species (especially those in pelagic systems). Water also dissolves and circulates nutrients. Even marine species that can be considered static as mature benthic forms (e.g. many molluscs

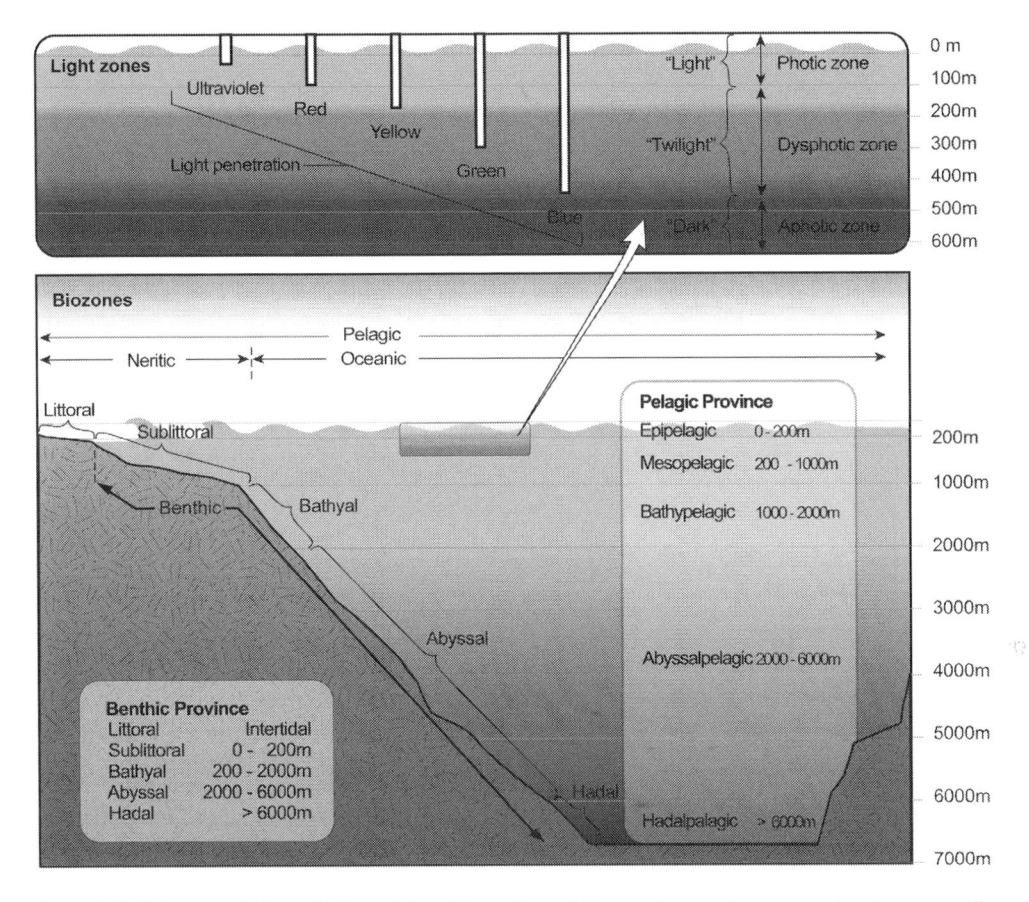

Figure 2.1 Diagram of the pelagic and benthic realms of the marine environment, showing generally recognized vertical depth and light zones

and seaweeds) usually have highly mobile larval or dispersive reproductive phases within the plankton of the pelagic realm. Seabed sediments in shallower depths and on slopes are often mobile to a degree that makes land erosion cycles seem slow and mild. On the other hand, in the deep-sea, tracks from scientific vehicles on the seafloor can still be seen decades later, making land cycles seem lightning fast!

Circulation differences

Although both terrestrial and aquatic environments exhibit patterns of circulation (i.e. circulation of the water in the oceans and of the atmosphere above the land and water), the two are not very similar. In the oceans, the medium – water – contains the organisms themselves; aquatic organisms live in the medium, flow with it and are subject to its physics and chemistry. In contrast, the presence of organisms in the atmosphere is strictly temporary. Thus, in the oceans, species and communities can be passively transported long distances, viably existing in the pelagos, whereas on land, colonization of distant areas generally requires active migration and passive dispersal through the atmosphere, which is not a long-term viable habitat (except for some microbes). For benthic communities, particularly 'suspension feeders', movements of the overlying waters are essential for the transport of their food resources, whereas this is generally not the function of the atmosphere for terrestrial communities.

Differences in primary production

Perhaps the most obvious biological difference between terrestrial and marine ecosystems is the types and source of primary production (i.e. photosynthesis, Steele, 1985). In terrestrial ecosystems, primary producers (mainly vascular plants such as trees and grass) constitute the great majority of biomass, and individual organisms are often large and long-lived. In contrast, the dominant primary producers in marine ecosystems – the phytoplankton – are generally microscopic and can reproduce rapidly before dying or being eaten. Consequently, they have much higher turnover rates compared to terrestrial ecosystems (e.g. forests or grasslands). In terrestrial ecosystems, the biomass of primary producers tends to be highly conserved (e.g. in woody plants). In contrast, in marine ecosystems, the biomass of primary producers is rapidly processed, either by consumers or reducers (e.g. bacteria). As a consequence, marine sediments are typically much lower in organic carbon than terrestrial soils and become progressively lower still with increasing depth. Even marine macrophytic plants (e.g. macroalgae and seagrass angiosperms), though often perennial, have short generation times compared to terrestrial plants. No marine plants have the longevity of terrestrial gymnosperms and angiosperms. At lower latitudes, the partially seawater-adapted mangroves (emergents) are recent re-invaders of marine waters, straddling both worlds.

Taxonomic differences

In the oceans, there is a high diversity of divisions among the algae, comprising many phyla of microscopic unicellular or multicellular forms (see below and Table 2.1). This in fact represents a much greater diversity at the level of division (phylum), and therefore genetic diversity, than in the terrestrial environment. This is often not recognized because of our bias and perception of the overwhelming diversity of flowering plants on land – which nevertheless all belong to a single division – the Angiosperms (or Anthophyta). Bryophytes, Pteridophytes and Gymnosperms are all absent in marine waters, and Angiosperms are significantly under-represented, mainly by a few species of seagrasses and mangroves.

Of the animal phyla, all are represented in the oceans in one way or another. On land, however, several of the major phyla are completely absent, having failed to adapt to the more demanding climate. Although the greatest species diversity undoubtedly lies within the terrestrial insects, the greatest phyletic (and therefore genetic) diversity is clearly among the flora and fauna of the oceans.

Temporal and spatial scales

Other important differences between terrestrial and marine ecosystems result from – or are associated with – the temporal and spatial scales of ecological responses to changes in the physical environment. In the oceans, these changes are related to the fundamental differences in the communities of primary producers. The high storage of biomass in terrestrial organisms and organic detritus tends to decouple biological and physical processes. In the oceans, the space and timescales of physical and biological processes are nearly coincident (at least in the pelagic realm), such that biological communities can respond rapidly to physical processes. Spatial variation and distributions of primary producers on land are largely related to topography and soil;

at the same scale, they change slowly compared to marine primary producers. On land, largely as a function of the longer generation times of larger organisms, population and community cycles may depend more on biological processes than on immediate physical processes.

Relatively short response times to environmental perturbations

As a result of the mobility, fluidity and interconnectedness of marine waters, environmental perturbations (e.g. oil spills, the introduction of toxic substances, toxic algal blooms) can be rapidly dispersed throughout the ocean environment. Depending on the properties of a pollutant, this can readily occur in two or three dimensions.

Boundary differences

Compared to their marine counterparts, terrestrial environments have more pronounced physical boundaries between ecosystems. Especially at finer scales, it may be difficult to identify distinct boundaries in marine systems. This is due to the dynamism of marine systems and because pelagic and benthic realms require separate consideration. This does not mean that there are no distinct marine ecosystems, but, generally, their boundaries are more 'fuzzy' or transitional. Thus, the concept is less useful in marine than in freshwater or terrestrial environments.

Longitudinal diversity gradients

In addition to the latitudinal diversity gradients observed in both terrestrial and marine communities, there is also a longitudinal diversity gradient in the marine environment. Species diversity decreases from west to east in both the Atlantic and Pacific Oceans. Furthermore, faunas in the Pacific Ocean (e.g. coral reefs) are on the whole more diverse than in the Atlantic Ocean (Thorne-Miller and Earle, 1999) (Figures 2.2 and 2.3). The greater diversity in the western part

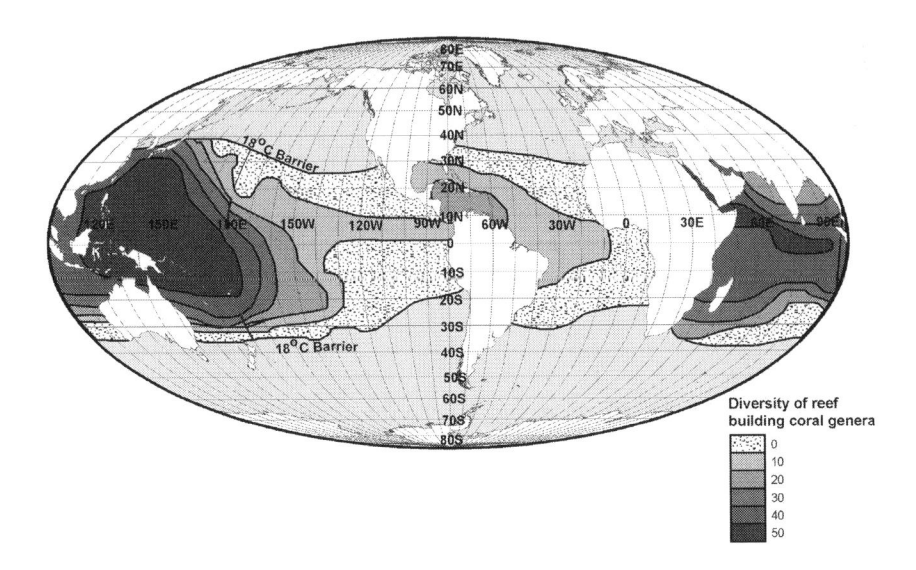

Figure 2.2 Global distribution of diversity of reef building coral genera

Source: Adapted from Stehli et al. (1967)

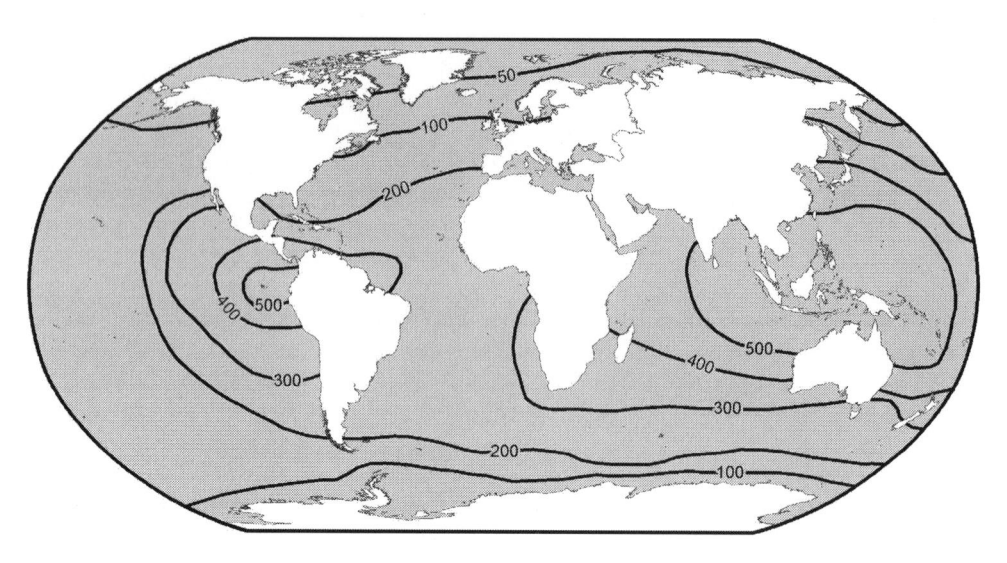

Figure 2.3 Global species richness of bivalve molluscs
Source: Adapted from Stehli et al. (1967)

of the basins can be related to the eccentric circulation pattern of the two oceans, whereby the west receives waters from low latitudes (with higher species diversity) at high velocity (e.g. the Gulf Stream and Kuroshio), while the eastern part of the basin receives lower velocity water from higher latitudes (with lower species diversity). The greater overall species diversity of the Pacific is generally explained in terms of its greater geological age.

Diversity at higher taxonomic levels

At higher taxonomic levels, the diversity of marine fauna is much greater than the diversity of terrestrial fauna. All phyla of animals are represented in the oceans. Some taxa, fishes for example, are extraordinarily diverse while others are less diverse (although in the 'lower' phyla many species remain to be described, particularly in the deep pelagos and benthos). Most marine communities are also highly patchy and variable in species composition.

Physiographic characteristics of the marine environment

Physiographic (also known as 'geomorphology') characteristics are those features broadly recognized as 'marine landform' – in essence, related to the topography and substrates of the seafloor. The physiography of the shoreline and seabed directs the flow of tides and currents and shapes the broad character of benthic biological communities (along with the overlying geological, biogeographical and oceanographic contexts). Primary physiographic characteristics of the marine environment are discussed below.

Horizontal divisions and bathymetry

In the horizontal dimension, the marine environment is generally regarded as divisible into several major provinces (Figure 2.1). Coastal waters comprising the coastal zone (generally defined as less than 30m in depth) extend seaward from the high water level and include the littoral zone. Near to shore are various kinds of inlets including estuaries, bays and coves and associated wetlands. Estuaries are the meeting place of freshwaters with the ocean, where the salinity is measurably diluted by freshwater runoff (see above). Bays and coves are shoreline concavities of the ocean where salinity may not be diluted (unless they are bays *within* estuaries). The coastal regions run into the neritic province of sub-littoral waters, which is the region of the ocean that lies above the continental shelf out to depths of 200m. Although (in some places) the edge of the continental shelf corresponds to the exclusive economic zones (EEZ) of states in some locales, there is no relationship between the two. The neritic province then merges into the oceanic province at the shelf edge or the shelf break. The oceanic province comprises those vast areas of the oceans that physically lie beyond the edge of the continental shelves and whose waters exceed depths greater than 200m.

Depth, light and pressure

The next major division of the oceans is made in terms of depth, where a variety of terms are used to describe the habitats and their biological communities. The conventional descriptive divisions of the oceans with respect to depth, which have long been recognized, are shown for both the pelagic and benthic realms in Figure 2.1. Depth is an important factor in both the pelagic and benthic realms. In combination with temperature, salinity, light and pressure (with which it co-varies) depth defines the distributions of major community types.

Somewhat arbitrarily, the oceans are vertically subdivided into the epipelagic (down to 200m), the mesopelagic (200 to 1,000m), the bathypelagic (1,000 to 2,000m) and the abyssal/ hadal zones (> 2,000m). Similar terms are applied to the benthic realm (see Figure 2.1). In the oceans, light intensity diminishes exponentially with depth. In the vertical dimension, and fundamentally for the photosynthetic organisms, the oceans can therefore be subdivided as follows: The euphotic (= photic, or well lighted) zone is the region in which sufficient light penetrates to allow net photosynthesis and plant growth to occur. Below this is the mesophotic or dysphotic (poorly lighted) zone where light is still present, but its intensity is too low to support plant growth, though it still fuels some microbes and symbiots. Below the dysphotic zone, the great majority of the oceans' depths lie within the aphotic zone, where no light penetrates (Figure 2.1).

The **compensation point** occurs at the bottom of the euphotic zone, a depth below which the rate of photosynthesis exceeds the rate of respiration. The actual depth of the euphotic zone increases with water depth itself, from the coast towards the edge of the shelf and into oceanic waters, and it also varies at different times of the year. For example, in estuaries, the euphotic zone may be less than 2m in depth, in average coastal waters it approximates 30–50m, while in oceanic waters it may exceed 200m. In contrast, in the Arctic Ocean, the euphotic zone may exceed 100m during the spring and suddenly decrease to only a few metres during the summer phytoplankton bloom.

Depth is also a surrogate for **pressure**. The increase in pressure with depth has a significant impact on organisms. With every 10 metres of depth, the water pressure increases by approximately one atmosphere (with the greatest change from 0 to 1 atmosphere occurring in the top ten metres). Additional physical, chemical and biological changes lead to a decrease of dissolved oxygen and increase of dissolved carbon dioxide (see below). Organisms that live in the deeper regions of the oceans are adapted to these physical conditions of high pressure, low temperatures and dilute resource concentrations and rarely move into the epipelagic region.

Temperature also decreases with depth, from ambient surface values to a nearly constant 0–4°C in the deepest oceanic waters. Conversely, salinity typically increases with depth. Concentrations of particulate organic carbon (the detrital flux from the euphotic zone) also decrease exponentially with depth, while oxygen concentrations decline and carbon dioxide concentrations increase with depth. Depth is therefore an index of a variety of concurrently changing physical and chemical conditions, which collectively influence the nature of biological communities.

Oceanographic characteristics of the marine environment

Oceanographic structures

Temperature

Temperature is often regarded as the most critical factor governing the distribution and behaviour of organisms on earth. There is considerable scientific literature on the effects of temperature on the distribution, physiology and behaviour of aquatic organisms. All marine organisms show some pattern of limitation of their distribution according to temperature, and very few species have geographical ranges covering the entire range of observed aquatic temperatures.

Temperatures are relatively stable within the oceans, and there is far less seasonal fluctuation than in terrestrial systems. Seasonal variation in temperature is generally low at both high and low latitudes, but fluctuations may be higher at mid-latitudes – e.g. from 0 to +20°C. The temperature in the oceans varies from −2°C in Polar Oceans to +32°C in tropical waters. The lower temperature is set by the freezing point of seawater. Due to its salt content, the freezing point of seawater is depressed below that of freshwater (by definition −0°C) and the temperature of freezing is proportional to the salinity. The temperature of maximum density lies below the freezing point for all open ocean seawaters; it is not +4°C as for freshwaters. This means that seawater of a given salinity will continue to sink below warmer waters of the same salinity as it cools.

The oceans are vertically thermally stratified by a 'permanent' thermocline. To a first approximation, ocean waters can be divided into an upper wind-mixed layer – from the surface to a depth of 600 to 1,000m. Here temperatures range from +8°C to +32°C. Temperatures of the lower layers – from 600 m to the bottom – are generally well below those at the surface, averaging between 0 and +4°C. The temperature of most marine waters therefore lies between 0 and +4°C. Water temperatures in the deeper layers are also more constant than in surface waters and may be significantly influenced by ocean currents. Bottom topography also has an influence on water temperature, with pockets of warmer (but more saline) waters often being trapped for periods of time in large basins. Deeper colder waters can also be transported towards the surface in upwellings along ridges and continental margins.

With its impact on growth and its complex relations to biodiversity and nutrient supply (see below), temperature is a major determinant of marine communities. However, unlike on land, it is generally much more constant, hence the communities are also more constant. Exceptions are discussed below.

Ice cover and scour

Ice cover may be permanent, seasonal or absent. Its extent and development are a substantial influence on community types over broad geographical areas. Not surprisingly, the presence of permanent ice greatly restricts marine productivity. Through a variety of mechanisms, seasonal ice has a major impact on seasonally eliminating or enhancing production. The physical effects of ice scour and glaciers have the greatest impact on shallow water marine communities, especially in intertidal regions, primarily by reducing community diversity or by restricting some organisms to crevices. Except for inshore areas – or in Polar Oceans where effects of rafted ice can reach to considerable depths – ice does not have an impact on the sub-tidal benthos.

Temperature gradients and anomalies

Temperature anomalies are well documented as being associated with several types of regions. Upwelling regions, carrying nutrient-rich waters to the surface, also carry a signature of surface temperature lower than surrounding waters. While the geographic location of upwelling events is predictable, the timing may not be and may depend on unpredictable meteorological events. Upwellings may be seasonally or annually variable in development; the water column is typically vertically mixed, and stratification is weak or absent.

Salinity

The salt content of virtually all parts of the open ocean lies between 32 and 39psu (**practical salinity units** or parts per thousand – also written as ‰, or ppt) and the chemical composition of the principal ions of seawater remains virtually constant. This is because the rate of input and export of salts to and from the oceans is low compared to the rate of global mixing. The salinity of surface waters around the globe correlates strongly to the difference between latitudinal variations in precipitation and evaporation and to the rate of mixing with sub-surface waters. Salinity variations with depth also contribute to vertical stratification and stability of the water column (see below).

Variations in salt content over the range of 32 to 39psu have little discernible effect on marine organisms. However, the majority of marine species are **stenohaline** (adapted to constant salinity) and are intolerant of salinity changes greater than this. Although, in the past, considerable attention has been devoted to the relationships between distributions of organisms and salinity, other factors such as temperature and the movements of water masses dictate the distributions of marine organisms.

Only in estuaries, estuarine bays, (semi-) enclosed seas (e.g. the Baltic) or sheltered mountainous coastal regions experiencing high rainfall (e.g. southern Chile) – where salinity varies from close to zero to over 33psu – does salinity have major effects on the distribution of aquatic species. This is because of its significance in osmotic and ionic regulation. The majority of aquatic species are adapted either to a life in freshwaters at low but relatively constant salt content or to the much higher, nearly constant salinities of the oceans. Relatively few species (**euryhaline**) are well adapted to life in estuaries at intermediate and fluctuating salt content. In estuaries, salinity exhibits complex behaviour, both physically and in its impact on the distribution and abundance of organisms (see below).

Composition of seawater and nutrients

Seawater is a complex solution containing virtually every element of the periodic table. This is a testament to the high dissolving power of water. Despite variations in total salt content (the salinity), the composition of seawater (that is the ratio of the major elements to one another) remains virtually constant. This is occasioned by the fact that the concentrations of most elements are high relative to their rate of use by the biota. The exception to the rule is for elements (or ionic compounds) called nutrients.

Nutrients in marine environments are those inorganic substances (PO_4, NO_3, NH_4, Fe and Si) that are required for, and can control the rate of, aquatic plant growth. Only nutrients within the euphotic zone are significant to our understanding of processes for marine management purposes. Nutrient concentrations below the euphotic are much higher than within it. Not until these nutrients reach the surface waters in upwellings can they be exploited by primary producers.

Nutrients in aquatic environments are derived from the atmosphere, from land drainage, by recirculation of deep waters to the surface, underwater volcanos and vents and from internal regeneration. Various components of the marine environment obtain their nutrients by different processes and from different predominant sources.

In the open water pelagic realm, nutrient inputs are predominantly a function of the mixing regime and/or internal regeneration by the food web. Atmospheric inputs are assumed to be uniform over macroclimatic regions. Terrestrial inputs to the oceans are low compared to internal recycling, except in estuaries and the nearshore coastal zone. Except in the immediate coastal zone, internal regeneration and recirculation of nutrients within the ocean thus far outweigh the significance of external inputs.

A major (internal) source of nutrients to coastal areas is upwelling or entrainment of subsurface water. While the geographic location of these events is largely predictable, the timing may depend on unpredictable meteorological events. Such water mixing regimes may be evident from local low-temperature anomalies.

Within water masses, the nutrient regime is a function of whether the water column is stratified or mixed, as described by the stratification parameter (see below).

Oxygen and other dissolved gases

Within the neritic and pelagic realms of the oceans (and larger lakes), dissolved gases are normally at or close to saturation in surface waters. This is because physical processes that cause exchange of gases between water and atmosphere typically overshadow in magnitude the biological processes of photosynthesis and respiration. The wide diel (daily) variations of gas content, typical of smaller bodies of water, are generally not observed. Even below a thermocline, oxygen rarely becomes naturally seasonally depleted within the water column. The exceptions to this are stratified inlets and estuaries where there are major inputs of organic matter and the occurrence of summer stratification. In oceanic tropical and subtropical waters, an oxygen minimum layer develops at a depth of between 600 and 1,000m in both the Atlantic and Pacific oceans (more highly developed in the latter).

Low oxygen concentrations – often engendered by human actions – are prompting considerable concern for several regions around the globe. These hypoxic (reduced oxygen) and anoxic areas (virtually no oxygen – also termed 'dead zones') are caused by the inputs of dissolved nutrients and suspended solids from land runoff (Diaz and Rosenberg, 2008). Perhaps the largest of these areas lies in the Gulf of Mexico. Here, large areas to the south and west of the Mississippi River have become anoxic following algal blooms caused by terrestrial and freshwater nutrient inputs into coastal waters, largely originating from upstream agriculture.

Water masses and density

The combination of temperature and salinity determines the density of seawater. Higher temperature reduces density and higher salinity increases it. Outside of estuarine environments, density itself is of little direct biological consequence in the oceans because it varies only within narrow limits. Density is, however, of major global significance for physical oceanographic processes as a determinant of ocean thermohaline circulation (see below).

Certain 'usual' or 'habitual' combinations of temperature and salinity (defining the water masses) persistently occur, and the distribution of these combinations can be traced over wide swaths of the oceans, frequently for thousands of square kilometres. Water masses can in some respects be considered as analogues of the major climatic regions of terrestrial environments, and they define the extent and influence of major ocean currents (Figures 2.4

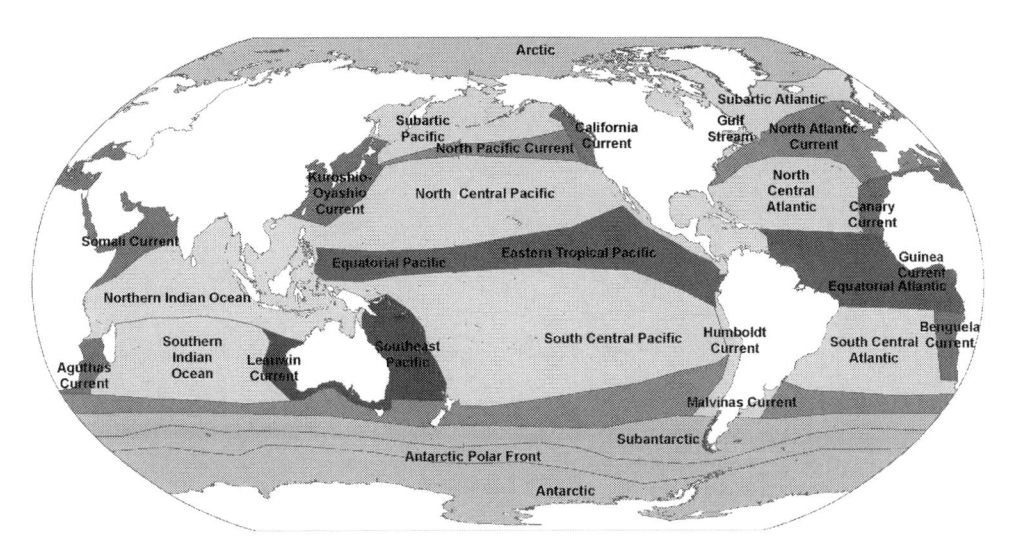

Figure 2.4 Global map of pelagic provinces
Source: UNESCO (2009)

and 2.5). Both temperature and salinity need to be quantified in order to define the origins, movements and extent of the distribution of water masses. A water mass will seasonally change its temperature (salinity is far less seasonally variable) because of atmospheric interactions. Temperature by itself, therefore, is not ideal as a measure or descriptor of water masses or origin of water masses.

Oceanographic processes

Water motions

Natural bodies of water show many types of movements at all spatial and temporal scales. Water motion is an essential feature of the oceans and is essential for all life. It follows that a proper perspective on water movements is imperative to understanding aquatic ecosystem function. Ultimately, three 'forces' cause water motion on the earth: the sun's radiation (causing heating and evaporation/precipitation); the earth's rotation on its own axis (a modifying action rather than a true force); and gravitational effects of the sun and moon. A minor fourth factor is geological (tectonic), which causes tsunamis. Proximately, water motion may be generated by winds, tides or density differences.

Water motion sustains life by replenishing and stimulating the production of resources for organisms of every trophic level. Motion provides the passive transport for the dispersal of organisms (or their larvae) in the broad-scale or in more local patterns of ocean connectivity. The junctions between the gyral systems and major ocean currents contribute to the barriers that separate species and higher taxa. Essentially, these intersections define the major biogeographic provinces and regions of the oceans.

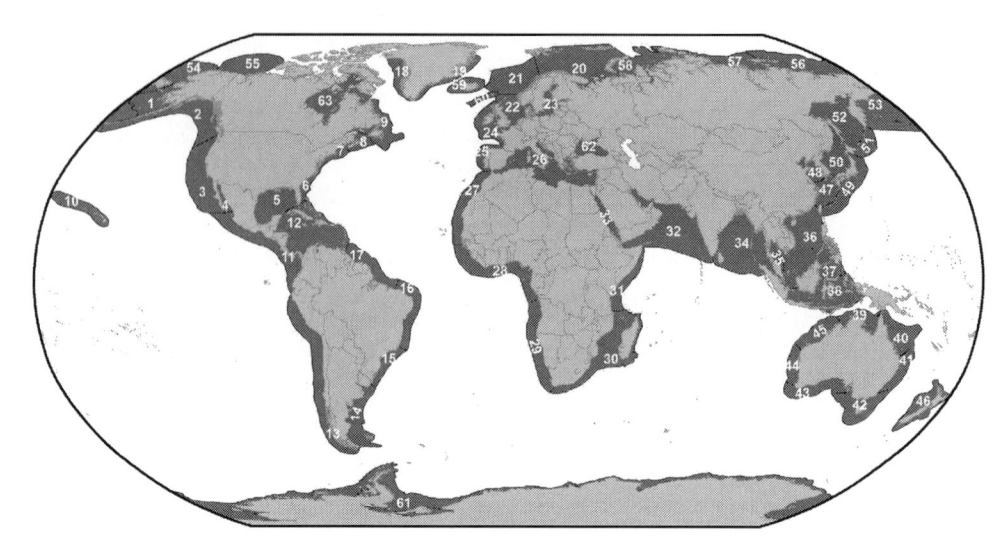

Figure 2.5 Global map of Large Marine Ecosystems (LMEs). LMEs are ocean areas approximately 200,000km² or greater, adjacent to the continents in coastal waters where primary productivity is generally higher than in open ocean areas. The 64 LMEs are delineated based on bathymetry, hydrography, productivity, trophic relationships. LME names are as follows:

1 East Bering Sea
2 Gulf of Alaska
3 California Current
4 Gulf of California
5 Gulf of Mexico
6 Southeast US Continental Shelf
7 Northeast US Continental Shelf
8 Scotian Shelf
9 Newfoundland-Labrador Shelf
10 Insular Pacific-Hawaiian
11 Pacific Central-American Coastal
12 Caribbean Sea
13 Humboldt Current
14 Patagonian Shelf
15 South Brazil Shelf
16 East Brazil Shelf
17 North Brazil Shelf
18 West Greenland Shelf
19 East Greenland Shelf
20 Barents Sea
21 Norwegian Shelf
22 North Sea
23 Baltic Sea
24 Celtic-Biscay Shelf
25 Iberian Coastal
26 Mediterranean Sea
27 Canary Current
28 Guinea Current
29 Benguela Current
30 Agulhas Current
31 Somali Coastal Current
32 Arabian Sea
33 Red Sea
34 Bay of Bengal
35 Gulf of Thailand
36 South China Sea
37 Sulu-Celebes Sea
38 Indonesian Sea
39 North Australian Shelf
40 Northeast Australian Shelf-Great Barrier Reef
41 East-Central Australian Shelf
42 Southeast Australian Shelf
43 Southwest Australian Shelf
44 West Central Australian Shelf
45 Northwest Australian Shelf
46 New Zealand Shelf
47 East China Sea
48 Yellow Sea
49 Kuroshio Current
50 Sea of Japan
51 Oyashio Current
52 Okhotsk Sea
53 West Bering Sea
54 Chukchi Sea
55 Beaufort Sea
56 East Siberian Sea
57 Laptev Sea
58 Kara Sea
59 Iceland Shelf
60 Faroe Plateau
61 Antarctic
62 Black Sea
63 Hudson Bay
64 Arctic Ocean

Sources: Sherman et al. (1992); lme.noaa.gov

Surface ocean circulation

Water motion in the surface waters of the oceans (down to a depth of about 600m) is driven primarily by winds and tides. Below approximately 600m, water motion and horizontal currents are much slower and driven primarily by density differences. The frequency and strength of water movements at the ocean's surface have profound effects on biological communities.

At the global scale, the basic patterns of surface ocean circulation are the same in the Pacific and Atlantic oceans. These patterns consist of sets of currents that together comprise a sub-polar gyre and a subtropical gyre (Figure 2.6). The counterparts of each of these currents in the North Atlantic and North Pacific and in the South Atlantic and South Pacific is very clear. For example, the counterparts of the sub-polar Labrador current and the subtropic Gulf Stream in the North Atlantic Ocean are the Kurile and Kuroshio currents in the North Pacific Ocean, respectively. In all oceans, circulation is eccentric, resulting in a westward intensified current and weaker eastern dispersion. It is the effect of the earth's rotation (described in terms of the Coriolis 'force' or parameter) that produces these various eccentric ocean gyres with their westward, intensified currents.

Each gyre revolves around a central but eccentric hub, simply referred to as a 'central gyre'. The only gyre named specifically is the Sargasso Sea of the North Atlantic (the only sea without a shoreline, but for Bermuda). Each subtropical gyre encloses a relatively warm, salty, clear blue sea, poor in nutrients and of low productivity. Production is dependent primarily on internally recycled NH_4. These are essentially the deep, desert areas of the ocean. The best-known example is the Sargasso Sea, known for its eponymous matts of *Sargassum* algae that offer unique floating habitats and ecological communities, providing protection to juvenile turtles and fish from larger predators. However, other analogous central gyre areas are found in the north and south Pacific Ocean.

In the tropical areas of the oceans, both a north equatorial current and a south equatorial current can be recognized flowing from east to west. In the Atlantic and Pacific Oceans, an equatorial countercurrent is found between these two, running from west to east. The Pacific equatorial countercurrent is 8,000nmi in length and about 250nmi in width. The Atlantic equatorial countercurrent is more variable; it becomes stronger in summer, extending its influence westward towards South America.

Of all the ocean currents, the most pronounced and persistent is the Antarctic circumpolar current (West Wind Drift). It subsumes the sub-polar gyres of the southern oceans and completely

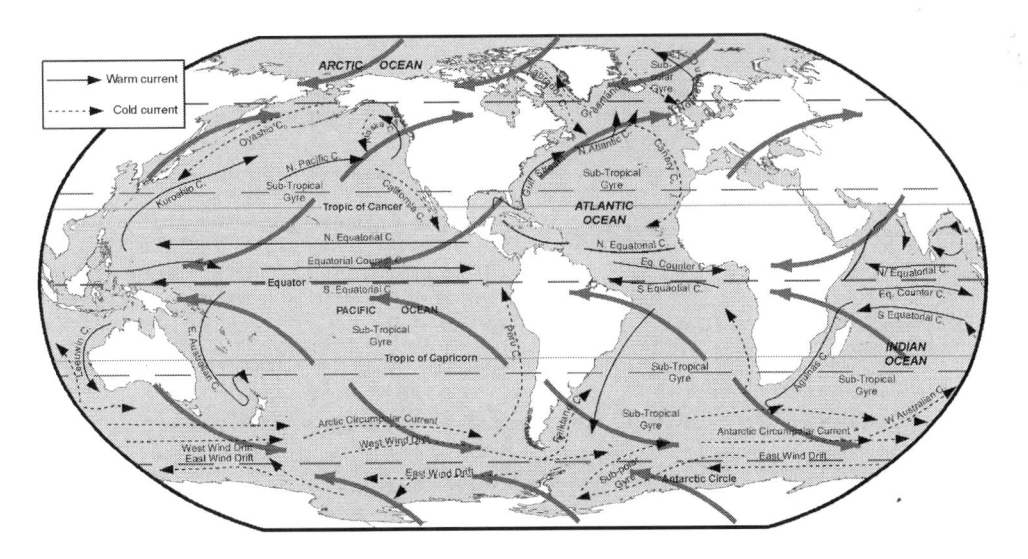

Figure 2.6 Surface ocean circulation and the major surface ocean currents

circles the globe running from west to east. Meanwhile, the most variable surface ocean currents are found in the Indian Ocean. Here, even the major ocean currents show great variability in velocity, direction and latitudinal and longitudinal position. For example, the Somali current flows south in winter but north in summer at high velocity. The north equatorial current reverses direction under monsoon influence; indeed, while flowing east in summer, it is dubbed the monsoon current. The equatorial countercurrent is only observed in winter and disappears in summer when the monsoon current occurs. The major force that causes these changes is the variable winds of the monsoon season, clearly showing the dependence of surface ocean circulation on the atmospheric winds.

Oceanic convergences and divergences

Where the various oceanic gyral systems meet, corresponding to regions of reversal of major atmospheric wind systems, water masses either converge or diverge. That is, at the surface, the circulation pattern induces a series of convergences and divergences where water either descends below the surface or is brought up from below the surface. Where oceanic water masses meet and descend, they are known as **convergences**; where they rise to the surface and disperse, they are known as **divergences**. They are present throughout the oceans but are typically restricted to oceanic waters and more pronounced in the southern oceans. These are typically regions of much higher than average production, at all trophic levels. However, these regions are much more clearly defined in the tropics, subtropics and southern oceans than at northern latitudes.

It is important to realize the differences in the character of oceanic convergences and divergences. Convergences are where surface waters meet and biota may accumulate – thus biomass is increased. They may be areas of rich life that has physically accumulated – that is, it has been advected to the region, not produced there. Rich fish feeding areas may nevertheless correspond to convergence zones. A divergence is not usually so well defined, though it carries a signature of reduced surface temperature. At divergences, nutrient-rich water is brought to the surface, stimulating new production. Perhaps the best-known example is the Antarctic divergence where extremely high levels of primary planktonic production are known.

Coastal upwelling

Production is also high in regions of near-coastal upwelling and divergence, where production is dependant primarily on inputs of NO_3 from deeper waters. If the coast lies to the left of the wind direction, then surface water is transported to the right in the Northern Hemisphere. This is a result of deflection cause by the Coriolis parameter. This necessitates a deep replacement current moving onshore and upwards – that is, upwelling. This upwelling water brings with it much higher concentrations of nutrients which have been biologically regenerated in deepwater. The Coriolis 'force' is thus directly linked to areas of high production. Examples include the Grand Banks of Newfoundland where the Gulf Stream has moved offshore.

In the Southern Hemisphere, the situation is reversed. The best-known example here is the northward flowing Peru-Chile current that flows into the Humbolt current (Figure 2.6). It progresses along the west coast of South America and leads to upwelling of cold deep nutrient-rich water, which, in turn, stimulates the planktonic production. Coastal upwellings may be relatively constant, periodically interrupted (e.g. as in the El Niño Southern Oscillation System off Peru and Chile), or more episodic as in the case of the Benguela current off southwestern Africa.

Deep ocean circulation

Deep ocean currents generally move at a slower pace than surface waters and are driven by small differences in density caused by both temperature and salinity variations (thermohaline circulation). Formation of the deep and bottom waters of the oceans occurs only in two ways. Saline water is transported to high latitudes and is subsequently cooled without significant dilution and then sinks. Cold surface water freezes; as it does, it forms ice (above the eutectic point where solvent alone freezes and eliminates the salt solutes) so that the salinity of the remaining water is increased and consequently sinks. The first process is of greatest importance in the North Atlantic between Iceland and Norway. The second is around Antarctica, primarily in the Weddell Sea region. These are the only two regions of the world where deep and bottom waters are formed in any quantities.

Water sinking in the North Atlantic makes its way south towards the Antarctic. Here it meets water sinking from the Weddell Sea where it is re-cooled and moves off to the Indian Ocean, the North Pacific and then the South Pacific. As deduced from carbon-14 dating techniques, this entire journey takes some 1,500 years. Despite the long time that these waters have been out of contact with the atmosphere, they still retain significant levels of oxygen. This indicates the sparse nature of life in the depths of the oceans, and its low metabolic rates at low temperatures.

Tidal amplitude and currents

Tides are caused by the gravitational effects of the moon and the sun. In the absence of land, the world ocean would experience semi-diurnal tides of a periodicity of 12.25 and 24.50 hours and a spring-neap cycle of variable amplitude depending on the conjunction of the moon and the sun. Regionally, both tidal periodicity and tidal amplitude depend on the physical dimensions of the ocean basin. Some locations in the world experience only one tide per day (diurnal) and others up to four per day (quadridiurnal). Which type occurs depends on the location of the amphidromic points (geographic points in the middle of ocean basins around which tides circulate) and which of the sun's and moon's gravitational influences most nearly reinforce the natural period of oscillation of the regional ocean basin.

Tidal amplitude is a consequence of the natural period of oscillation of a regional ocean basin, sea or bay. Canada's Bay of Fundy and its extension, the Minas Basin, experience the highest tides in the world because the natural period of oscillation of the Bay of Fundy is approximately 13 hours. This corresponds closely to the moon's gravitational period (12.25 hours), creating a natural resonance with the waters in the Bay of Fundy that induces spring tide amplitudes up to 16m. In contrast, the natural periods of oscillation of the basins in the Mediterranean Sea are about 9.30 hours. Here, there is interference between gravitational period and natural period of oscillation. The result is tides of only a few centimetres.

Currents in the oceans can originate from several processes, including general ocean circulation, density-driven origins of water masses, tides and winds. In coastal waters, currents due to tides are generally the most predictable and best documented. Tidal amplitude, together with bottom slope, determine the vertical and horizontal extent of the intertidal zone. Tidal currents and bottom slope determine, in part, the substrate characteristics, which, in turn, determine the extent and type of benthic communities. A major distinction between substrate types and associated benthic communities can be recognized as follows: erosional areas, where current speeds are high and material is removed from the substrate resulting in hard bottoms (rock, boulders, gravel etc.), and depositional areas, where water movement is sufficiently slow that particles sediment out, resulting in soft bottoms (mud, silt, etc.).

Stratification and mixing regime

All temperate marine and freshwaters typically mix vertically in spring and autumn and may or may not become stratified in the summer. The seasonal cycle and timing of stratification has major implications for the productivity regime of a region. Two forces cause near-surface water mixing: tides and winds. The effects of wind mixing are difficult to formulate, but they are more uniformly distributed than those of tides. In the oceans, when the water column depth exceeds ~ 50m, winds cannot prevent the water column from stratifying during the annual heating cycle; however, tidal action can. In freshwaters, tides are insignificant, and temperate lakes over approximately 50m in depth always stratify.

Polar waters typically mix in early summer and then become stratified. Mainly, this results from vertical salinity differences. In sub-Arctic waters, stratification can be due to a combination of salinity and temperature effects. In temperate and subtropical waters, stratification of the water column is primarily due to temperature; however, it may also result from a change of salinity. In coastal waters, this may be associated with seasonal freshwater runoff from land, and it can also be related to the cycle of primary production.

Tsunamis, storm surges, hurricanes and water spouts

These massively destructive atmospheric events are now more-or-less predictable in terms of location of impact and scale of effect, at least on the time scale of days. Although considerable local destruction to coastal marine communities may result from these phenomena, over a period of years (for most intertidal communities) to decades (for coral reefs, mangroves) the local biological communities are generally re-established. Such phenomena, although with profound local effects in shaping marine communities, should be thought of as resetting the ecological 'clock of succession'. Climate change, however, has added an extra degree of variability, and some locations unaccustomed to such high-energy events are now experiencing them for the first time in recorded history. The consequences for these ecosystems are still unclear but will likely be longer term than for those places that have over millennia adapted to such events.

The marine biological environment

The flora and fauna of the oceans range in taxonomy and size from minute marine viruses to large marine mammals and are generally organized into a series of more-or-less definable communities outlined in Table 2.1. Overall, the great majority of phyla of multicellular animals are found only in the marine environment.

This section discusses the major biological community types within the pelagic and benthic realms.

Marine biology of the pelagic realm

The pelagic realm supports a diversity of organisms and life forms that have no dimensional or community counterparts on land. Pelagic organisms inhabit a fully three-dimensional realm; most of its species inhabit it permanently or some temporarily as larvae. Within the realm, conditions vary dramatically with depth, changing the array of organisms present and community structure. Pelagic species may also occupy different trophic levels and different stages

Table 2.1 Summary of some major marine communities

Biological realm	Major community 'units'	Community type
Fringing communities	Estuaries	
	Non-biogenic communities	Rocky shores
		Tidal flats
		Sub-tidal soft bottoms
		Shorebirds
	Biogenic communities	Biogenic reefs
		Corals
		Kelp bed Macrophytic algae
		Mangroves
		Seagrasses
		Salt marshes
Pelagic	Pelagic communities	Phytoplankton
		Holozooplankton
		Meroplankton
		Fish
		Diadromous fish
		Seabirds/Waterfowl
		Marine reptiles
		Marine mammals
Benthic	Benthic communities	Demersal fish
		Territorial fish
		Epi-benthic zoobenthos
		In fauna burrowing zoobenthos
		Cold seeps
Deep-Sea		Deep-sea corals
		Sponge beds
		Soft corals (Gorgonians)
	Seamounts	
	Hydrothermal vents	
	Abyssal plains	
	Trenches	

of development. The 'fringing communities' of the benthic realm – intertidal and sub-tidal communities within the euphotic zone – that contain various photosynthetic plants could be considered the functional equivalent of terrestrial communities, where species either live on the ground or are dependent on it for habitat and resources. However, a significant difference between the benthic realm and terrestrial environments is that most of the benthic realm lies below the photic zone and is therefore devoid of its own primary producers (except at hydrothermal vents). It must therefore rely on detrital resources settling from the photic zone and hence has no terrestrial community counterpart.

For descriptive purposes, the pelagic realm can be subdivided into the **planktonic** community of smaller organisms and the **nektonic** assemblage of larger, more mobile species. The plankton comprises all those aquatic organisms that are 'suspended' freely in the water mass. They drift passively in the water currents, and their powers of locomotion are insufficient to enable them to move against the horizontal motion of the water. However, many planktonic species may make extensive vertical movements. One specialized component of the plankton is the **neuston**, which comprises a diversity of organisms inhabiting the immediate water surface.

The nekton is composed of those actively swimming consumers that constitute the middle and upper trophic levels of marine ecosystems. Among these, the dominant organisms are many types of fish (e.g. herring, tuna). They increase in size up to, and including, marine reptiles and mammals. Some species of fish, reptiles and marine mammals make extensive seasonal migrations, while other fish species make regional movements. The smaller components of the nekton, which may also make seasonal migrations, comprise invertebrates such as euphausiids. The chief feature separating nekton from plankton is their relative mobility and their ability to swim or migrate contrary to ocean currents. The distinction between plankton and nekton is by no means absolute. For example, while being nektonic as adults, fish larvae are members of the plankton.

Because of their ubiquity, plankton has generally not been considered worthy of management attention until recently. Nevertheless, an appreciation of their global significance is warranted. As in other communities, the plankton comprises primary producers, primary and secondary consumers and decomposers. The primary producers, the **phytoplankton**, are unicellular or chain-forming micro-algae, ranging from $< 1\mu m$ to $> 1mm$ in diameter. Functionally, they are often divided into **picoplankton** ($< 2\mu m$), **nannoplankton** (2–$20\mu m$) and **netplankton** ($> 20\mu m$). Phytoplankton are the dominant photosynthetic organisms on the planet, yet they remain virtually unknown by most humans.

Within the oceans themselves, the phytoplankton are responsible for > 95 per cent of the annual primary production. In temperate waters, the seasonal cycle of primary production within the plankton is typically highly pulsed. It typically peaks in the spring with an outburst of diatom growth, as the water column becomes **stratified**. Seasonal cycles are generally more extreme at high latitudes (e.g. in the Arctic) and in coastal waters. Cycles are less modulated in the tropics and in offshore oceanic waters. Within the pelagic realm, the rate of primary production is controlled by the availability of light and nutrients (typically: nitrogen, phosphorus, iron, silica).

The presence of floating ice at high latitudes also affords a new habitat for micro-algae. A major new community of primary producers – the **epontic** or ice-algae community develops seasonally in the bottom few centimetres of ice.

The zooplankton comprises a wide diversity of phyla, body forms and modes of life. **Holoplankton**, the predominant members, spend their entire life cycle within the water column. They are distinguished from the **meroplankton**, the larval forms of benthos, which recolonize the benthos after a period in the plankton. A category of **ichthyoplankton**, comprising the larvae of many fish species, is also often recognized. Individuals may be herbivores, omnivores or strict carnivores or function ontogenically at different 'trophic levels'. Every phylum of the animal world is represented somewhere in the marine plankton; however, they typically do not exhibit a high diversity of species, compared to benthic or terrestrial communities. **Zooplankton** are the most abundant animals (metazoa) in the world, yet like phytoplankton, they are largely unknown to the public. The most abundant zooplankton are the copepods. Their nauplius larva is the most common and abundant type of animal body plan on our planet.

Zooplankton range in size from $< 50\mu m$ (nauplii) to over $2m$ (lion's mane jellyfish *Cyanea spp*). Despite being at the mercy of horizontal water motions, many zooplankton can make extensive vertical migrations. The extent of these migrations generally increases with increasing body size and with depth. In neritic and oceanic waters, this typically leads to larger species inhabiting deeper levels of the water column and overlapping in their migrations with smaller species living closer to the surface.

The phytoplankton are grazed by organisms of two major food webs. The **classical food web** consists of the larger phytoplankton (nanno- and net-plankton), grazed by zooplankton, which, in turn, are consumed by the nekton (euphausiids, fish, marine mammals etc.). This food web has been recognized for many years since the major commercial fisheries of the world depend on it. The classical food web, fuelled largely by nutrients from seasonally mixed or upwelling waters, is most highly developed in coastal (neritic) regions; it becomes progressively diluted into oceanic waters. A second pelagic food web, the **microbial food web**, has more recently been recognized (e.g. Azam, 1998), though in evolutionary terms it is much more ancient. It is spatially less variable from neritic to oceanic waters, and ecologically it dominates in oceanic waters. It consists of the smaller primary producers (pico- and nanno-plankton), bacteria, flagellates and ciliates. The ubiquitous nauplii may be an important interface between these two food webs (Turner and Roff, 1993).

The planktonic community is pre-eminently a region where organic production is directly available to grazing consumers. Since there is no equivalent to the resistant woody tissues of terrestrial plants a large proportion of the production of the phytoplankton is consumed immediately. Ungrazed organic material and faecal pellets sink towards the ocean floor. Depending on depth, seasonal timing of cycles and so on, this material may be efficiently recycled several times within the water column before the residue finally becomes available to the zoobenthos.

Both pelagic and benthic communities are strongly related to depth in the water column, as a function of light and pressure (Figure 2.1). The epipelagic lies within the uppermost part of the water column, with characteristic communities of plankton and fish. The biomass of the upper mesopelagic has been found to be greater than previously thought, with swaths of organisms that mainly rely on food from the overlaying layers. Organisms that live in the lower mesopelagic and bathy-pelagic regions are less dense and adapted to the sparser resource concentrations, having evolved in response to the enduring factor of pressure. However, some fish and invertebrates in the bathy-pelagic zone (such as the sergestid prawns and myctophid fishes) migrate into the upper two layers at night. Where bathy-pelagic zones are occupied by specially adapted fish or invertebrates, the actual populations may be low (except in areas where food is more abundant such as around seamounts). Vertical changes in these biological communities with depth are real, but, especially in the pelagic realm where organisms form a series of overlapping distributions and vertical migration ranges, divisions should not be regarded as discrete.

Marine biology of the benthic realm

The benthic realm is sometimes conceived as a fluid extension of the terrestrial environment. In fact, in evolutionary terms and from a global view, the opposite perspective is more appropriate; the terrestrial environment is really an extension of the benthic realm – above water.

For practical purposes, the benthic realm can be considered two-dimensional. While it can be argued that benthic organisms use their space three-dimensionally (e.g. plants extend well into the water column, animals burrow into the substrate), their mode of life and physical adaptations lie in sharp contrast to organisms of the pelagic realm. Benthic plants and animals are bottom referenced, and their distributions are also strongly related to water depth.

Fringing communities

In the shallowest euphotic regions of aquatic habitats, photosynthetic fixation of carbon is performed by several types of fringing floras. Despite their diversity and their potential high biomasses, compared to phytoplankton, their overall contribution to primary production of marine ecosystems is low, except in shallow coastal waters.

Several community types, many of them familiar to us because of superficial similarities to terrestrial forms, can be recognized:

- Dense stands of sub-tidal (and therefore submergent) macrophytic angiosperm vegetation (basically of terrestrial ancestry) are found here. In marine habitats, these are represented by species such as the truly marine seagrasses *Spartina* and *Thalassia*.
- Salt marshes around the ocean margins are intertidal (rarely extending below mean tide level) and essentially semi-terrestrial in nature.
- In wetlands and marshes, both submergent and emergent macrophyte plant communities develop. These communities are a mixture of truly aquatic and semi-terrestrial species.
- The algal (seaweed) communities are a dominant feature of rocky marine shores. High biomasses of these macrophytic algae may develop both intertidally where they are temporarily exposed to the air during the phases of the tide and sub-tidally within the euphotic zone.
- The intertidal and sub-tidal micro-algae are often forgotten in comparison with the more visually striking macrophytic algae of the shorelines. These microscopic algal communities (single cells, chains or mats) are often dominated by the benthic diatoms that form mats on the bottom sediments or rock or may form epiphytic growths on the macroalgae themselves.
- The hermatypic corals forming reef structures are found here. Although the corals themselves are of course animals, their symbiotic micro-algae (zooxanthellae) are responsible for high rates of photosynthesis. Reef forming and other hard corals are not represented at higher latitudes, although non-photosynthetic 'soft corals' extend into the Polar Regions.
- Mangrove swamps typically develop in hot wet regions of the ocean margins where tidal amplitude is low.

The rate of primary production in these fringing communities may be extremely high. It is usually much higher per unit area per unit time than that of the planktonic algae. However, the greater vertical extent and horizontal coverage of the phytoplankton more than compensates in terms of overall production. A major distinction between planktonic and benthic producers is that the macrophytic plants tend to conserve their biomass. The carbon fixed by them tends to enter detritus pathways (as is generally the case in terrestrial ecosystems) rather than being directly grazed.

Associated with the plants of the fringing communities is a wide diversity of bottom-dwelling invertebrates. These are considered next under zoobenthos communities.

Zoobenthos communities

Marine life extends from the highest limit of the high tides to the abyssal depths of the oceans. Yet, unlike the pelagic realm, life is either attached to or roams the substrate, or is buried only a few centimetres within it.

In the shallow waters of the euphotic zone, benthic and nektonic grazers can feed directly on a diet of macrophytic or microphytic algae. Within the intertidal zone, the upper limits to distributions are set by tolerance to atmospheric desiccation, while lower extents are often set by less physically tolerant predators. There are major differences between the communities that develop on rocky shores versus those consisting of silts and muds.

In the immediate sub-tidal, plants are still present within the euphotic zone. However, in deeper waters, the living plants are absent below the photic zone, and crucial changes occur in the animal communities. Below the euphotic zone, the substrate has generally changed to a relatively uniform sand/silt/mud composition. The benthos in these regions can only be supported by the rain of detrital material, which is derived from the other photosynthetic aquatic communities. The benthos in the dysphotic and aphotic zones is essentially a region of detritus-based food webs and of heterotrophic bacterial activity. It is also a dominant site of nutrient regeneration. Overall, detritus-based food chains are extremely significant; in fact, as mentioned earlier, they are the dominant food webs of the ocean.

In the deeper benthos (bathyal and greater), sparse density is complemented by surprisingly high diversity. In areas where polymetallic nodules are found (discussed in Chapter 11 under seabed mining) or other hard substrates, these support a range of suspension feeders that would not otherwise survive on a purely muddy bottom.

Specialized members of the fish community also exist at various depths within the benthos. For example, flatfish, skates, dogfish and many other species are essentially bottom dependent or at least bottom referenced (demersal). They may be relatively static or highly localized and territorial. Other bottom-dwelling species are periodically mobile or migratory (e.g. crabs and lobsters). Others are either attached permanently to a substrate or burrow within it.

Discussion questions

- What marine environments are the most vulnerable to human activities? Which types of environments would recover quickly from human impacts?
- What marine biological communities are the most vulnerable to human activities? Which types of biological communities would recover quickly from human impacts?
- Marine food webs have more trophic levels and are more complex than terrestrial food webs. What are the implications of this for their conservation and management?

Further reading

Azam, F. (1998) 'Microbial control of oceanic carbon flux: The plot thickens', *Science*, vol 280, no 5364, pp. 694–696.

Barnes, R. S. K. and Hughes, R. N. (1999) *An Introduction to Marine Ecology*, Blackwell Science, Oxford.

Bertnes, M. D., Gaines, S. D. and Hay, M. E. (2001) *Marine Community Ecology*, Sinauer Associates, Sunderland, MA.

Charette, M. A. and Smith, W. H. F. (2010) 'The volume of earth's ocean', *Oceanography*, vol 23, no 2, pp. 112–114.

Day, J. C. and Roff, J. C. (2000) *Planning for Representative Marine Protected Areas: A Framework for Canada's Oceans*, World Wildlife Fund Canada, Toronto.

Diaz, R. J. and Rosenberg, R. (2008) 'Spreading dead zones and consequences for marine ecosystems', *Science*, vol 321, no 5891, pp. 926–929.

Knox, G. A. (2001) *The Ecology of Seashores*, CRC Press, London.

Mann, K. H. and Lazier, J. R. N. (2005) *Dynamics of Marine Ecosystems: Biological-Physical Interactions in the Oceans*, Blackwell Science, Cambridge, MA.

Sherman, K., Alexander, L. M. and Gold, B. D. (1992) *Large Marine Ecosystems: Patterns, Processes and Yields*, AAAS Press, Washington, DC.

Steele, J. H. (1985) 'A comparison of terrestrial and marine ecological systems', *Nature*, vol 313, pp. 355–358.

Stehli, F. G., McAlester, A. L. and Heisley, C. E. (1967) 'Taxonomic diversity of recent bivalves and some implications for geology', *Geological Society of America Bulletin*, no 78, pp. 455–466.

Sverdrup, K. A., Duxbury, A. C. and Duxbury, A. B. (2003) *An Introduction to the World's Oceans*, McGraw-Hill, New York.

Thorne-Miller, B. and Earle, S. A. (1999) *The Living Ocean: Understanding and Protecting Marine Biodiversity*, Island Press, Washington, DC.

Turner, J. T. and Roff, J. C. (1993) 'Trophic levels and trophospecies in marine plankton: Lessons from the microbial food web', *Marine Microbial Food Webs*, vol 7, pp. 225–248.

UNESCO (2009) *Global Open Oceans and Deep Seabed (GOODS): Biogeographic Classification*, UNESCO, Paris.

Valiela, I. (2015) *Marine Ecological Processes*, Springer-Verlag, New York.

Chapter 3

An introduction to international law and international marine law

How nations cooperate to govern the oceans

Key topics

- For most of human history, oceans have been managed as *res nullis*, meaning 'nobody's property'.
- International law is the framework for how states conduct themselves at the national level.
- The objective of international law is to establish acceptable practices or behavioural norms between states to regularize and make predictable international interactions.
- International law is established through agreements between states, customs, general principles of law and judicial rulings.
- The central differences between domestic (national) and international law are: who makes it (legislatures vs. states); to whom does it apply (persons vs. states); who interprets it (judges/bureaucracies vs. states); the near absence of sanctions in the international system; and courts (largely irrelevant in international law).

Introduction

While human use of the marine environment began approximately 164,000 years ago (Marean et al., 2007), it was not until the rise of secular sovereign states in 17th-Century Western Europe that modern ocean governance regimes and jurisprudence began to develop to address the socio-economic, political and environmental challenges created by population growth and technological advancement. The 20th Century saw the creation of the majority of the more than 500 major multilateral instruments that have been deposited with the Secretary-General of the United Nations (UN), covering a range of subject matter, including human rights, conflict, trade, refugees, the environment and the law of the sea. Considering only the environmental treaties (terrestrial and marine) between two or more countries, there are currently almost 500 such international agreements of varying scale and importance.

Many comprehensive marine law texts dissect the various international agreements and the cases that arise from them (e.g. Anand, 1983; Brown, 1994; Rothwell and Stephens, 2010; Tanaka, 2012), and it is not the intent of this chapter to duplicate these efforts. Instead, this chapter will introduce in the foundations of marine law and policy for the purposes of gaining a better understanding of how marine environments are governed. This chapter begins by tracing the development of international law and international marine law through an exploration of types of ownership. This is followed by a summary of the primary international treaties, conventions and agreements and their associated institutions.

Ownership regimes in the ocean

Before discussing the legal frameworks that govern the oceans, an understanding of the ways marine environments can be appropriated is necessary, for much of international marine law and policy is concerned with concepts of appropriation and ownership. There are four types of ownership regimes applicable to coastal and marine environments. The first is **open access** (*res nullis* meaning 'nobody's property'), where the oceans and resources within them are neither exclusive nor transferable and, as such, available to all. The concept, developed under Roman law, applies to things that cannot be appropriated, such as wild animals or fish on the high seas and also in situations where there is an absence of the right to exclude. Under pure open access, regime users have no responsibility for the maintenance of the asset. Marine capture fisheries before the end of the 19th Century were, for all intents and purposes, open to all, and 'ownership' only occurred once a resource was appropriated – a caught fish for example. The 1893 Bering Sea Arbitration between Great Britain and the United States on the harvest of seals was perhaps the first example of the repudiation of *res nullis* as a foundation of marine management. Another component of open access is the concept of *terra nullis* (land belonging to no one), which is a Roman law concept that describes territory (terrestrial or marine) over which no state has declared or exerted sovereignty. The concept has particular relevance to the Antarctic continent, which still has unclaimed (and unratified claims to) terrestrial and marine environments (discussed in Chapter 10).

Private ownership (*res privata*) confers exclusive, transferable rights (*abusus* in legal terms) to a resource with an associated duty to refrain from socially unacceptable uses. While relatively rare in marine environments, private ownership of access rights to certain amounts of fish within certain fish stocks, termed *individual transferrable quotas*, is increasing (this is discussed in Chapters 8 and 12). Private ownership also occurs when a non-privately-owned resource is captured or killed.

Common ownership (*res communes*) is a blended model between pure property rights and pure regulation where a defined group of independent users exert a collective right to a jointly held resource or geographic area. The model uses explicitly or implicitly understood rules about how the resource is allocated. Common ownership resources are rarely exclusive and transferable. Common ownership in marine environments is usually associated with smaller coastal fisheries in traditional societies, including Aboriginal-controlled fisheries that coexist alongside industrial fisheries in developed countries.

Lastly, **public ownership** (*res publicae*) is where governments, on behalf of their citizens, control marine areas either through international agreements/treaties or through outright appropriation through force. States make laws or rules to control access and management of resources. Public ownership is exerted by nearly every coastal nation, and a key objective of the various Law of the Sea conventions (discussed below) was to establish a common agreement on the geographical extent of public (state) ownership. Under public ownership, use rights in marine environments are fundamentally a function of the relationship between states and persons (in the legal sense, including citizens and legal business entities).

An introduction to law and types of legal systems

The term 'law' most likely originates from 'lagu', an Anglo-Saxon word generally meaning the rules that bind a community. In the contemporary context, 'law' refers to the principles and rules that govern the behaviour of individuals and groups within a society as well as the institutions that oversee these rules. Law is typically administered through a state system comprised of three components: political institutions (composed of politicians selected, appointed or otherwise)

Table 3.1 Distribution of legal systems among United Nations member states

Legal system	Number of member states	Percentage of member states
Civil law	77	40.1
Civil law/customary law	25	13.0
Common law	23	12.0
Common law/customary law	14	7.3
Civil law/Muslim law	11	5.7
Civil law/common law	10	5.2
Civil law/common law/Muslim law	7	3.7
Common law/Muslim law/customary law	6	3.1
Civil law/common law/customary law	5	2.6
Muslim law/common law/civil law/customary law	4	2.1
Muslim law	3	1.6
Customary law	1	0.5
Civil law/common law/Jewish law/Muslim law	1	0.5
Muslim law/customary law	1	0.5

Source: www.juriglobe.ca/eng/syst-onu/index-alpha.php

in senior positions of power that make laws; the bureaucracy, which administers laws (and may add regulations that support implementation of them); and the 'judiciary' (judicial system) comprised of courts where judges interpret laws and rule on questions and disputes.

In order to inform a discussion on the evolution of the development and application of international law, an understanding of domestic (national) law is useful. While there continues to be disagreement on how to categorize the world's legal systems, five types of legal systems are in common use throughout the world (Table 3.1). Importantly, these types are not mutually exclusive, and many states utilize a combination of two or more systems. In addition, many subnational entities (e.g. French territories) have different legal systems than their parent nation. With respect to religious law, Muslim law is the only recurring religious-based legal system in use across many states. Many other religion-based legal systems have been subsumed into customary law with the exception of Jewish law, used only in Israel.

Beginning with the most popular law system, **civil law** – used in some form by approximately 70 per cent of UN member states – is based on the Roman and Germanic law heritage where all laws are codified (written down). In civil law systems, statutes (written laws) take precedence over case law (law made by judges based on past precedent). Judges have limited authority to interpret law and use of juries is limited to serious criminal matters.

Common law – used in some form by approximately 40 per cent of UN member states – is the primary contrasting legal system to civil law. Common law systems, originally developed in England, are based on case law, or precedent, where law is based on prior court decisions rather than written statutes. Common law is based on the principle of *stare decisis*, which states that judges are obligated to respect past judicial decisions. This entails that case law is produced in common law jurisdictions. However, case law may be overturned by higher (appellate) courts and may not be recognized between sub-national jurisdictions.

Customary law, used in some form by approximately 23 per cent of UN member states, is a legal system based on accepted practice and behaviour (termed 'norms') where a legal practice (although not written down or developed by the judiciary) becomes accepted as law by all parties affected by the practice. Customary law is a foundational part of international law (see below) and is part of the legal systems of nearly 60 states, including China, Japan and states in Africa and the Persian Gulf.

Islamic law, also known as Muslim law or Sharia ('a path to the watering hole') law, is a religious-based legal system used in approximately 33 states. Islamic law is a religious code that governs both the secular (e.g. crime, governance, finance) as well as the spiritual (e.g. diet, prayer). Islamic law is based on the Qur'ān and the rulings of Muhammad (the Sunnah). Many Muslim states have adopted Islamic law as a component of their legal systems, and the law is interpreted by judges and religious leaders.

A primer on international law

International law is not new: Evidence exists of treaties between states as far back as the Babylonian and Sumerian empires (> 2,200 years ago). Greek city states and the ancient kingdoms of India were known to have well-developed relationships with other political bodies that were codified in what we would term today as international law. Since its background is so rich and complex, this section provides only a brief overview of the history and development of international law, mainly focussing on the structure and processes that form international marine law and policy agreements (those interested in a more thorough treatment of this topic are encouraged to review Higgins (1994), Malanczuk (1997) and others).

International law, also known as the law of nations and language of international relations, is the framework for how states conduct themselves at the national level. It comprises both a system for securing values (e.g. freedom, security, material goods) and a method to resolve disputes between states (Higgins, 1994). The objective of international law is to establish acceptable practices or behavioural norms between states to regularize and make predictable international interactions.

The central differences between domestic and international law are: who makes it (legislatures vs. state parties); to whom does it apply (persons vs. states); who interprets it (courts/bureaucracies vs. states and occasionally international courts); the near absence of sanctions in the international system; and courts (used infrequently in international law). Given the absence of a central governing authority and judicial system, international law is often perceived as gently compelling good behaviours at best and non-existent at worst. The 'law' in international law is therefore more a collection of rules and customary practices rather than a codified system of enforceable statutes.

States generally have legislatures that enact laws, a judiciary that enforces laws and an executive branch that oversees the legislature and judiciary. International law, on the other hand, has no central governing authority; however, states use their executive and legislative branches – often unilaterally – to establish rules on how the state will interact with other states. The state may also unilaterally ignore the rules that other states create for themselves, without the threat of international sanctions (except from the other state).

International law is sometimes further subdivided into public and private international law. **Public law** applies to the interactions between states and may be defined by agreements, customs, principles and rules. **Private law** applies to individuals and corporations who interact with and between states.

The terms **soft law** and **hard law** are also used to further differentiate international law. While many jurists and scholars consider all international law a type of soft law, soft law, in the international law context, generally consists of non-binding multilateral agreements. These include decisions/resolutions adopted at conferences, recommendations, codes of conduct, principles, action plans, declarations and resolutions such as those from the UN General Assembly (Juda, 2002). Soft law can be an alternative to creating treaties, and, while non-binding, states often pay careful attention to the development of soft law as it can signal future developments of hard law (Boyle, 1999). Hard law and customary law consist of binding agreements between states in the form of treaties (conventions). Certain treaties may be appealed to the International Court of Justice or other convention-specific tribunals, such as the International Tribunal for the Law of the Sea (ITLOS, Chapter 4).

The state in international law

In international law, the state (see Introduction) is the fundamental entity. While international law affects sub-sovereigns, nations, corporations and occasionally individuals (e.g. pirates), the state is the primary agent in international law. It is important also to note that governments are different from states. To be considered a state under international law (articulated in the Montevideo Convention on the Rights and Duties of States), the state must:

- Demonstrate sovereignty (control) over a defined territory and its nationals.
- Support a permanent human population.
- Have a functioning government.
- Have the capacity to engage with other states.

While recognition by other states is not a requirement for statehood, recognition is a key goal for newly formed states (e.g. South Sudan), proposed states (e.g. Palestine), and states where it is unclear who is the government (e.g. Somalia). Unrecognized states (e.g. Macedonia, a member of the UN) may act as states and undertake the duties of a recognized state. More recently, after the dissolution of the former Yugoslavia in the early 1990s, the European Economic Community (EEC) added a number of other criteria (e.g. support for democracy, human rights, rule of law) before the EEC would recognize the newly formed Balkan states.

Other key aspects of statehood and international law include:

- The political existence of a state is independent of recognition by the other states.
- It is possible to recognize a state without recognizing its government.
- It is also possible to recognize a government but elect not to have any diplomatic ties to the government.
- A government in exile may not exert control of a state, but the government can be recognized.
- An unrecognized state may function in the same manner as a recognized state.
- People have the right of self-determination, but states may only be established to end colonialism or foreign subjugation.

While these criteria are straightforward for most states, many new states, including South Sudan, Palestine and those states in the former Yugoslavia, have difficulty meeting them.

International law also governs any entity that possesses 'international personality'. These 'non-state' entities include governmental organizations (e.g. the UN). Very rarely will multinational companies, individuals and non-governmental organizations be recognized as having 'international personality'. It is important to note that international law governs states, which in turn regulate the activities of their nationals and businesses. Thus, many types of international laws – including human rights or trade – are enacted through domestic legislation.

Sources of international law

Unlike domestic law that is established by a legislative body, international law is much more fluid and draws on a range of sources to inform its development and application. While treaties and customary law are generally viewed as the primary sources of international law, the International Court of Justice (ICJ – Article 38.1) defines the sources of international law, which include:

- International conventions, whether general or particular, establishing rules expressly recognized by the contesting states.
- International custom, as evidence of a general practice accepted as law.
- The general principles of law recognized by civilized nations.
- Judicial decisions and the teachings of the most highly qualified publicists of the various nations, as subsidiary means for the determination of rules of law.

Treaties and other bilateral agreements to which states are parties that directly address the issue are the primary source of international law. These are often referred to as 'express international agreements'. Agreements – if given the authority by a treaty – may be between international organizations (e.g. World Bank) and states. It is important that while the ICJ statute refers to 'conventions' the terms 'treaties', 'agreements', 'declarations', 'accords', 'protocols' and so on are synonymous with 'conventions' although there are certain exceptions.

The second principal source is customary international law, which is best understood as being where a practice of states becomes so accepted that it becomes legally binding. Any action by any state that is adopted (or accepted) by other states can become part of international law, which is also known as *opinion juris sive necessitates* (the belief by states that it is an international legal obligation). Stated differently, there are two elements that demonstrate whether a practice/action is part of customary international law: consistent action taken by numerous states over a period of time and the belief by a state that a practice has become binding. Statements by government officials may be evidence of customary international law or demonstrate what a state considers as customary. However, customary international law is normally predicated on the actions of a state rather than the verbal or written intentions of states. No written agreement is required to develop customary international law, and the routes to establishing customary international law vary; however, it is important to note that international law may be created by actions of a single state that are adopted by other states.

Third, the general principles of mature legal systems can be used to create international law (i.e. the 'law recognized by civilized nations'). Although there has been much discussion on how to define mature legal systems, it has come to be understood that international law may be formed by 'borrowing' aspects of national law common across states and, where appropriate, expanding national law into the international realm.

Lastly, international law may be formed by accepting and incorporating rulings of state's legal systems (e.g. Supreme Court) into international law as well as the 'teachings' of highly qualified jurists.

There are other sources of international law not contemplated by the ICJ in the 1940s. These include the acts or resolutions of international organizations such as the UN General Assembly (UNGA), which produce frequent resolutions that may become accepted as customary international law. The growth of international organizations in the late 20th Century has vastly expanded the number of agreements between nations that could be considered as informing customary international law.

Operation of international law

A foundation of international law is the Vienna Convention on the Law of Treaties (signed 23 May 1969 and entered into force 17 January 1980), which is generally accepted as the 'rulebook' related to treaties (UN, 2005). For the Vienna Convention to be applicable, parties must be states (Article 1), parties must have agreed (Articles 11–13) and agreements must be in writing and intended to be binding (Article 2). Moreover, the agreement must state that the governing law will be international law (Box 3.1).

Under the Vienna Convention, states have to agree to be bound to any aspect of international law. The only exception to this rule is if an action violates a 'peremptory norm' defined as 'a

---

Box 3.1 Key terms as defined by the Vienna Convention on the Law of Treaties

Treaty: An international agreement in written form forged between states and governed by international law, whether embodied in a single instrument or in two or more related instruments and whatever its particular designation (Article 2, paragraph 1[f]).

Party: A state which has consented to be bound by the treaty and for which the treaty is in force (Article 2, paragraph 1[g]).

Plenipotentiary: A person (usually a diplomat) invested with the full power to represent their government in negotiations.

Date of acceptance: The date at which a state becomes a party to a treaty. Treaties normally identify whether treaties are as 'signature subject to acceptance' (analogous to ratification) or by acceptance without prior signature (analogous to accession).

Date of accession: The date at which a state becomes party to a treaty of which it is not a signatory. The right of accession is independent of the entry into force of the treaty; that is, a state may accede to a treaty which has not yet entered into force.

Date of adoption: The date when states participating in the negotiation of a treaty agree on its final form and content. This usually occurs before signature.

Date of denunciation: When a state announces that it is no longer willing to be bound by a treaty.

Date of entry into force: The date when a treaty binds the parties that have agreed to be bound by it. It is often triggered when a certain number of states have ratified the treaty.

Date of ratification: The date when states obtain internal approvals to be bound by a treaty. This usually occurs after signature.

Date of reservations: The date when a state excludes or modifies the legal effect of certain provisions of a treaty.

Date of signature: The date when a state consents to be bound by a treaty but prior to ratification when internal approvals have been obtained.

Date of succession: The date when newly formed states (e.g. Russia) agree to be bound by treaties signed by predecessor states (e.g. Soviet Union).

Source: UN (2005)

norm accepted and recognized by the international community as a whole as a norm from which no derogation is permitted and which can be modified only by a subsequent norm of general international law having the same character' (UN, 2005). Peremptory norms include prohibitions against slavery, genocide, piracy, aggressive war and torture.

States may make agreements between themselves, but the accords are not considered treaties if a peremptory norm is violated. Additional states may enter into an existing treaty between two or more states if provided for in the treaty. Termed 'accession', the additional state may or may not require (depending on the treaty) the consent of the existing signatories to the treaty to be able enter into the treaty.

Eight steps are common in the development of multilateral treaties:

1. Negotiation: A pledge to address a shared issue through a treaty, agreement or convention.
2. Adoption: The conclusion of negotiation after the form and content of an agreement is reached. Adoption may be secured via meetings of parties in their capacity as participating in international organizations or an international conference convened specifically to address the shared issue.
3. Signature: Parties agree to abide by the conditions of the agreement while the approval process is underway. Signature does not bind a state to a treaty, but states are expected not to defeat the purpose of the treaty while it proceeds through ratification and enters into force.
4. Ratification: A third party or depository holds an agreement while parties secure the necessary domestic approvals that may include domestic legislative change.
5. Acceptance (approval): Binds a state to conform with the treaty while it proceeds through the ratification process.
6. Confirmation: Agreement by a state to be bound by the treaty prior to it entering into force.
7. Entry into force: The date when the treaty enters into force and binds all states that have ratified the treaty.
8. Accession: Request by a state to become party to a treaty already negotiated.

Source: UN (2005)

Certain treaties allow 'reservations', which permit states to reject or amend certain provisions of a treaty with which they disagree. However, the United Nations Convention on the Law of the Sea (discussed below) expressly prohibits reservations; it is an all-or-nothing treaty.

The United Nations Charter (the UN Charter) was developed after World War II to 'save succeeding generations from the scourge of war'. The purpose of the Charter is to prohibit acts of aggression and unlawful use of force between states. Under the UN Charter, states with disputes over points of law or facts are obligated to settle disputes by peaceful means. The Charter sets out a continuum of increasingly interventionist dispute reconciliation mechanisms starting with negotiation between states/parties with disputes. If negotiation fails, the Charter directs parties to use non-binding tactics including inquiry, conciliation and mediation. This promotes dialogue, collection of the facts to an issue, and use of a third party to resolve disputes. The next step, if non-binding approaches fail and parties agree to further action, is arbitration, where an independent party evaluates proposals from the affected parties and chooses one. In arbitration, the arbitrator cannot develop an alternative option; one of the proposals from the affected parties must be selected. Lastly, parties are directed to the courts for decision. All states that are signatories to the Charter also agree to the rules of the ICJ. The ICJ is final arbiter of 240+ international treaties, and the purpose of the ICJ is to settle international disputes and provide advisory opinions. States cannot be forced to settle disputes before the ICJ. Headquartered in the Hague, the ICJ is composed of 15 judges and has no powers of enforcement, which is left to the UN Security Council. Only states may appear before the ICJ, and its rulings only apply to the disputed issue brought forward by the states; however, certain ICJ rulings have come to be recognized as important in determining international legal principles. The Court is not bound by the principle of *stare decisis* so that it is not obligated to consider past precedents in their deliberations. However, the ICJ carefully considers past decisions, and they are regularly cited by both the Court and in argument.

The relationship between domestic and international law

Congruence between domestic and international law offers a state a number of efficiencies. These include: cost saving as a result of harmonized rules and procedures; access to international markets; standardized compliance and enforcement; and positive international reputation and optics. For many states, such as those in the European Union, the number of international laws developed in Brussels far exceeds the number of domestic statutes passed. However, states – including those in the European Union – have different practices, often based on constitutional frameworks, and the extent to which international law supersedes domestic law. As a consequence, there is little consistency in how states treat international law in their domestic legal systems. Most states accept that international treaties or conventions (any type of agreement that is legally binding) to which they are a party cannot be casually overridden by domestic legislation. However, states have different perspectives with respect to the process by which international law is accepted as part of domestic law. 'Monist' states, such as the Netherlands, are those that accept international agreements to which they are a party as having the force of law akin to domestic laws. 'Dualist' states, such as Australia and Canada, intentionally separate international and domestic laws and carefully choose which international treaties, once ratified, evolve into domestic laws. A number of states, including Germany and the United States, occupy the middle ground. Here, international treaties must meet domestic legal principles (e.g. Germany reserves the right to evaluate all international treaties against German *Basic Law*) or have checks and balances to ensure that treaties represent the public interest (e.g. the United States uses primarily its Senate, but also its House, in interpreting international law) (de Mestral and Fox-Decent, 2008; Law Society of New South Wales, 2010).

A murkier issue with which many states wrestle is whether customary international law can be superseded by domestic law. The United Kingdom, for example, accepts customary international law as part of domestic (common) law, except where it conflicts with domestic laws. Other states, including Canada and New Zealand, generally follow the UK practice. However, they use a general principle of statutory interpretation that statutes should be interpreted to the extent possible not to be inconsistent with their international legal obligations, with the obvious proviso that if such interpretation is not possible, the domestic statute prevails.

Why international law works

Often the public perception of international law is that, due to its lack of sanctions and voluntary nature, states routinely flout international conventions. This perception is fed by high-profile media coverage of trade wars, border disputes, non-compliance with UN directives, illegal fishing and specific nations (e.g. North Korea) refusing to recognize international agreements.

The reality, however, is that compliance with international law is similar to that of domestic laws. It is rarely ignored and very infrequently challenged. The reason for high compliance rates with international law is not the threat of sanctions. International compliance might be compared with national taxes: most citizens pay their taxes knowing full well that the chance of being audited is low. However, compliance with tax codes is high, and the reason for this is that citizens are raised to understand the societal importance of legal behaviour. So too with international law. Rather than fearing sanctions, states recognize the benefits of compliance and cooperation.

There are many reasons for the success of international law, which are summarized from Malanczuk (1997):

- States create laws for themselves and in their interests. As such, there is little to be gained from ignoring or contravening their own laws.
- Most international law begins as customary practice and, therefore, even in the absence of international law, these customs would often still be followed. As such, international law is often merely the formalization and expansion of the current practices between states.
- States are, for the most part, geographically anchored to specific territory, meaning that they must coexist with neighbouring states for long periods of time. As such, cooperation is necessary for social and economic prosperity and for avoiding groups of other states from imposing sanctions for breaking international laws.

International marine law

Unlike the world's terrestrial land masses and islands where, with the exception of places such as Antarctica, states have established and acknowledged claims of ownership, most of the ocean continues to be free for the use of all. Consequently, the ocean has traditionally been poorly governed. The result has been little to no concern for ocean health, significant loss of human life due to poor maritime safety and political friction among coastal and seafaring states.

Traces of international marine law date back to the Roman Empire, which declared the oceans as a commons, free to all. In 529 AD, Emperor Justinian declared that 'the sea is common to all, both as to ownership and as to use. It is owned by no one; it is incapable of appropriation, just as is the air. And its use is open freely to all men' (Watson, 2009). Throughout the Middle Ages, coastal states began to exert their sovereignty over their adjacent territorial waters. This culminated in a 1494 Papal Bull from Pope Alexander VI resulting in the Treaty of Tordesillas, where it was agreed that Spain would own all marine areas in the West Atlantic while Portugal would possess everything in the East Atlantic. Furthermore, Spain was given the Pacific and the Gulf of Mexico and Portugal the south Atlantic and Indian Ocean.

As other European nations built merchant and naval vessels to participate in colonization and international trade, the Treaty of Tordesillas was rendered obsolete. The next phase of the development of international marine law consisted of the writings of Hugo Grotius, a Dutch jurist who authored the 1608 tome *Mare Liberum* (Free Sea) (Grotius, 1633). Grotius argued on behalf of the Dutch East India Company, which was seeking free passage for trade, that since oceans cannot be 'occupied' or defended in the same manner as terrestrial environments, the sea therefore should be free to all. This doctrine resulted in the 'three-mile rule', whereupon states 'occupied' (held jurisdiction) areas from the coastline out to three miles, or the maximum distance a cannon ball could be fired in the 18th Century. In contrast, at the request of King Charles I, John Selden, an English jurist, authored *Mare Clausum* (Closed Sea) in 1635 that argued that oceans can be appropriated by states without physical occupation.

Ultimately, Grotius's proposition became the accepted practice of most states, and the free seas doctrine became one of the first recognized principles of international marine law. This conviction endured until the early 20th Century when a number of states raised the need to codify international marine law. In 1930, the League of Nations convened a Codification Conference to formalize the rules of international law and identify outstanding areas of international law that had not been adequately addressed. Forty-four states attended the conference, and, though territorial waters was one of the three Conference themes, the only agreement that was reached was the Convention on Certain Questions Relating to the Conflict of Nationality Laws (League of Nations, 1930).

The next major development in international marine law was the 1945 decision by the United States to unilaterally declare its sovereignty over its continental shelves, including the natural resources on and within the seabed in these areas. Termed the 'Truman Proclamation', this action established a precedent that was soon followed by other nations.

In 1949, The United Kingdom sued Albania in the ICJ for compensation resulting from damage to two British destroyers that had struck mines in Albanian waters in 1946. This was the first case (the *Corfu* Channel case) of the newly formed ICJ, and Albania was ordered to pay £843,947 in compensation. This case is significant as it provided the Court the opportunity to consider issues related to territorial seas, including the coastal state's rights within these areas and the rights of other states to peaceful navigation (ICJ, 1949).

The final significant legal case prior to convening the various United Nations Law of the Sea Conferences was the 1951 *Fisheries Case* where the United Kingdom requested that the ICJ rule on the extent of Norwegian territorial waters. Norway had delineated its territorial seas using 'straight baselines' (discussed below) from headland to headland and therefore claimed control of fishing rights in these areas (ICJ, 1951). This excluded the UK from fishing in what it considered were 'high seas' areas. The ICJ ruled in favour of Norway, however, soon prompting other nations to connect headlands.

The International Law Commission (ILC) was established by the UN General Assembly in 1948 and tasked with the development and codification of international law. In particular, the 'regime of the high seas' was identified as a priority area. Over the next decade, the ILC made a number of recommendations including freedom of the high seas, flagging for ships, the right to safe passage, acts of piracy, baselines, high seas fishing and the 12nmi territorial sea (www.un.org/law/ilc/).

To date, over 100 international conventions with significance to the marine environment have been adopted and are summarized in Table 3.2.

Differences between marine law and other legal systems

As far back as Roman times, marine law was differentiated from Roman (land) law. It was recognized that, regardless of nationality, culture or geographic location, maritime legal issues were predominantly international in nature and concerned with addressing safety, navigation, ship registration, contracts, liability and insurance.

Marine (or 'maritime') law originates from civil law tradition. While influenced by common law, international maritime law exhibits the key tenets of civil law, which include codification modified by aspects of customary law and past practices and traditions. There are, however, a number of key differences between marine law and other legal systems. These are discussed in Box 3.2.

Table 3.2 List of international marine conventions and non-binding agreements

Convention name	Short name	Year of adoption	Year of entry into force	Secretariat/relevant authority	Ratifying parties as of 2012	Relevant chapter
Global non-marine conventions with importance to the marine environment						
1959 Antarctic Treaty		1959	1961	Secretariat of the Antarctic Treaty	50	8
1963 Treaty Banning Nuclear Weapon Tests in the Atmosphere, in Outer Space and Under Water	Partial Nuclear Test Ban Treaty (PTBT)	1963	1963	United Nations: UNODA	136	
1985 Vienna Convention for the Protection of the Ozone Layer		1985	1988	United Nations: UNEP	197	
1987 Montreal Protocol on Substances that Deplete the Ozone Layer	Montreal Protocol	1987	1989	United Nations: UNEP	197	
1991 Protocol on Environmental Protection to the Antarctic Treaty	Madrid Protocol	1991	1998	Secretariat of the Antarctic Treaty	33	9
1992 Agenda 21 Preamble and Chapter 17	Agenda 21	1992	1992	United Nations: Dept of Economic and Social Affairs		5
1992 Rio Declaration on Environment and Development	Rio Declaration	1992	1992	United Nations: UNEP		5
1992 Convention on Biological Diversity	CBD	1992	1993	Conference of the Parties	193	5
1992 United Nations Framework Convention on Climate Change	UNFCCC	1992	1994	UNFCCC Secretariat	194	5
1997 Kyoto Protocol to the UN Framework Convention on Climate Change (UNFCCC)	Kyoto Protocol	1997	2005	UNFCCC Secretariat	190	5
1997 Montreal Amendment to Montreal Protocol		1997	1999	United Nations: UNEP	179	

1999 Basel Protocol on Liability and Compensation for Damage Resulting from Transboundary Movements of Hazardous Wastes and Their Disposal	The Basel Protocol on Liability and Compensation	1999	Not yet in force	United Nations: UNEP	10
1999 Beijing Amendment to the Montreal Protocol on Substances that Deplete the Ozone Layer	Beijing Amendment	1999	2002	United Nations: UNEP	156
2001 Stockholm Convention on Persistent Organic Pollutants	Stockholm Convention 2001	2001	2004	United Nations: UNEP	168
Global marine conventions					
1884 Convention for the Protection of Submarine Telegraph Cables	1884 Cable Convention	1884	1888	France	41
1910 International Convention for the Unification of Certain Rules of Law related to Assistance and Salvage at Sea and Protocol of Signature	Assistance and Salvage Convention 1910	1910	1913	Comité Maritime International	86
1910 International Convention for the Unification of Certain Rules of Law Related to Collision between Vessels and Protocol of Signature	Collision Convention 1910	1910	1913	Comité Maritime International	95
1914 International Convention for the Safety of Life at Sea	SOLAS 1914	1914	1919	International Maritime Organization	162
1923 Convention and Statute of the International Regime of Maritime Ports	1923 Ports Convention	1923	1926	UN Secretary-General	43
1924 International Convention for the Unification of Certain Rules of Law relating to Bills of Lading and Protocol of Signature	Hague Rules 1924	1924	1931	Comité Maritime International	80

5
7
7
7

(Continued)

Table 3.2 (Continued)

Convention name	Short name	Year of adoption	Year of entry into force	Secretariat/relevant authority	Ratifying parties as of 2012	Relevant chapter
1926 International Convention for the Unification of Certain Rules concerning the Immunity of State-owned Vessels and 1934 Additional Protocol	Immunity of State-Owned Ships Convention 1926/1934	1926, 1934 (Additional Protocol)	1937	Comité Maritime International	30	6
1946 International Convention for the Regulation of Whaling	Whaling Convention	1946	1948	International Whaling Commission	88	7
1948 Convention on the International Maritime Organization	IMO Convention	1948	1958	International Maritime Organization	170	7
1952 International Convention for the Unification of Certain Rules relating to Civil Jurisdiction in Matters of Collision	Collision/ Civil Jurisdiction Convention 1952	1952	1955	Comité Maritime International	85	7
1952 International Convention for the Unification of Certain Rules relating to Penal Jurisdiction in Matters of Collision and Other Incidents of Navigation	Collision/Penal Jurisdiction Convention 1952	1952	1955	Comité Maritime International	91	7
1952 International Convention for the Unification of Certain Rules relating to the Arrest of Sea-going Ships	Arrest Convention 1952	1952	1956	Comité Maritime International	94	7
1957 International Convention relating to the Limitation of the Liability of Owners of Sea-going Ships and Protocol of Signature	Limitation of Shipowners' Liability (Revised) Convention 1957	1957	1968	Comité Maritime International		7

1958 Convention on the Continental Shelf	UNCLOS 1958	1958	1964	UN Division for Ocean Affairs & Law of the Sea	58	2
1958 Convention on Fishing and Conservation of the Living Resources of the High Seas	UNCLOS 1958	1958	1966	United Nations	38	2
1958 Convention on the Territorial Sea and the Contiguous Zone	UNCLOS 1958	1958	1964	United Nations	52	2
1958 Convention on the High Seas	Geneva Convention on the High Seas	1958	1962	United Nations	63	2
1961 International Convention for the Unification of Certain Rules relating to Carriage of Passengers by Sea and Protocol	Carriage of Passengers Convention 1961	1961	1965	Comité Maritime International	11	7
1966 International Convention on Load Lines	CLL 66	1966	1968	International Maritime Organization		7
1967 Protocol to Amend the International Convention for the Unification of Certain Rules of Law relating to Assistance and Salvage at Sea	Assistance and Salvage Protocol 1967	1967	1977	Comité Maritime International	9	7
1968 Protocol to Amend the 1924 International Convention for the Unification of Certain Rules of Law Relating to Bills of Lading	Visby Rules	1968	1977	Comité Maritime International	30	7
1969 International Convention Relating to Intervention on the High Seas in Cases of Oil Pollution Casualties	Intervention Convention	1969	1975	International Maritime Organization	87	7
1969 International Convention on Civil Liability for Oil Pollution Damage	1969 CLC/1969 Civil Liability Convention	1969	1975	International Maritime Organization	36	

(Continued)

Table 3.2 (Continued)

Convention name	Short name	Year of adoption	Year of entry into force	Secretariat/relevant authority	Ratifying parties as of 2012	Relevant chapter
1969 UN General Assembly Resolution 2574 D (XXIV): Question of the Reservation Exclusively for Peaceful Purposes of the SeaBed and the Ocean Floor, and the Subsoil Thereof, Underlying the High Seas Beyond the Limits of Present National Jurisdiction, and the Use of Their Resources in the Interests of Mankind	UNGAR 2574	1969		United Nations		9
1970 UN General Assembly Resolution 2749 (XXV) on Declaration of Principles Governing the SeaBed and the Ocean Floor, and the Subsoil Thereof, Beyond the Limits of National Jurisdiction (GAR 2749)	UNGAR 2749/ Declaration of Principles	1970		United Nations		9
1972 Convention for the Prevention of Marine Pollution by Dumping from Ships and Aircraft	Oslo Convention	1972	1974	OSPAR Commission	13	7
1972 Convention on the International Regulations for Preventing Collisions at Sea	COLREG 1972	1972	1977	International Maritime Organization	155	7
1973 Protocol Relating to Intervention on the High Seas in Cases of Pollution by Substances Other Than Oil	Intervention Protocol	1973	1983	International Maritime Organization	54	7
1974 United Nations Convention on a Code of Conduct for Liner Conferences	Liner Code 1974	06 April 1974, Geneva	06 October 1983	United Nations	78	7

1974 International Convention for the Safety of Life at Sea	SOLAS 1974	1 November 1974, London, United Kingdom	25 May 1980	International Maritime Organization	162	7
1976 Convention on Limitation of Liability for Maritime Claims	LLMC	19 November 1976, London, United Kingdom	1 December 1986	International Maritime Organization	52	7
1977 Convention on Civil Liability for Oil Pollution Damage Resulting from Exploration for and Exploitation of Seabed Mineral Resources	CLEE 1977	1 May 1977, London, United Kingdom		United Kingdom	0	7
1978 Protocol Relating to the 1973 International Convention for the Prevention of Pollution from Ships (including Annexes, Final Act and 1973 International Convention)	MARPOL 73/78	17 February 1978	2 October 1983	International Maritime Organization	151	7
1979 Protocol Amending the International Convention for the Unification of Certain Rules relating to Bills of Lading as modified by the Amending Protocol of 23rd February 1968	SDR Protocol	21 December 1979, Brussels, Belgium	14 February 1984	Comité Maritime International	33	7
1979 International Convention on Maritime Search and Rescue	SAR Convention	27 April 1979, Hamburg, Germany	22 June 1985	International Maritime Organization	96	7
1982 Paris Memorandum of Understanding on Port State Control	Paris MOU	26 January 1982, Paris, France	1 July 1982	The Paris Memorandum of Understanding on Port State Control	27	7
1982 United Nations Convention on the Law of the Sea	Law of the Sea Convention (LOSC or UNCLOS)	10 December 1982, Montego Bay, Jamaica	16 November 1994	United Nations	162	2

(Continued)

Table 3.2 (Continued)

Convention name	Short name	Year of adoption	Year of entry into force	Secretariat/relevant authority	Ratifying parties as of 2012	Relevant chapter
1988 Convention for the Suppression of Unlawful Acts against the Safety of Maritime Navigation	SUA 1988/1988 SUA Convention/Maritime Convention	10 March 1988, Rome, Italy	01 March 1992	International Maritime Organization	156	7
1988 Protocol for the Suppression of Unlawful Acts against the Safety of Fixed Platforms Located on the Continental Shelf	1988 SUA Protocol/Fixed Platform Protocol/SUA PROT	10 March 1988, Rome, Italy	01 March 1992	International Maritime Organization	145	7
1989 Protocol Concerning Marine Pollution Resulting from Exploration and Exploitation of the Continental Shelf	1989 Protocol	29 March 1989, Kuwait	17 February 1990	Regional Organization for the Protection of the Marine Environment	8	7
1989 International Convention on Salvage	Salvage Convention	28 Apr 1989, London, United Kingdom	14 Jul 1996	International Maritime Organization	58	7
1991 UN General Assembly Resolution 46/215 (1991): Large-Scale Pelagic DriftNet Fishing and Its Impact on the Living Marine Resources of the World Ocean and Seas	United Nations Resolution 46/215	20 December 1991		United Nations		6
1992 Protocol to Amend the 1969 International Convention on Civil Liability for Oil Pollution Damage	1992 Protocol/CLC Protocol 1992	27 November 1992, London, United Kingdom	30 May 1996	International Maritime Organization	128	7
1992 International Convention on Civil Liability for Oil Pollution Damage (Consolidated text of the 1969 Convention, incorporating the amendments of 1976, 1992 and 2000)	1992 CLC/1992 Civil Liability Convention	27 November 1992, London, United Kingdom	30 May 1996	International Maritime Organization	128	7

1993 International Convention on Maritime Liens and Mortgages	Maritime Liens and Mortgages Convention	06 May 1993, Geneva, Switzerland	05 September 2004	United Nations	13	7
1993 Agreement to Promote Compliance with International Conservation and Management Measures by Fishing Vessels on the High Seas	FAO Compliance Agreement	24 Nov 1993, Rome, Italy	24 Apr 2003	Food and Agriculture Organization	39	5
Part XI of the 1982 United Nations Convention on the Law of the Sea as Amended by the 1994 Agreement on the Implementation of Part XI of the 1982 United Nations Convention on the Law of the Sea	Installation/Marine Environment/ Decommission/ Decommissioning/ Mining/Seabed	28 July 1994, New York, USA	28 July 1996	United Nations	141	2
1994 Agreement Relating to the Implementation of Part XI of the United Nations Convention on the Law of the Sea of 10 December 1982		28 July 1994, New York	28 July 1996	United Nations	141	2
1995 Agreement for the Implementation of the Provisions of the United Nations Convention on the Law of the Sea of 10 December 1982 relating to the Conservation and Management of Straddling Fish Stocks and Highly Migratory Fish Stocks	Fish Stocks Agreement 1995	4 December 1995, New York	11 December 2001	United Nations	77	6
1996 Protocol to Amend the 1976 Convention on Limitation of Liability for Maritime Claims	LLMC PROT 1996	2 May 1996, London, United Kingdom	13 May 2004	International Maritime Organization	35	7
1996 Protocol to the 1972 Convention on the Prevention of Marine Pollution by Dumping of Wastes and Other Matter	LC PROT 1996	7 November 1996, London, UK	24 Mar 2006	International Maritime Organization	42	7

(Continued)

Table 3.2 (Continued)

Convention name	Short name	Year of adoption	Year of entry into force	Secretariat/relevant authority	Ratifying parties as of 2012	Relevant chapter
1997 Agreement on the Privileges and Immunities of the International Tribunal for the Law of the Sea		23 May 1997 in New York, USA	30 December 2001	International Tribunal for the Law of the Sea	39	2
1998 Protocol on the Privileges and Immunities of the International Seabed Authority		26 March 1998 in Kingston, Jamaica	31 May 2003	International Seabed Authority	32	2
1999 International Convention on Arrest of Ships	Arrest of Ships Convention/ ARREST 1999	12 March 1999, Geneva, Switzerland	Not yet in force	United Nations	7	7
2000 Specific Guidelines for Assessment of Platforms or Other Man-Made Structures at Sea		18–22 September 2000		International Maritime Organization		
2001 International Convention on the Control of Harmful Anti-fouling Systems on Ships	The HAFS Convention/AFS 2001	5 October 2001, London, United Kingdom	17 September 2008	International Maritime Organization	60	5
2001 International Convention on Civil Liability for Bunker Oil Pollution Damage	Bunkers Convention/ BUNKERS 2001	23 Mar 2001, London, UK	21 Nov 2008	International Maritime Organization	46	7
2004 International Convention for the Control and Management of Ships' Ballast Water and Sediments	BWM 2004/Ballast Water Convention	13 February 2004	Not yet in force	International Maritime Organization	35	5
2005 Protocol to the 1988 Convention for the Suppression of Unlawful Acts against the Safety of Maritime Navigation	2005 Protocol to the 1988 SUA Convention/2005 SUA Protocol	14 October 2005, London, UK	28 July 2010	International Maritime Organization	17	7
2005 Protocol to the 1988 Protocol for the Suppression of Unlawful Acts against the Safety of Fixed Platforms Located on the Continental Shelf	2005 SUA Platforms Protocol/ SUA PROT 2005	14 Oct 2005, London, UK	28 July 2010	International Maritime Organization	13	7

2005 Protocol for the Suppression of Unlawful Acts against the Safety of Fixed Platforms Located on the Continental Shelf	2005 SUA Platforms Protocol/ SUA PROT 2005	14 October 2005, London, UK	28 July 2010	International Maritime Organization	13 7
2005 Convention for the Suppression of Unlawful Acts Against the Safety of Maritime Navigation	2005 SUA Convention	14 October 2005, London, UK	28 July 2010	International Maritime Organization	17 7
2006 Maritime Labour Convention	MLC	7 February 2006 in Geneva, Switzerland	20 August 2013	International Labour Organization	15 7
2007 Nairobi International Convention on the Removal of Wrecks	WRC 2007/ NAIROBI WR Convention 2007/ Nairobi WRC 2007	18 May 2007, Nairobi, Kenya	14 April 2015	International Maritime Organization	5 7
2009 London Convention and Protocol/UNEP Guidelines for the Placement of Artificial Reefs		27–31 October 2008		United Nations	
2009 Hong Kong International Convention for the Safe and Environmentally Sound Recycling of Ships	Hong Kong Convention	15 May 2009, Hong Kong, People's Republic of China	Not yet in force	International Maritime Organization	0 7
2009 Agreement on Port State Measures to Prevent, Deter and Eliminate Illegal, Unreported and Unregulated Fishing	Port State Measures Agreement	22 November 2009, Rome, Italy	5 June 2016	Food and Agriculture Organization	16 5
2010 United Nations General Assembly Resolution A/65/37A on Oceans and the Law of the Sea	UNGAR A/65/37A	7 December 2010			

(Continued)

Table 3.2 (Continued)

Convention name	Short name	Year of adoption	Year of entry into force	Secretariat/relevant authority	Ratifying parties as of 2012	Relevant chapter
Important multilateral regional marine conventions and national proclamations						
1839 Convention between France and Great Britain for Defining the Limits of Exclusive Fishing Rights		2 August 1839	17 August 1839	United Nations	2	
1882 Convention regulating the Police of the North Sea Fisheries	North Sea Fisheries Convention	6 May 1882	15 March 1884	Government of Netherlands	6	
1884 Convention for the Protection of Submarine Telegraph Cables	1884 Cable Convention	14 March 1884, Paris, France	1 May 1888	Government of France	41	
1936 Convention Regarding the Regime of the Straits	Montreux Convention	20 Jul 1936, Montreux, Switzerland	9 Nov 1936	Government of Turkey	9	2
1942 Treaty between Great Britain and Venezuela Relating to the Submarine Areas of the Gulf of Paria		26 Feb 1942, Caracas, Venezuela	22 Sept 1942	United Nations	3	
1945 US Presidential Proclamation No. 2668, Policy of the United States with Respect to Coastal Fisheries in Certain Areas of the High Seas	Truman Proclamation	28 Sep 1945, Washington, DC USA		Government of the USA		2
1945 US Presidential Proclamation No. 2667, Policy of the United States with Respect to the Natural Resources of the Subsoil and SeaBed of the Continental Shelf	Truman Proclamation	28 Sep 1945, Washington, DC USA		Government of the USA		2
1972 Convention for the Conservation of Antarctic Seals	CCAS	1 June 1972, London, UK	11 March 1978	British Antarctic Survey	17	8

1980 Convention on the Conservation of Antarctic Marine Living Resources	CCAMLR	20 May 1980, Canberra, Australia	7 April 1982	Commission for the Conservation of Antarctic Marine Living Resources	30	8
1980 Protocol for the Protection of the Mediterranean Sea Against Pollution from Land-Based Sources and Activities	LBS Protocol	17 May 1980, Athens, Greece	17 June 1983	United Nations: UNEP	22	5
1982 Agreement Concerning Interim Arrangements Relating to Polymetallic Nodules on the Deep SeaBed between France, the Federal Republic of Germany, the United Kingdom and the United States		2 Sep 1982, Washington DC, USA	2 Sep 1982	United States	4	9
1984 Provisional Understanding Regarding Deep Seabed Matters between Belgium, France, the Federal Republic of Germany, Italy, Japan, the Netherlands, the United Kingdom and the United States	Provisional Understanding Regarding Deep Seabed Matters	3 Aug 1984, Geneva, Switzerland	2 Sep 1984	United States	6	
1989 Guidelines and Standards for the Removal of Offshore Installations and Structures on the Continental Shelf and in the Exclusive Economic Zone (IMO Resolution A.672 (16))	Offshore Installations Resolution	19 October 1989		International Maritime Organization		
1994 Protocol for the Protection of the Mediterranean Sea against Pollution Resulting from Exploration and Exploitation of the Continental Shelf and the Seabed and its Subsoil	Offshore Protocol	14 October 1994, Madrid, Spain	24 March 2011	United Nations: UNEP	6	

(Continued)

Table 3.2 (Continued)

Convention name	Short name	Year of adoption	Year of entry into force	Secretariat/relevant authority	Ratifying parties as of 2012	Relevant chapter
1995 Protocol Concerning Specially Protected Areas and Biological Diversity in the Mediterranean	SPA and Biodiversity Protocol	10 June 1995, Barcelona, Spain	12 December 1999	United Nations: UNEP	18	
1996 Protocol on the Prevention of Pollution of the Mediterranean Sea by Transboundary Movements of Hazardous Wastes and Their Disposal	Hazardous Wastes Protocol	1 Oct 1996, Izmir, Turkey	19 December 2008	United Nations: UNEP	6	
1998 OSPAR Decision 98/3 on the Disposal of Disused Offshore Installations		23 July 1998, Sintra, Portugal	9 February 1999	OSPAR Commission		
2001 Convention on the Protection of the Underwater Cultural Heritage	2001 Convention	2 November 2001, Paris, France	2 January 2009	United Nations: UNESCO	41	
2002 Protocol Concerning Cooperation in Preventing Pollution from Ships, and in Cases of Emergency, Combating Pollution of the Mediterranean Sea	Prevention and Emergency Protocol	25 Jan 2002, Valetta, Malta	17 Mar 2004	International Maritime Organization	12	
2004 Regional Cooperation Agreement on Combating Piracy and Armed Robbery against Ships in Asia	ReCAAP	11 November 2004, Tokyo, Japan	4 September 2006	ReCAAP Information Sharing Centre	15	
2008 Protocol on Integrated Coastal Zone Management in the Mediterranean	ICZM Protocol	21 January 2008, Madrid, Spain	24 March 2011	United Nations: UNEP	7	
2008 United Nations Security Council Resolutions 1816, 1848, and 1851 on the Situation in Somalia	UNSCR 1816, 1848 and 1851	2 June 2008 2 December 2008 16 December 2008		United Nations Security Council		7

2010 United Nations Security Council Resolution 1918 on the Situation in Somalia	UNSCR 1918	27 April 2010	United Nations Security Council	10	5
UNEP Regional Seas Programme Conventions				
Convention on the Protection of the Marine Environment of the Baltic Sea Area	Helsinki Convention	1974, amended 1992	1980, and amendments in 2000	Helsinki Commission	18	5
Convention for the Protection of the Marine Environment and the Coastal Region of the Mediterranean	Barcelona Convention	1976, amended 1995	1978, and amendments in 2004	Mediterranean Action Plan	18	5
Kuwait Regional Convention for Co-operation on the Protection of the Marine Environment from Pollution	Kuwait Convention	1978	1978	ROPME Secretariat	8	5
Abidjan Convention for Co-operation in the protection and Development of the Marine and Coastal Environment of the West and Central African Region	Abidjan Convention	1981	1984	Abidjan Convention Secretariat	33	5
Convention for the Protection of the Marine Environment and Coastal Area of the Southeast Pacific	Lima Convention	12 November 1981	1986	Permanent Commission on the South Pacific (CPPS)	5	5
Regional Convention for the Conservation of the Red Sea and Gulf of Aden Environment	Jeddah Convention	1982	1985	Programme Secretariat	7	5
Convention for the Protection and Development of the Marine Environment of the Wider Caribbean Region	Cartagena Convention	1983	1986	Caribbean Environment Programme	28	5

(*Continued*)

Table 3.2 (Continued)

Convention name	Short name	Year of adoption	Year of entry into force	Secretariat/relevant authority	Ratifying parties as of 2012	Relevant chapter
Convention for the Protection, Management and Development of the Marine and Coastal Environment of the Eastern African Region	Nairobi Convention	1985, amended 2010	1996, amendments not yet entered into force	United Nations Environment Program	10	5
Convention for the Protection of Natural Resources and Environment of the South Pacific Region	Noumea Convention	1986	1990	Secretariat of the Pacific Regional Environment Programme	24	5
Convention on the Protection of the Black Sea Against Pollution	Bucharest Convention	21 April 1992	15 January 1994	Black Sea Commission	5	5
Convention for the Protection of the Marine Environment of the Northeast Atlantic	OSPAR Convention	1992	1998	OSPAR Commission	16	5
Convention for Cooperation in the Protection and Sustainable Development of the Marine and Coastal Environment of the Northeast Pacific	Antigua Convention	1992	Not yet in force	United States	8	5

Source: treaties.un.org/Pages/MSDatabase.aspx

Box 3.2 Differences between marine law and other legal systems

Jurisdiction: Jurisdiction is generally less important in marine law than other domestic legal systems. In most international agreements and conventions, a number of jurisdictions may be involved in a marine dispute. Imagine a ship that is registered (flagged) in a state different than its ownership. Now imagine it to be involved in an incident during transit in the territory of one or more coastal states while travelling between two different port states. As such, multiple states may have some jurisdiction over a matter.

Dispute resolution: Marine law utilizes alternative dispute resolution mechanisms as a means to arbitrate disputes and an international body of jurisprudence regarding arbitration is developing. In particular, the UN Law of the Sea Convention outlines specific dispute resolution mechanisms that are mandatory for state parties to the Convention, including the creation of International Tribunal for Law of the Sea (ITLOS).

Self-regulation: Marine law is increasingly self-regulatory, and protocols are established independent of government organizations.

Financial caps: Tort (harms) law operates on the principle of *restitutio in integrum*, meaning that full compensation is due to the victim. However, from as early as the 11th Century, shipping law has operated under the principle of limitation of liability (financial caps), which caps damages at certain, agreed-upon limits. Limitation of liability applies across a number of international conventions and agreements pertaining to the transport of goods, passengers, noxious substances and oil.

Source: Cho (2010)

Tetley (2000) defines the following three types of maritime law:

1. Public International Maritime Law: Encompasses the legal relationships between states in respect of marine matters and consists of agreements, conventions and treaties negotiated between states or supranational organizations (e.g. European Union).
2. Conflict of Maritime Laws: The collection of rules used to resolve maritime disputes between private parties subject to the laws of different states.
3. International Private Maritime Law: Encompasses the legal maritime relationships between private parties of different States. While negotiated between states, private international maritime law primarily applies to the actions of non-state actors such as shipping companies.

In addition, the collection of international marine laws that govern how states may use the ocean is sometimes referred to as 'admiralty law' or the 'law of the seas' (not to be confused with the UN Law of the Sea Convention discussed in Chapter 4). 'Maritime law' often refers to the 'International Private Marine Law' discussed above that primarily applies to private parties operating in the marine environment.

Discussion questions

- What are the central differences between domestic (state) and international law?
- Define 'hard' and 'soft' law. How might an international convention be construed to be 'harder' or 'softer' than others?
- Why does international law work if there are few penalties and sanctions?

Further reading

Anand, R. P. (1983) *Origin and Development of the Law of the Sea*, Martinus Nijhoff, The Hague.
Boyle, A. (1999) 'Some reflections on the relationship of treaties and soft law', *International and Comparative Law Quarterly*, vol 48, pp. 901–913.
Brown, E. D. (1994) *The International Law of the Sea*, vol 1–2, Dartmouth, Aldershot, UK.
Cho, D. (2010) 'Limitations of 1992 CLC/FC and enactment of the special law on M/V Hebei Spirit incident in Korea', *Marine Policy*, vol 34, no 3, pp. 447–452.
De Mestral, A. and Fox-Decent, E. (2008) 'Rethinking the relationship between international and domestic law', *McGill Law Journal*, vol 53, pp. 573–648.
Grotius, H. (1633) *Freedom of the Seas or the Right which Belongs to the Dutch to Take Part in the East Indian Trade*, Oxford University Press, New York.
Higgins, R. (1994) *Problems and Process: International Law and How We Use It*, Oxford University Press, Oxford.
International Court of Justice (1949) *Corfu Channel Case (United Kingdom of Great Britain and Northern Ireland v. Albania)*, ICJ Reports, www.icj-cij.org/en/case/1, accessed 04 February 2019.
International Court of Justice (1951) *Fisheries Case (United Kingdom v Norway)*, ICJ Reports, www.icj-cij.org/en/case/5, accessed 04 February 2019.
Juda, L. (2002) 'Rio plus ten: The evolution of international marine fisheries governance', *Ocean Development & International Law*, vol 33, no 2, 109–144.
Law Society of New South Wales (2010) *The Practitioners Guide to International Law*, www.lawsociety.com.au/sites/default/files/2018-05/NSWYL%20-%20The%20Practitioner%27s%20Guide%20to%20International%20Law%2C%20Second%20Edition-ilovepdf-compressed.pdf, accessed 04 February 2019.
League of Nations (1930) 'Convention on certain questions relating to the conflict of nationality law', *Treaty Series*, vol 179, no 4137, p. 89.
Malanczuk, P. (1997) *Akehurst's Modern Introduction to International Law*, Routledge, Oxford, UK.
Marean, C. W., Bar-Matthews, M., Bernatchez, J., Fisher, E., Goldberg, P., Herries, A. I. R., Jacobs, Z., Jerardino, A., Karkanas, P., Minichillo, T., Nilssen, P. J., Thompson, E., Watts, I., and Williams, H. M. (2007) 'Early human use of marine resources and pigment in South Africa during the Middle Pleistocene', *Nature*, vol 449, no 905.
Rothwell, D. R. and Stephens, T. (2010) *The International Law of the Sea*, Hart Publishing, Oxford, Oregon.
Tanaka, Y. (2012) *The International Law of the Sea*, Cambridge University Press, Cambridge, UK.
Tetley, W. (2000) 'Uniformity of international private maritime law: The pros, cons and alternatives to international conventions: How to adopt an international convention', *Tulane Maritime Law Journal*, vol 24, pp. 775–856.
United Nations (2005) *Vienna Convention on the Law of Treaties 1969*, United Nations, Treaty Series, vol 1155, p. 331, legal.un.org/ilc/texts/instruments/english/conventions/1_1_1969.pdf, accessed 04 February 2019.
Watson, A. (2009) *The Digest of Justinian*, vol 1, University of Pennsylvania Press, Philadelphia, PA.

Chapter 4

The United Nations Convention on the Law of the Sea and related agreements

Moving from 'free seas' to codifying jurisdiction

Key topics

- The United Nations Convention on the Law of the Sea is the primary legal agreement governing the conduct between states in the world ocean. It is one of the most complex international agreements ever drafted and also one of the most fully subscribed (168 parties).
- The Convention defines the rights of states to waters adjacent to their territory; defines the seabed beyond national jurisdictions and its resources as the 'common heritage of mankind'; and obliges states to protect and preserve the ocean and accommodate the needs of other states.
- The Convention establishes the International Seabed Authority and International Tribunal for the Law of the Sea.

UN conventions on the Law of the Sea

By the 1950s, human uses of the oceans were changing significantly, driven by globalization, technological advances and human population growth. Offshore hydrocarbon development, seabed mining, container-based shipping, high seas fishing and militarization of the oceans highlighted the limitations of the existing rules of the ocean and the vulnerabilities of the relationships between nations that use the ocean.

In particular, a number of jurisdictions, led by the United States, had unilaterally increased their jurisdiction beyond the 12nmi territorial seas. This precipitated friction between coastal states and seafaring nations with respect to shipping and other coastal economic activities. In the 1960s–1970s, there was also considerable interest in the potential of seabed mining, yet little agreement on its governance, whether on or off the continental shelves. States at this time were also positioning themselves to their best advantage based on geography, foreign policy and shipping to their markets. **Coastal states** blessed with significant coastal areas adopted national policies exerting control and exclusive rights over their marine areas. Also termed 'territorialists', states such as Canada and Russia considered marine resource development and conservation a national right. **Maritime states**, including the United States and other geographically smaller seafaring nations (e.g. the United Kingdom), advocated for the freedom of the seas. Also termed 'internationalists', maritime states were increasingly concerned that a patchwork of national regulations would complicate shipping and other resource development (McDorman, 2008). States' positions on marine sovereignty were further articulated by how they viewed uses of their marine territory by other states. **Functionalist states** believed that states should only regulate their marine areas for certain activities such as vessel-source pollution. **Unilateralist states** believed that *all* marine activities in their waters should be regulated.

The first United Nations Conference on the Law of the Sea (UNCLOS I) was held in Geneva, Switzerland, in 1958. Attended by 86 states, UNCLOS I produced the following four conventions:

1. Convention on the Territorial Sea and Contiguous Zone (entered into force in 1964). While this Convention acknowledged the existence of the territorial sea (including the airspace above, and seabed below) and defined how the landward boundary is to be measured, the breadth of the territorial sea provoked controversy. The Convention also codified the rights of innocent passage through territorial waters and international straits. Lastly, the Convention defined the contiguous zone as 12nmi seaward of the baseline and permitted coastal states to intervene in these waters. This was hoped to prevent and punish infringement of its customs, fiscal, immigration or sanitary regulations within its territory or territorial sea (UN, 2005a).
2. Convention on the Continental Shelf (entered into force in 1964). This Convention gave coastal states rights over the natural resources on their continental shelves. The shelves were defined as 'to the seabed and subsoil of the submarine areas adjacent to the coast but outside the area of the territorial sea, to a depth of 200 metres or, beyond that limit, to where the depth of the superjacent waters admits of the exploitation of the natural resources of the said areas; (b) to the seabed and subsoil of similar submarine areas adjacent to the coasts of islands'. Within the continental shelf areas the Convention provided direction on permissible activities and their relation to safe navigation (UN, 2005b).
3. Convention on the High Seas (entered into force in 1962). This Convention defined the 'high seas' as any non-territorial or non-internal waters and defined the permissible activities on the high seas along with the requirements for flagging of ships and navigational safety (UN, 2005c).
4. Convention on Fishing and Conservation of the Living Resources of the High Seas (entered into force in 1966). This Convention allows all states to fish on the high seas and outlines measures to ensure that stocks are not overfished (UN, 2005d).

The second United Nations Conference on the Law of the Sea (UNCLOS II) was again held in Geneva in 1960. The Conference was intended to focus on the breadth of the territorial sea and fishery limits. However, it failed to reach agreement on any new conventions.

The Third United Nations Conference on the Law of the Sea (UNCLOS III and hereafter referred to as the 'LOSC') was held in New York in 1973. While the original intent of the Conference was to regulate seabed mining, the Conference became a nine-year exercise amongst 160 states to rewrite the rules of the ocean. Eleven negotiating sessions were held over 585 days during this protracted period. The final Convention, adopted in 1982, established navigational rights, territorial sea limits, economic jurisdiction, rights to seabed resources, management of shipping through narrow straits and the security of the marine environment (UN, 1982).

The LOSC is perhaps the most complex and comprehensive international agreement ever completed. It rivals the International Declaration of Human Rights and Charter of the United Nations (both non-binding) for its scope and impact on humanity. Indeed, there is no terrestrial agreement that approaches the comprehensiveness of the LOSC.

The LOSC can be summarized in three principles (after LeGresley, 1993):

1   States own and have the geographically defined rights to certain resources in their adjacent marine areas.
2   Parts of the ocean are the 'common heritage of mankind'.
3   States are obliged to preserve the seas and accommodate the needs of other states.

The LOSC consists of 17 Parts, 400 Articles and 9 Annexes and currently has 168 parties. The LOSC entered into force for those States a party to it 12 months after the signature of the 60th state, in November 1994.

The LOSC enabled the establishment of four institutions (later discussed in this Chapter):

1   International Tribunal for the Law of the Sea.
2   Commission on the Limits of the Continental Shelf.
3   International Seabed Authority.
4   Meeting of States Parties to the Convention.

The LOSC also led to, or informed, a number of other agreements that will be discussed later in this chapter.

The LOSC also includes a number of provisions for dispute resolution. Unlike most treaties where mechanisms for dispute resolution are contained in a separate protocol, the LOSC is unique in that dispute resolution is part of the Convention and signatories to the agreement must abide by the LOSC's dispute resolution apparatus. Unique to international law, the LOSC contains both negotiation and arbitration approaches to dispute resolution.

Under the LOSC (Article 287, Part XV), arbitration (binding) approaches include submission to:

- The International Tribunal for the Law of the Sea (ITLOS), which includes the Seabed Disputes Chamber.
- The International Court of Justice.
- Ad hoc arbitration (in accordance with Annex VII).
- Special arbitration tribunals (established under Annex VIII).

For disputes related to maritime boundaries, military matters and issues under discussion by the United Nations Security Council (UNSC), the LOSC allows states at any time ('optional exceptions') to opt out of compulsory adjudication. In addition, for issues regarding national sovereignty, the LOSC permits states to submit disputes to a conciliation commission, which is non-binding but uses moral suasion to pressure states into compliance with the commission's judgments.

The LOSC establishes no global forum or process where states meet to discuss the implementation of the Convention or other ocean matters. As such, addressing imminent ocean issues has fallen to the UN General Assembly (UNGA) which has passed a number of non-binding resolutions on contentious topics such as high seas driftnet fishing and piracy (McDorman, 2008). For less urgent matters, the UN Secretary-General, using authority provided for in the LOSC, has convened annual meetings of the State Parties to the LOS Convention (SPLOS). These assemblies address administrative matters such as the election of judges for ITLOS and members of the Commission on the Limits of the Continental Shelf.

Marine jurisdiction under the LOSC

In the years following Hugo Grotius' *Mare Liberum* proposal that oceans that could not be continuously occupied were free to all, many states claimed jurisdiction out to 3nmi seaward of their coasts (the 'cannon shot rule'). Despite the absence of global agreement, this distance gradually increased to 12nmi by the 1960s as many countries expanded their marine sovereignty. However, it was noted that if every state increased their jurisdiction 12nmi, this would effectively bring under national jurisdiction more than 100 straits and channels, many of which, including the Straits of Gibraltar (Mediterranean) and Hormuz (Persian Gulf), are important global shipping routes. This situation created tension between the major ocean-faring states and smaller coastal states with territories adjacent to straights and narrows.

While many states were taking sides on the 3 and 12nmi jurisdictional issue, other states had more ambitious claims of marine sovereignty. The 1945 Truman Proclamation unilaterally claimed jurisdiction to natural resources on the continental shelves of the United States. Argentina, Chile, Peru and Ecuador soon followed suit, but, given their very narrow continental shelves, they claimed jurisdiction out to 200nmi. Other nations, such as Canada, adopted a middle position, claiming a 100nmi jurisdiction over navigation in an effort to protect its Arctic areas from pollution.

The following jurisdictional arrangements were made under the LOSC.

The 'baseline'

All the LOSC jurisdictional agreements originate from what is termed in the agreement as 'the baseline'. The baseline is defined as the low tide mark; all other jurisdictional zones are measured seaward from this line (Article 5). While the baseline concept works well for certain types of coastlines (i.e. ones with easily established low tide lines), other types of coastlines, including island archipelagos, estuarine/lagoon systems, and Antarctic ice shelves, are more difficult to ascertain. Accordingly, Article 7 permits straight lines to be drawn between points on a coastline where the coastline is convoluted or low tide marks difficult to determine.

Internal waters

All waters (marine, estuarine, freshwater) on the landward side of the baseline fall under the jurisdiction of the state. In internal waters, states may establish and enforce domestic (national) laws, regulate use and access any resource (Figure 4.1). Foreign vessels have no right of passage within internal waters unless authorized by the respective state (Article 8). In instances such as bays, where the width of the mouth of the bay is greater than 24nmi from headland to headland, internal waters will be indented towards the baseline.

Territorial waters: the 12-mile territorial sea

From the baseline to 12nmi, states are free to establish laws, regulate use and use any resource within this zone. Delineating territorial waters is optional, and states are not required to exert jurisdiction to the full 12nmi. The right to designate territorial waters extends even to rocks and islets above the water line. Within this zone, vessels have the right to 'innocent passage', defined as passage that is 'not prejudicial to the peace, good order or the security' of the coastal state. A vessel may not fish, pollute, discharge weapons for practice or spy on the coastal state.

Coastal states may suspend the right of innocent passage in strategic areas of their coastal seas. Warships are immune from coastal states' jurisdiction within the territorial sea, and the coastal

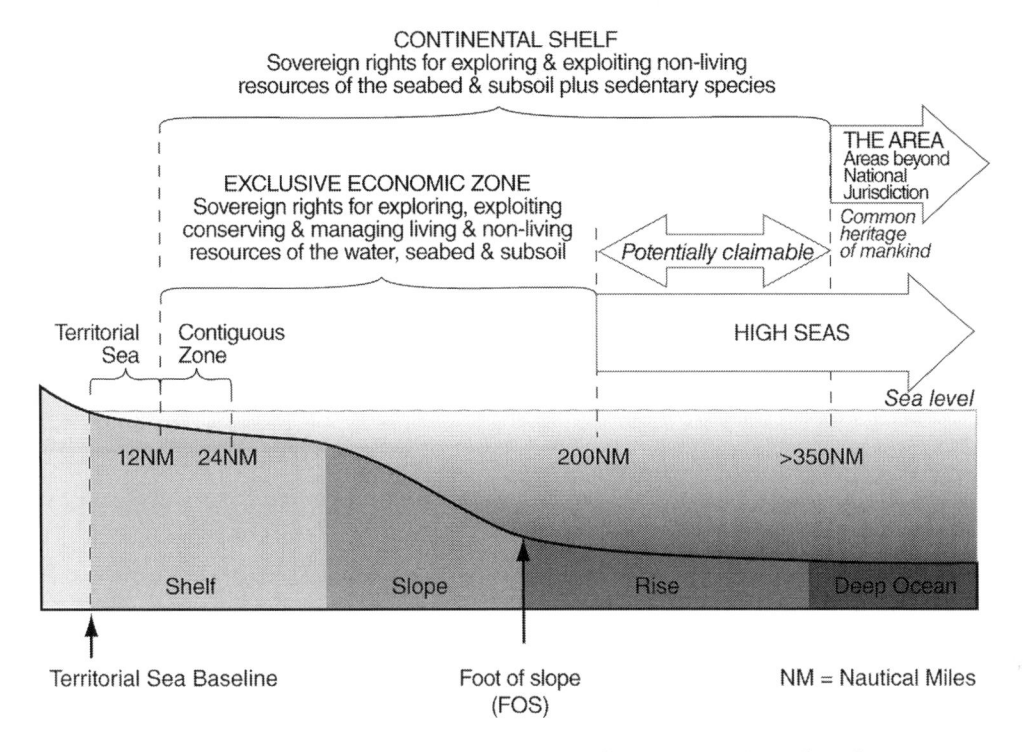

Figure 4.1 Maritime jurisdiction under the United Nations Convention on Law of the Sea

state cannot board a warship. This is even if a crime is committed on board while in the port of the coastal state (a function of the sovereign immunity of the vessel).

For those states with strategic straits (necessary for global trade), the concept of 'transit passage' was agreed to in the LOSC. In this instance, naval vessels, including submarines and ship-born associated aircraft, are allowed to maintain configurations (e.g. carrying weapons) or postures (readiness) that would not be permitted under the rights of innocent passage. Ships in transit passage must adhere to international civilian regulations (e.g. navigation safety, pollution) and must traverse through the strait without stopping except in the case of emergencies. In particular, under transit passage, submarines must surface (and display their flag – although debated by the United States), aircraft may not be launched from ships, and research or surveying must not be performed.

The LOSC provides consideration for archipelagic states such as Indonesia where the 12nmi zone is defined using a line joining the outermost points of the outermost islands. In the archipelagic zone, ships of all states have the right of innocent passage; however, the state has the right to designate shipping lanes to ensure unobstructed passage.

Territorial waters: beyond the 12nmi territorial sea to 24nmi

Beyond the 12nmi territorial waters seaward to 24nmi lies the 'contiguous zone' where a state can enforce jurisdiction related to customs, taxation, immigration and pollution. In addition, the contiguous zone and beyond is a zone of 'hot pursuit'. In these areas, states may pursue those who have violated, or are about to violate, a state's laws in the territorial sea. A vessel can be hotly pursued up to the territorial sea of another state, and hot pursuit also applies to violations of domestic laws.

The Exclusive Economic Zone

The Exclusive Economic Zone (EEZ) extends from the baseline out to 200nmi. The EEZ extends outward to 200nmi even if the coastal state's continental shelf is less than 200nmi in width. Under this definition, the total area of the world's EEZs is approximately 30 million square nmi (Figure 4.2). EEZs were established primarily to address fishing by foreign fleets and to extract offshore oil and gas in certain areas such as the Gulf of Mexico. EEZs are becoming increasingly important, however, as technology to extract hydrocarbons and other non-renewable resources improves.

The LOSC confers on the coastal state the rights to explore, exploit, conserve and manage the resources in the water column and on and under the seabed (Article 56). Included in these rights is the authority to construct (or dismantle) and regulate the operation of installations, structures and artificial islands in accordance with international regulations (Article 60). Coastal states are also required to promote the 'optimum utilization' of living resources within their EEZs. Based on the best available science, states are obliged to establish catch limits for living resources and to ensure that fisheries do not negatively affect other species in their EEZs (Article 61). Any surplus catch below optimum utilization is to be made available to other states based on historical use and the needs of less developed states. In addition, the coastal state is required to publish their domestic fishing laws and regulations (Article 62). Coastal states may not transfer rights to resources in their EEZs to other states or their nationals (Article 72). A coastal state has the right to inspect non-military ships within its EEZ and make arrests if the ship contravenes the coastal state's laws. Coastal states, however, must release arrested vessels and their crews upon posting of a reasonable bond, and crews may not be imprisoned. In cases of arrest or detention of foreign vessels, the coastal state shall promptly notify the flag state (Article 73).

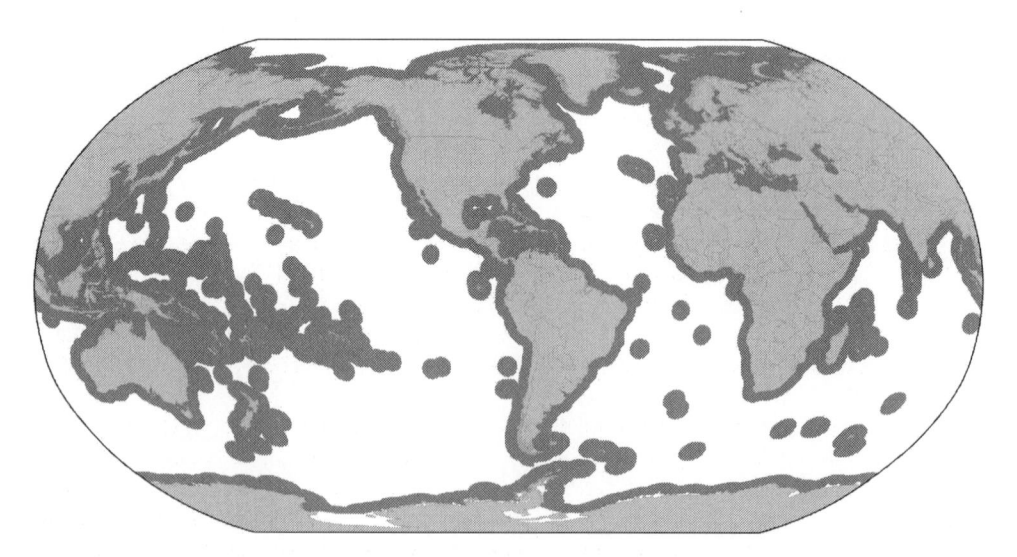

Figure 4.2 Exclusive Economic Zones of the world

Within another state's EEZ, foreign ships or aircraft (used for spotting schools of fish or marine mammals) retain the right to fish, conduct research, construct seabed cables or pipelines and explore and develop natural resources with the permission, and within the regulations, of the coastal state (Article 58).

Continental shelves

The EEZ and a state's continental shelf are not the same. The continental shelf is the submerged portion of a continent (seabed) that extends from the baseline to the 200nmi EEZ or until the submerged continent ends, whichever is greater. The LOSC refers to the 'natural prolongation' of the land territory as the 'extended continental shelf', including the seabed and subsoil of the shelf (Article 76). The LOSC provides two constraints on the maximum extents of the extended continental shelf: The first is that the shelf may not be more than 350nmi from a state's coastal baselines. The second constraint limits the extended continental shelf 100nmi seaward from the 2,500 metre isobath (depth contour). Within the two above constraints, continental shelves can be delineated by the following means, also termed formula lines:

- Sediment thickness: A line, termed the **Gardiner line**, is defined by the points where the thickness of the sediment is at least 1 per cent of the distance to the foot of the slope.
- Bathymetry: A formula, termed the **Hedberg formula**, defines a line by points 60nmi seaward from the foot of the continental slope where the foot of the slope is defined as the point of maximum change in the gradient at its base.

Once the Gardiner/Hedberg lines are established, then the constraint lines are applied to determine the outer limit of the continental shelf.

Approximately 85 coastal states have continental shelves beyond the 200nmi limit (e.g. Canada's east coast extends to 400nmi and beyond). In instances where states have extended continental shelves, they are required to submit information to the Commission on the Limits of the Continental Shelf, which makes recommendations that the submitting states considers and may make a resubmission. Ultimately, it is the submitting state that decides where the outer limit of its shelf is and submits the coordinates and maps to the UN Secretary-General. At this time, the outer limit is final and binding on the submitting state (subject to bilateral resolution of overlapping claims of other States) (Article 76).

Coastal states have rights to the non-biological resources on and within continental shelves as well as living sedentary species (species that do not leave the seabed) that extend beyond 200nmi but not to resources (i.e. fish) within the water column in these areas (Article 77). Also, coastal states have the exclusive right to authorize and regulate drilling on the continental shelves. However, they do not have rights to the water column or airspace above the extended continental shelf, nor can coastal states' rights interfere with navigation (Article 78). States may set conditions and establish locations for the laying of seabed cables and pipelines, but they may not prohibit other states from installing seabed cables and pipelines (Article 79). The rights of coastal states to the extended continental shelf are not conditional on occupation, proclamation or use (Article 77).

Regarding revenue produced from the extended continental shelf, coastal states are required to make royalty-like payments to the International Seabed Authority (see below), which will redistribute to other states, based on equitable sharing, taking account of the needs and interests of less developed states. Payments are to be made annually after the first five years of production on a sliding royalty scale starting at 1 per cent of the value or volume of production and increasing by 1 per cent annually until year 12, after which it remains at 7 per cent (Article 82). However, how exactly production/volume is to be calculated is not made clear. To date, this clause has not yet been implemented, and planned offshore petroleum on eastern Canada's extended continental shelf may be the first testing of this article.

High seas

The high seas are best understood as the water column beyond the 200nmi EEZ where domestic law, outside of those states with accepted extended continental shelves, no longer applies. The LOSC affirms the reservation of the high seas for peaceful purposes (Article 88), prohibits high seas sovereignty claims (Article 89) and affirms the right of all states to operate ships under their national flags (Article 90). The LOSC further declares that the high seas are open to all states, including those that are landlocked and/or developing (Article 87). High seas rights conferred on all states include freedom of navigation, fishing, scientific research, overflight, laying submarine cables or pipelines and the construction of artificial islands or structures (Article 87).

The LOSC provides the administrative framework for shipping on the high seas (Articles 90–99). Specific provisions include direction on the duties of flag states (Articles 90–94), including the requirement for states to set the conditions (construction, training, seaworthiness, communications) under which ships may sail under their flags (Articles 91 and 94). Ships are therefore bound to the national laws of their flag (Article 92) while flag states are required to maintain a register of ships flying their flags (Article 94). Flag states are also obliged to investigate any reports by other states of non-compliance with national or international laws by flag state ships (Article 94).

The LOSC provides immunity to warships and ships (e.g. survey ships) on non-commercial government business (Article 96). For those ships involved in collisions or other navigational incidents, the arrest and detention of a ship may only be undertaken by the flag state. Furthermore, disciplinary proceedings against masters and crew may be undertaken only by the flag state or state(s) of which the master and crew are nationals. Lastly, only the state that issued the certificates or licenses of a disciplined master or crew member may revoke these privileges (Article 97).

All ships must render assistance to persons and vessels in distress, and coastal states must establish and maintain effective search and rescue services (Article 98). No ship may transport slaves (Article 99), and states shall cooperate to intercept ships carrying illegal drugs (Article 108).

Articles 100–107 address piracy on the high seas. The LOSC binds all states to cooperate in the elimination of piracy (Article 100) and defines piracy as any illegal act of violence, detention or depredation against a ship or aircraft and/or its crew or cargo. Piracy can transpire in ABNJ (Article 101), and it includes the mutiny of a warship, government ship or aircraft by its crew (Article 102). The LOSC defines a pirate ship as any ship or aircraft controlled by persons with the intent as outlined in Article 101 and allows any state to seize or arrest a pirate ship (or aircraft) using government ships or aircraft and prosecute pirates under the state's national laws (Articles 105 and 107).

The LOSC also deals with broadcasting. With the exception of distress calls, no ship may broadcast to the general public from the high seas. Persons engaged in unauthorized broadcasting may be prosecuted by the flag state, coastal states (receiving the transmissions), state of which the broadcasters are nationals or states where the broadcasting equipment was installed (Article 109).

Warships may not board ships of another flag state unless the ship is involved in piracy, slave trade or unauthorized broadcasting or is suspected to be without nationality (Article 110). To discourage this practice, ship owners may receive compensation for any loss or damage from states whose warships or aircraft wrongly board ships believed to be contravening Articles 100–109.

Foreign ships believed to be contravening national or international laws in territorial waters may be pursued by domestic warships or aircraft into the high seas. This right of 'hot pursuit' begins once the offending ship has been warned by visual and auditory means to stop. It ends when the ship that is being pursued enters the territorial sea of its flag state (Article 111). Ship owners may receive compensation for any loss or damage from states whose warships or aircraft incorrectly seize or arrest ships believed to be contravening national laws.

As mentioned previously, all states have the right to lay pipelines and submarine cables in ABNJ (Article 112). On the other hand, states have several obligations regarding pipelines and cables. For instance, damaging or interfering with them is necessarily a punishable offence. They are also required to pass domestic laws to indemnify ship owners for damages that are caused by attempts to avoid harm to pipelines (Article 113).

The LOSC affirms the rights of nationals from all states to fish on the high seas (see Chapter 6). This is subject to regional fisheries agreements and treaties, as well as the rights of coastal states regarding straddling stocks and highly migratory species (Article 116). All states, however, must cooperate on the conservation of living resources (fish, marine mammals, seabirds) on the high seas and, where one or more states use the same resources, negotiate agreements (or establish regional fisheries management organizations) to ensure resource management produces optimum sustained yield (Articles 117–119). States are also required to base decisions on the best available science. This science is to be shared between states through sub-regional, regional and international organizations (Article 119).

Islands

Under the LOSC, islands have the same jurisdictional regime (territorial sea, contiguous zone, EEZ and continental shelf) as other land masses. The exception is what the LOSC terms 'rocks', which are emergent lands that are unable to support human habitation or economic activity (Article 121).

Enclosed or semi-enclosed seas

The LOSC defines enclosed or semi-enclosed seas as a:

> gulf, basin or sea surrounded by two or more States and connected to another sea or the ocean by a narrow outlet or consisting entirely or primarily of the territorial seas and exclusive economic zones of two or more coastal States.
>
> (UN, 1982)

Examples of enclosed or semi-enclosed seas include some of the world's most important water bodies and include the Black, Baltic, Caribbean and Mediterranean Seas and the Persian Gulf, Gulf of Mexico and Gulf of Thailand. The management of enclosed or semi-enclosed seas is convoluted by a number of problematic facts: some states have not ratified the LOSC (e.g. Turkey, the United States); other states have not delineated their marine boundaries (e.g. Caribbean states); some states have strategically important seaways (e.g. South China Sea); and some states have interests in seabed petroleum resources (e.g. North Sea). Moreover, some places with these kinds of seas are not recognized as states (e.g. Taiwan, Northern Cyprus); (CEPAL, 2004).

The LOSC dictates that states bordering on enclosed or semi-enclosed seas cooperate on the management of living marine resources; protection and preservation of the marine environment; scientific research; and working with other states and international organizations (Article 123).

The Area

The 'Area' is formally defined in the LOSC as those non-biological resources on or under the seabed outside of the continental shelf areas. An example of such a resource came from the H.M.S. *Challenger* expeditions from 1872–1876 when nodular polymetallic structures in the deep seabed that contained high amounts of manganese were discovered (Bezpalko, 2004). By the 1960s, the commercial potential of these manganese nodules was recognized. As the resource was 3,600–5,500 metres below the surface, the deep seabed outside of the continental shelves became a region of interest during the LOSC negotiations. The LOSC defines these non-biological resources as 'minerals' (Article 133), regardless of whether or not they are metallic in nature. The Area does not apply to the water column or airspace above the seabed outside of continental shelves (Article 135).

Under the LOSC, the Area is the 'common heritage of mankind', to be used exclusively for peaceful, beneficial purposes, and states should cooperate to achieve these ends (Articles 136 and 138). The LOSC provides for the economic development of the Area under the conditions that development contributes to the world economy and growth in international trade (Article 150). Specifically, all states, regardless of geographic location, degree of economic development or whether landlocked or not, have equal rights to participation and access to the Area (Article 141). According to the LOSC, a portion of the economic benefits derived from the Area should be shared among all states (Article 139).

Activities in the Area are overseen by the International Seabed Authority (the 'Authority' – see below), which reports to the UNGA (Articles 140, 143–146 – see below). No state is permitted to claim any territory or the natural resources within the Area. However, mineral resources from the Area may be 'alienated' (taken) under the rules of the Authority (Article 137).

In addition, archaeological artefacts or discoveries of historical significance must be managed for the benefit of humanity all the while recognizing the importance of the culture of origin (Article 149). States bear the responsibility for actions of their nationals (individuals, corporations or government) and are liable for their actions (Article 139).

Rights of passage through straits under the LOSC

Straits are marine corridors frequently used by ships to transit between, and occasionally within, states. Straits are generally narrow bodies of water geographically constrained by land (or ice) and may connect EEZs, high seas, national waters or any combination of these areas. Given the importance of global trade, straits frequently have significant economic or military importance to both the coastal and maritime states (Figure 4.3).

Examples of well-known straits include:

- The Strait of Hormuz bordered by Iran, the United Arab Emirates and the Omani enclave of Musandam (14.3km at its narrowest point). Twenty per cent of the world's petroleum is transported through the Strait of Homuz, making the Strait a strategically important site for ensuring the uninterrupted flow of petroleum products.
- The Strait of Gibraltar, bordered by Morocco and Gibraltar (13km at its narrowest point), connects the Mediterranean Sea from the Atlantic Ocean.
- The Strait of Malacca, bordered by Malaysia and Indonesia (2.7km at its narrowest point), is traversed by more than 50,000 ships annually carrying approximately 25 per cent of the world's traded goods and is the shortest route between Asia and the Middle East.

As seaborne trade increased in the 1960s and 1970s, states realized the risk of moving goods through straits and therefore encouraged the codification of the rights and duties of coastal and

Panama Canal 2016
0.9M barrels/day

Danish Straits 2016
3.2M barrels/day

Turkish Straits 2016
2.4M barrels/day

Strait of Hormuz 2016
18.5M barrels/day

Strait of Malacca

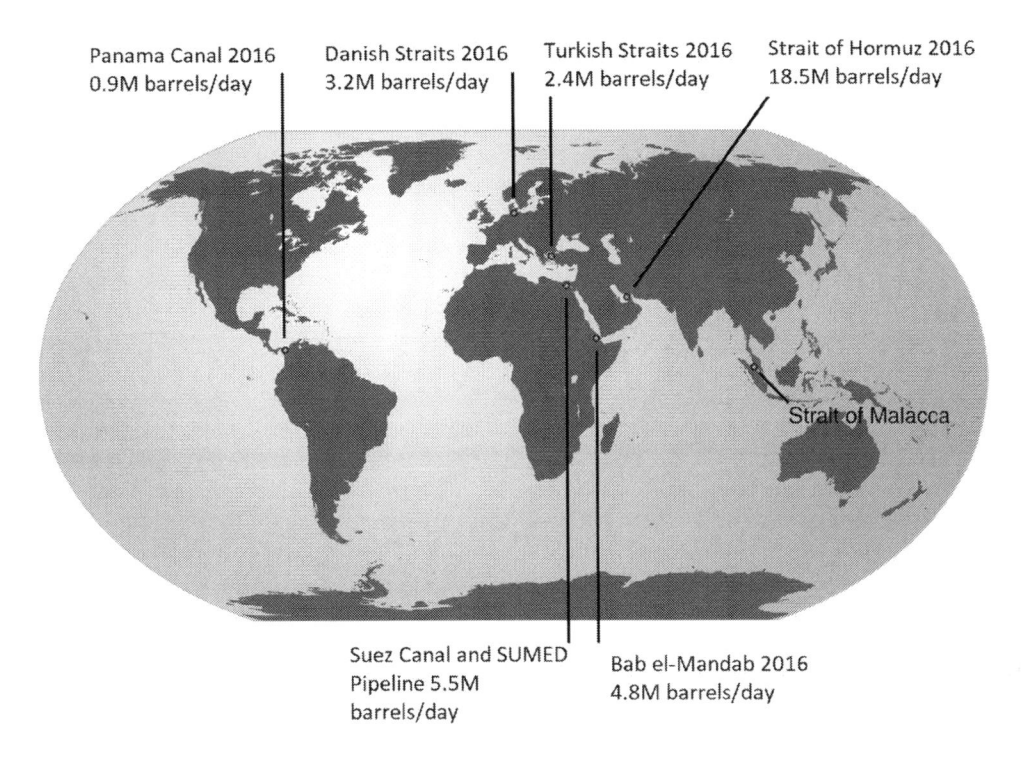

Suez Canal and SUMED
Pipeline 5.5M
barrels/day

Bab el-Mandab 2016
4.8M barrels/day

Figure 4.3 Globally important shipping straits and volume of crude oil and petroleum products shipped in 2010 for comparison purposes

Source: EIA (2017)

maritime states operating in straits. The LOSC articulates the rights of passage through straits in Part III of the Convention.

First and foremost, the LOSC mandated that all ships and aircraft enjoy the right of transit passage between one part of the high seas or EEZ and another part of the high seas or EEZ (Article 38). However, if an alternative high seas route exists to the strait, there is no right of passage (Article 36). No rights of passage were conferred in internal waters unless the establishment of strait baselines encompassed waters historically used for transit (Article 35). Rights of passage in no way affect marine jurisdiction of coastal states bordering straits as outlined in Part II of the Convention (Article 34). No non-transit related activity, including marine scientific research and hydrographic studies, is permitted without the permission of the coastal state(s). Where these activities are permitted by the coastal state, they are subject to the rules of the Convention (Articles 38, 40). Ships during transit passage are required to proceed without delay, avoid any non-transit related activities, refrain from any threat or use of force against the bordering coastal state(s) (Article 39) and comply with international navigation and pollution regulations. Bordering states may establish sea lanes and traffic separation schemes so long as they conform to international regulations and are communicated to seafarers (Article 41). States bordering straits may establish regulations governing the safety of navigation, protection from pollution, fishing and commerce in the strait so long as these regulations do not impede the right of innocent passage (Article 42). Bordering states may communicate known dangers to passage within straits but may not suspend transit passage (Article 44).

Rights of archipelagic states under the LOSC

Part IV of the LOSC governs the unique nature of archipelagic states, which are defined in the LOSC as 'states constituted wholly by one or more archipelagos and may include other islands' (Article 46). The LOSC further defines archipelagos as a group of islands that are socially, politically and geographically interconnected. Many states, including Indonesia (17,500 islands), Philippines (6,800 islands), and Fiji (300 islands), are entirely archipelagic in nature and, when EEZs are mapped, the territorial waters of archipelagic states are extensive and complex. This had led to overlapping jurisdictions, difficulties in interpreting internal and territorial waters and creating political uncertainties for shipping.

The rights and responsibilities of archipelagic states were raised as far back as 1888. It was then that it was suggested that islands constituting a state should have a separate method of determining territorial waters (IDI, 1891). At the first two UNCLOS Conventions in 1958 and 1960, Indonesia and the Philippines championed the notion of recognizing archipelagic states as differentiated from other coastal (e.g. Canada) and marine states (e.g. Germany) (Coquia, 1983).

Nevertheless, it was not until the LOSC finished that the rights of archipelagic states were codified into international law. Archipelagic states may identify archipelagic waters (mostly equivalent to internal waters) by drawing strait baselines that join the outmost points of the outermost islands (and reefs exposed at low tide) (Article 47) so long as:

- The main islands included in this area and the ratio of the area of water to land (including atolls) does not exceed 9:1.
- Baselines do not exceed 100nmi; however, up to 3 per cent of the baselines may exceed 100nmi to a distance of 125nmi.
- Baselines will generally configure to the geographic configuration of the archipelago.
- Baselines do not exclude access to the high seas of another state.
- Baselines do not infringe on or remove access to another state's territorial waters.
- Baselines must be identified on charts or their geographic locations provided to the Secretary-General of the United Nations.

Territorial seas, contiguous zones and EEZs will be measured from the baselines drawn as per Article 47 (Article 48). Archipelagic states have authority over the airspace over their archipelagic waters, as well as the water column, seabed and subsoil resources (Article 49). However, unlike internal waters, the LOSC provides for the right of innocent passage in archipelagic waters (Article 50). This right may only be suspended for protection of the state's security and only after proper notification to other states. Furthermore, archipelagic states must recognize traditional fishing rights and existing agreements with bordering states. As in the case with other coastal states, they too must respect the rights of undersea cables transiting archipelagic waters (Article 52). Archipelagic states may designate sea lanes and air routes through their archipelagic waters consistent with international regulations (Article 53).

No non-transit related activity, including marine scientific research and hydrographic studies, is permitted without the permission of the archipelagic state. Where these activities are permitted by the archipelagic state, they are subject to the rules of the Convention (Articles 38, 40). Ships during transit passage are required to proceed without delay, avoid any non-transit related activities and refrain from any threat or use of force against the archipelagic state (Article 39). Archipelagic states may communicate known dangers to passage within straits but may not suspend transit passage (Article 44).

Rights of landlocked, geographically disadvantaged and developing states under the LOSC

The LOSC is unique among international legal instruments in its approach to states without the means or access to capitalize on oceanic opportunities, whether these opportunities are through trade, fishing or the development of non-renewable ocean resources (Figures 4.4, and 4.5). The

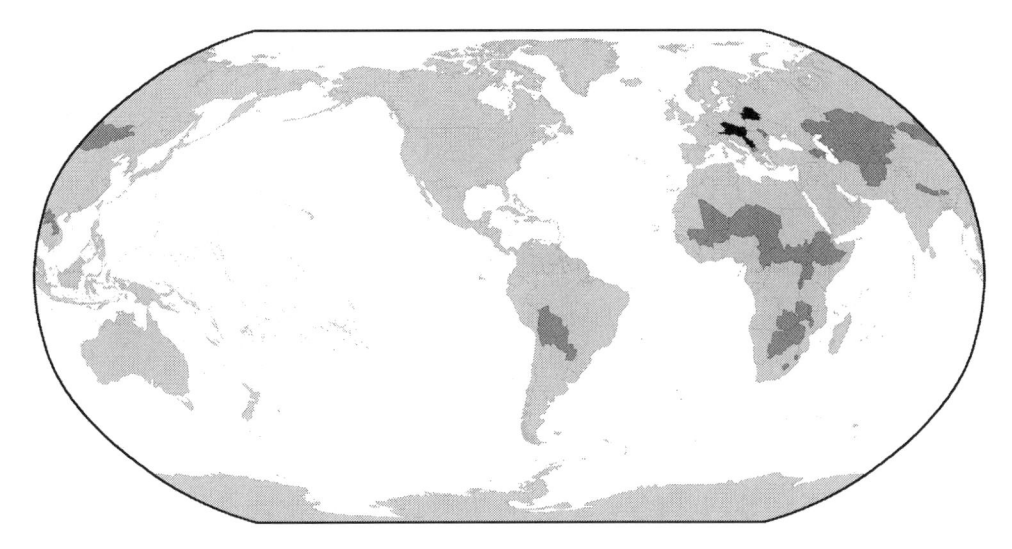

Figure 4.4 Landlocked developed states (darker shading) and landlocked developing countries (LLDCs – lighter shading)

Source: www.un.org/special-rep/ohrlls/lldc/list.htm

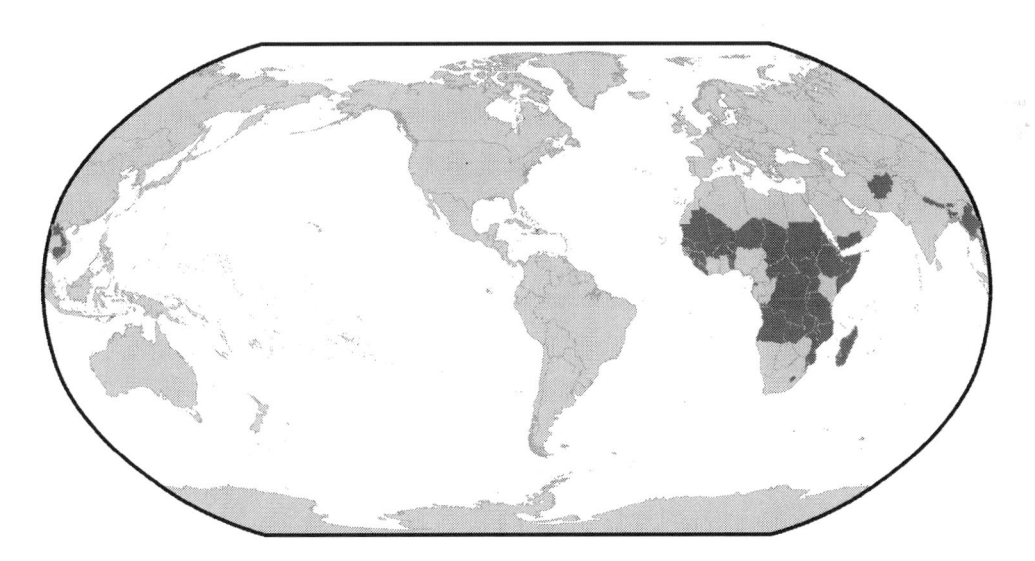

Figure 4.5 Least developed countries (LDCs – shaded states)

Source: www.unohrlls.org/en/ldc/25/

LOSC affirms the rights of these states to participate in the ocean economy as well as mandating the means (e.g. wealth redistribution for economic activity generated in the Area, sharing of surplus fisheries capacity) and mechanisms (e.g. International Seabed Authority to oversee the redistribution of wealth). Specifically, the LOSC addresses the rights of landlocked, geographically disadvantaged and developing states. It also addresses the obligations of other nations to support these rights, which are discussed below (Vasciannie, 1990).

Rights of landlocked states under the LOSC

Landlocked states are those states having no coastline fronting a sea or having no ready access to the sea. There are currently 48 landlocked states, mostly composed of less developed states in Africa and Asia (termed 'landlocked developing states' or LDS by the UN), and 22 of these states are party to the LOSC (UN, 2013b) (Figure 4.3). With the exception of European nations such as Luxembourg and Switzerland, most landlocked states are economically disadvantaged by cumbersome transit costs, protracted transit times and the need to negotiate with coastal states for port access (Bowen, 1986). Landlocked nations, and particularly the 31 landlocked developing countries (LLDCs), seek to participate in ocean development, particularly with respect to transportation, fisheries and seabed mining.

In response to these desires, the LOSC confers certain rights on landlocked states. Landlocked states have the right to participate in the exploitation of the oceans, including activities in the Area (Article 140). Coastal states are directed to allocate a portion of the surplus fish catch within EEZs (based on availability, environmental concerns and nutritional needs) to LLDCs. This is negotiated through bilateral and multilateral agreements with coastal states (Article 69). Article 87 affirms that the high seas are open to all states and that no special rights or privileges are conferred upon coastal or maritime states. Article 125 asserts the right of landlocked states to access the sea using all manner of transportation; however, these rights must be secured through agreements with coastal states. Articles 148 and 160 also commit coastal and developed states to assist LLDCs in participating in activities in the Area, including participation in marine science and research (Articles 143 and 256). Article 82 affirms that wealth produced from the development of states' extended continental shelves will be shared, taking account of the needs and interests of less developed states. Articles 266–278 commit more developed states to cooperate with landlocked states to develop and transfer marine technology.

For more developed landlocked states (Andorra, the Czech Republic, Hungary, Luxembourg, Serbia, Switzerland), the LOSC permits their participation in the exploitation of marine resources only within the EEZs of more developed states in the same geographic region (Article 69).

Rights of geographically disadvantaged states under the LOSC

Geographically disadvantaged states (GDS) are those coastal states – including states bordering enclosed or semi-enclosed seas – that require access to the EEZs of other adjacent states to satisfy their nutritional needs (Article 70).

Using bi- or multilateral agreements with coastal states, the LOSC affirms the rights of GDS to utilize the surplus catch of other coastal nations in the same geographical region or sub-region. The LOSC also promotes the participation of GDS in the Area, recognizing the unique challenges of these states to participate in the economic benefits of oceans' vast expanse. In addition, the LOSC commits to technology transfer to enable developing states to participate in activities in the Area (Article 150). Furthermore, GDS have a right of notification of, as well as participation in, marine research undertaken in an adjacent coastal state (Article 254).

Rights of less developed states under the LOSC

Less developed states (termed 'least developed countries' or LDCs by the UN) are not defined under the LOSC but are defined by the UN as those states with a three-year gross national income (GNI) under $US 1025 that also fall below thresholds as defined by the Human Assets Index (HAI) and Economic Vulnerability Index (EVI) (UN, 2004) (Figure 4.4). In 2018, 47 states met the UN test of LDCs, of which 16 states were landlocked and 11 states consisted of small island states (UN, 2018, 2012). Collectively, these 47 states comprised a total population of 832.3 million people with an average age under 25 years old (UN, 2018, 2012).

In addition, while not included in the LOSC, a companion set of states to the LDCs, termed Small Island Developing States (SIDS), is a particularly useful concept in applying the LOSC (Table 4.1). SIDSs were first recognized as a distinct group of states at the United Nations

Table 4.1 List of small island developing states (SIDS)

UN Members	
Antigua and Barbuda	Federated States of Micronesia
Bahamas	Mauritius
Bahrain	Nauru
Barbados	Palau
Belize	Papua New Guinea
Cape Verde*	Samoa*
Comoros*	São Tomé and Principe*
Cuba	Singapore
Dominica	St. Kitts and Nevis
Dominican Republic	St. Lucia
Fiji	St. Vincent and the Grenadines
Grenada	Seychelles
Guinea-Bissau*	Solomon Islands*
Guyana	Suriname
Haiti*	Timor-Lesté*
Jamaica	Tonga
Kiribati*	Trinidad and Tobago
Maldives*	Tuvalu*
Marshall Islands	Vanuatu*

Non-UN Members	
American Samoa	Guam
Anguilla	Montserrat
Aruba	Netherlands Antilles
British Virgin Islands	New Caledonia
Commonwealth of Northern Marianas	Niue
Cook Islands	Puerto Rico
French Polynesia	US Virgin Islands

Source: www.unohrlls.org/en/sids/46/

Note:
* Also less developed countries

Conference on Environment and Development in June 1992 (Chapter 9) based on the shared challenges of small resource bases; undiversified economies; difficulties accessing international markets; susceptibility to national disasters; high population growth; and high costs of energy, infrastructure and communications (UN, 2013a). There are currently 41 SIDS identified by the UN Office of the High Representative for the Least Developed Countries, Landlocked Developing Countries and the Small Island Developing States (UN, 2013b).

The LOSC affirms the same rights to developing states as geographically disadvantaged states. These rights include: right to utilize the surplus catch of other coastal states (Article 62); right of navigation and passage through straights (Article 38) in the high seas (Article 87); and right to the economic benefits to the Area, including technology transfer to participate in the development of the Area (Article 148).

In addition, more developed states are obligated to assist less developed states in the development of scientific and research capacity related to the protection of the marine environment (including ensuring developing states receive priority in funding for these programs). They are also obligated to make less developed states aware of the effects of their activities on marine environmental quality (Articles 202–203). Articles 266–278 commit developing states to cooperate with landlocked states to develop and transfer marine technology.

Marine scientific research

Part XIII of the LOSC articulates how marine research activities are to be undertaken both inside and outside of territorial waters. The Convention first affirms that all states have the right to undertake scientific research, including LDCs, LLDCs and GDSs (Article 238), that research is to be for peaceful purposes (Article 240) and that marine research cannot be used as a means to assert territorial claims (Article 241). States are obliged to cooperate on marine research, including assisting LDCs (Articles 242–243), and to disseminate the results of this work (Article 244). All states may undertake research in the Area or in the water column beyond the EEZ (Article 256).

Coastal states have complete control over research activities in their territorial seas, EEZs and continental shelves (Articles 246–246). Other states must secure consent of the coastal state before undertaking research in these areas; however, coastal states are expected to allow marine research to occur in their EEZs and continental shelves so long as the activities benefit mankind and conform to national procedures. The marine research is prohibited also from creating structures, drilling into the continental shelf, or exploiting living and non-living natural resources (Article 246).

Successes and challenges of the LOSC

Overall, the LOSC is viewed as one of the most successful international conventions, and the term 'Constitution for the Oceans' adequately represents the magnitude of the Convention. Given the scope and scale of issues, there were no expectations from states that were involved in drafting the Convention that they would 'get it right' the first time. However, to date, most states respect and adhere to the Convention, and the processes that were outlined in the LOSC have, for the most part, progressed in an orderly fashion (Freestone, 2013).

Nonetheless, with the hindsight of 30 years, one can see that there are a number of issues that were neglected during drafting of the Convention, as well as problems that surfaced after the completion of the LOSC. Summarized by Freestone (2012), they include:

- Underwater cultural heritage resources were not addressed but have since been managed under the UNESCO Convention on Underwater Cultural Heritage.

- The concept of managing to maximum sustainable yield (MSY) is a central tenant of the LOSC (Chapter 9) yet has been subsequently questioned as a workable fishery management tool.
- The Convention fails to provide an adequate definition of marine scientific research resulting in disagreement and conflict between states who conduct research-related activities (e.g. hydrographic surveys, ocean fertilization) within other states EEZs.
- The Convention is concerned with the relations between states rather than consideration of individual-related issues such as human rights.
- Climate change (Chapter 8) was never envisioned as a future issue when the Convention was drafted.
- While the LOSC addressed the high seas and deep seabed areas, the Convention was not a comprehensive treatise on ABNJ addressing topics such as bioprospecting and marine protected areas (Chapter 9).

In addition, Churchill (2012) determined that at least one third of the 160+ state parties to the LOSC are in breach of one or more of the Convention's provisions in the following ways:

- A number of states have established baselines, territorial seas and EEZs contrary to the LOSC direction.
- Not all states are in compliance with Article 94 of the Convention related to seaworthiness of ships flying their flag.
- Many states are in breach of their obligations under Article 61 (2) to maintain living resources within their EEZs as well as Article 117–19 to conserve living resources of the high seas.
- A number of states are failing to preserve or protect rare or fragile marine ecosystems (Article 194 [5]).

International organizations that establish and oversee international marine law

Permanent court of arbitration

The Permanent Court of Arbitration (PCA) was established by the Convention for the Pacific Settlement of International Disputes in 1899. The PCA's original mandate was to facilitate dispute resolution between states and has evolved to provide dispute resolution services to states, state entities, inter-governmental organizations and private parties. There are currently over 100 member states, and the PCA is based in The Hague, Netherlands.

The first marine arbitration undertaken by the PCA was the 1910 North Atlantic Coast Fisheries dispute between the United States and Great Britain. A fundamental part of this dispute was how the 3nmi boundary for territorial waters was measured for bays. Great Britain (representing what is now Canada) contended that 'bays' were defined from headland to headland regardless of size, thus excluding American fishers from most Canadian coastal nearshore waters. The United States retorted that only smaller 'bays' less than 6nmi from the entrance should be included as territorial waters. The PCA ruled that, for bays larger than 3nmi from headland to headland at the entrance, territorial waters should be defined by the 3nmi boundary following the geography of the coastline.

The PCA plays an important role in dispute resolution under the LOSC. Part XV of the Convention sets out the options for dispute resolutions. However, should a state not express a preference for a particular means of dispute resolution or states disagree on what means to use, an arbitral tribunal (constituted under Annex VII of the LOSC) becomes the default means of dispute resolution. The majority of dispute resolutions to date have involved disputes over maritime boundaries; since the LOSC came into force in 1994, five of the six of the Annex VII dispute resolutions have been arbitrated under the PCA. They are as follows:

- Bangladesh v. India (marine boundary delineation), instituted in October 2009 and decided in 2016.
- Ireland v. United Kingdom ('MOX Plant Case' concerning radioactive contamination of the Irish Sea by the United Kingdom), instituted in November 2001 and terminated through a tribunal order issued on 6 June 2008.
- Guyana v. Suriname (marine boundary delineation), instituted in February 2004 and decided by a final award rendered on 17 September 2007.
- Barbados v. Trinidad and Tobago (marine boundary delineation), instituted in February 2004 and decided by a final award rendered on 11 April 2006.
- Malaysia v. Singapore (land reclamation activities by Singapore in the Straits of Johor), instituted in July 2003 and terminated by an award on agreed terms rendered on 1 September 2005.
- Bangladesh v. India (marine boundary delineation), instituted in October 2009 and decided by a final award rendered on 7 July 2014.
- Mauritius v. United Kingdom (marine protected area delineation around the Chagos Archipelago), instituted December 2010 and decided by final award on 18 March 2015.
- Croatia v. Slovenia (territorial and maritime dispute), instituted in November 2009 and decided by final award on 29 June 2017.
- Philippines v. China (marine boundary dispute), instituted January 2013 and decided by a final award on 12 July 2016.
- Denmark v. the European Union (dispute over herring fishery allocation), instituted August 2013 and decided by a final award on 23 September 2014.
- Timor-Leste v. Australia (arbitration under the Timor Sea Treaty), instituted 2013 and decided 2016.

Source: www.pca-cpa.org

International court of Justice

The ICJ was established in 1945 by the Charter of the United Nations and began operating in 1946. The ICJ is one of the six principle organs of the UN and, like the PCA, is headquartered in The Hague. The court is composed of 15 judges elected for nine-year terms by the UNGA and UNSC. The Court is responsible for settling legal disputes submitted to it by states (termed 'Contentious cases') and providing advisory opinions on legal questions referred to it by authorized United Nations organs and specialized agencies (termed 'Advisory proceedings'). The ICJ's first case was the Corfu Channel Case; since then, the ICJ has adjudicated a number of marine disputes (discussed in Box 4.1).

Box 4.1 International Court of Justice marine-related cases

- Access to the Pacific Ocean (Bolivia v. Chile). Initiated 2013 and concluded 2018.
- Two cases reviewing the Judgment of 23 May 2008 in the case concerning Sovereignty over Pedra Branca/Pulau Batu Puteh, Middle Rocks and South Ledge (Malaysia v. Singapore). Initiated 2017 and concluded 2018.

- Maritime Delimitation in the Caribbean Sea and the Pacific Ocean (Costa Rica v. Nicaragua). Initiated 2014 and concluded 2018.
- Territorial and Maritime Dispute (Nicaragua v. Colombia). Nicaragua claims ownership of the maritime features (cays) off her Caribbean coast and instituted proceedings against Columbia in 2001. Honduras and Costa Rica submitted applications for permission to intervene in the dispute, but permission was not granted by the ICJ.
- Maritime Dispute (Peru v. Chile). Peru instituted proceedings against Chile in 2008 to delineate marine boundaries between the two nations. The ICJ issued a ruling on the boundary in January 2014.
- Whaling in the Antarctic (Australia v. Japan). Australia initiated proceedings against Japan in 2010, contending that Japan continues to ignore the global moratorium on commercial whaling under the International Convention for the Regulation of Whaling. In March 2014, the ICJ ruled in favour of Australia.
- Maritime Delimitation in the Black Sea (Romania v. Ukraine). Romania initiated proceedings against the Ukraine in 2004 to delineate marine boundaries between the two nations.
- Territorial and Maritime Dispute between Nicaragua and Honduras in the Caribbean Sea (Nicaragua v. Honduras). Nicaragua initiated proceedings against the Honduras in 1999 to delineate marine boundaries between the two nations.
- Fisheries Jurisdiction (Spain v. Canada). Spain initiated proceedings against Canada in 1995 following the seizure of a Spanish vessel fishing outside of Canada's EEZ, but on Canada's continental shelf. The Kingdom of Spain contended that no international law permits the application of domestic law and the seizure of vessels on the high seas. The ICJ ruled that it lacked jurisdiction to adjudicate the dispute in 1998.
- Oil Platforms (Islamic Republic of Iran v. United States of America). Iran initiated proceedings against the United States for violating the 1957 'Treaty of Amity' and for destroying offshore oil and gas platforms owned by the National Iranian Oil Company in 1987 and 1988. The ICJ rejected the claims of both states in 2003.
- Maritime Delimitation and Territorial Questions between Qatar and Bahrain (Qatar v. Bahrain). Qatar initiated proceedings against Bahrain in 1991 to resolve sovereignty over the Hawar islands, sovereign rights over the shoals of Dibal and Qit'at Jaradah, and the delimitation of the maritime areas of the two States. The ICJ ruled that Qatar has sovereignty over Zubarah and Janan Island and that the low tide elevation of Fasht ad Dibal falls under the sovereignty of Qatar. It also found that Bahrain has sovereignty over the Hawar Islands and the island of Qit'at Jaradah and the court drew a single maritime boundary between the two states.
- Maritime Delimitation between Guinea-Bissau and Senegal (Guinea-Bissau v. Senegal). Guinea-Bissau initiated proceedings against Senegal in 1991 to delineate marine boundaries between the two states. The ICJ removed the case from its docket when the two states reached agreement on the management of the disputed area.
- Maritime Delimitation in the Area between Greenland and Jan Mayen (Denmark v. Norway). Denmark initiated proceedings against Norway in 1988 to delineate marine boundaries between the two states. The ICJ established fixed a delimitation line for both the continental shelf and the fishery zones in the area between Greenland and Jan Mayen.
- Delimitation of the Maritime Boundary in the Gulf of Maine Area (Canada/United States of America). In 1979, Canada and the United States requested the ICJ to delineate a single line to delimit both the continental shelf and the 200-mile exclusive fishery zone in the Gulf of Maine. The ICJ issued its ruling in 1984.

- Continental Shelf (Tunisia/Libyan Arab Jamahiriya). In 1977, Tunisia and Libya requested the ICJ determine the principles and rules of international law as applied to the ownership of their respective continental shelves. The ICJ issued its ruling in 1982 using the principles of an 'equitable solution' to establish a segmented boundary delineating continental shelves.
- Aegean Sea Continental Shelf (Greece v. Turkey). Greece initiated proceedings against Turkey in 1976 to delineate marine boundaries between the two states. The ICJ determined it was without jurisdiction to entertain the application.
- Fisheries Jurisdiction (Federal Republic of Germany v. Iceland). Germany initiated proceedings against Iceland in 1972 after Iceland extended its jurisdiction over fisheries from 12nmi to 50nmi. The ICJ ruled that Iceland could not unilaterally exclude other states from its fisheries beyond 12nmi; however, Iceland was entitled to preferential rights to the fish in its coastal waters due to the dependence of Iceland on fishing.
- Fisheries Jurisdiction (United Kingdom of Great Britain and Northern Ireland v. Iceland). Great Britain initiated proceedings against Iceland in 1972 after Iceland extended its jurisdiction over fisheries from 12nmi to 50nmi. The ICJ ruled that Iceland could not unilaterally exclude other states from its fisheries beyond 12nmi; however, Iceland was entitled to preferential rights to the fish in its coastal waters due to the dependence of Iceland on fishing.
- North Sea Continental Shelf (Federal Republic of Germany/Netherlands and Federal Republic of Germany/Denmark). Germany and the Netherlands, as well as Germany and Denmark, requested the ICJ to state the principles and rules of international law applicable to resolving delimiting boundaries of their respective continental shelf in 1967. In 1969, the ICJ simultaneously ruled in both cases that there was no basis in customary law for delineating continental shelf areas based on equidistance from land. They directed the three states to carry out the delimitations on the basis of equitable principles.
- Constitution of the Maritime Safety Committee of the Inter-Governmental Maritime Consultative Organization. In 1959, The Assembly of the Inter-Governmental Maritime Consultative Organization requested the ICJ to provide an opinion on whether the Maritime Safety Committee of the Inter-Governmental Maritime Consultative Organization was constituted in accordance with the Convention for the Establishment of the Organization. In 1961, the ICJ ruled that that the Maritime Safety Committee was not properly constituted since the Convention required the Assembly to elect the eight states with the largest amount of registered shipping tonnage and this had not been done.
- Antarctica (United Kingdom v. Chile and United Kingdom v. Argentina). The United Kingdom initiated proceedings against Chile and Argentina in 1955 concerning disputed territory in the Antarctic. Both Chile and Argentina refused to submit to the jurisdiction of the ICJ and the case was removed from the ICJ docket in 1956.
- Fisheries (United Kingdom v. Norway). The United Kingdom initiated proceedings against Norway for claiming exclusive fishing rights beyond a 10nmi baseline in 1949. The ICJ ruled that the 10nmi baseline was not an established custom in international law and that Norway's claims were not inconsistent with international law.
- Corfu Channel (United Kingdom of Great Britain and Northern Ireland v. Albania). The United Kingdom initiated proceedings against Albania for compensation resulting from damage to two British destroyers that had struck mines in Albanian waters. The ICJ ordered Albania to pay £843,947 in 1949.

Source: www.icj-cij.org/en/list-of-all-cases

Organizations under the LOSC that establish and oversee international marine law

International Seabed Authority

The International Seabed Authority (the Authority or ISA) was established by Part XI of the LOSC in combination with the 1994 Agreement relating to the Implementation of Part XI of the United Nations Convention on the Law of the Sea of 10 December 1982. This Agreement and Part XI of the Convention are to be interpreted and applied together as a single instrument.

The Authority was established 16 November 1994 when the LOSC came into force. Headquartered in Kingston, Jamaica, it became operational in June 1996 and operates as an autonomous international organization within the United Nations system (www.isa.org.jm). The powers of the Authority are multifarious; within its dominion is the coordination of marine scientific research (Article 143), the transfer of technology (Article 144) and the protection of the marine environment and human life (Articles 145–146).

The ISA Council is responsible for implementing the seabed provision in the LOSC, including assigning rights to explore and develop geographic areas, approving work plans and setting environmental standards. The Legal and Technical Commission (LTC) advises Council on applications, technical reviews and environmental assessments. The LTC has developed *Regulations on Prospecting and Exploration for Polymetallic Nodules in the Area* and the *Regulations on Prospecting and Exploration for Polymetallic Sulphides and Cobalt-Rich Ferromanganese Crusts in the Area* and is currently engaged in developing exploitation regulations for the same (www.isa.org.jm).

The International Seabed Authority has approved and signed 29 15-year contracts for exploration for polymetallic nodules, polymetallic sulphides and cobalt-rich ferromanganese crusts. Seventeen of these contracts are for exploration for polymetallic nodules in the Clarion-Clipperton Fracture Zone (16) and Central Indian Ocean Basin (1). There are seven contracts for exploration for polymetallic sulphides in the Southwest Indian Ridge, Central Indian Ridge and the Mid-Atlantic Ridge and five contracts for exploration for cobalt-rich crusts in the Western Pacific Ocean.

Currently, ambiguity exists over whether the mineral resources of the Southern Ocean are regulated by the Authority under Part XI of the LOSC and the 1994 Implementation Agreement for Part XI or the Antarctic Treaty Consultative Parties under Article 7 of the Protocol on Environmental Protection to the Antarctic Treaty (Vidas, 2000). With the exception of scientific research, the Protocol prohibits any activity related to mineral resources and therefore conflicts with the 1994 Implementation Agreement. However, the incongruity remains moot because the 29 state parties to the Antarctic Treaty have to date agreed not to explore for deep-sea mineral resources in the Southern Ocean.

International Tribunal for the Law of the Sea

The International Tribunal for the Law of the Sea (ITLOS) was established under Annex VI of the LOSC to adjudicate disputes arising from the Convention. The panel is composed of 21 judges elected by member states and is headquartered in Hamburg, Germany. However, the Tribunal may sit and exercise its functions wherever necessary. Each state is limited to one member (judge) on the Tribunal, and at least three members are required from each geographic group as defined by the UN. Members are elected for nine years and can be re-elected. States party to a dispute who are not represented on the Tribunal are permitted to nominate an ad hoc judge with full powers of participation for the duration of the dispute they are selected to participate in. Parts XI and XV of the LOSC govern how disputes are referred to the Tribunal.

While tribunal decisions are final, they are only binding on parties to the dispute. Key Tribunal rulings are discussed in Box 4.2.

> **Box 4.2 International Tribunal for the Law of the Sea rulings**
>
> - The M/V 'SAIGA' Case (Saint Vincent and the Grenadines v. Guinea – case 1). On 28 October 1997, Guinea seized the oil tanker *SAIGA*, registered in the Saint Vincent and the Grenadines, on the basis the ship was involved in smuggling activities. The ship was arrested in international waters after a pursuit from Guinean waters. Saint Vincent and the Grenadines submitted their application on this matter 13 November 1997. Guinea contended that the Tribunal lacked jurisdiction in the matter. The Tribunal (divided 12 to 9) ordered the prompt release of the vessel and its crew from detention in Conakry, Guinea.
> - The M/V 'SAIGA' (No. 2) Case (Saint Vincent and the Grenadines v. Guinea – case 2). Following the Tribunal order to release the *SAIGA* on 4 December 1997, Guinea prosecuted the Master of the vessel resulting in a fine of US $15,000,000, confiscation of the vessel and six months imprisonment for the Master (all convictions suspended). The Tribunal ordered Guinea to refrain from carrying out its national court decision.
> - Southern Bluefin Tuna Cases (New Zealand v. Japan; Australia v. Japan – cases 3 and 4). On 30 July 1999, Australia and New Zealand submitted requests to the Tribunal for provisional measures (an injunction) against Japan to immediately halt an experimental fishery for southern bluefin tuna. On 9 August 1999, Japan filed a request to the Tribunal to deny any provisional measures for the fishery. Australia and New Zealand requested an arbitration procedure under Annex VII of the LOSC. On 27 August 1999, the Tribunal ordered the three parties to resume negotiations with the goal of reaching an agreement on measures to conserve and manage southern bluefin tuna and that the parties should restrict their catches. The Tribunal also ordered a halt to the experimental fishery unless the experimental fishery catch was included in the total allowable catch.
> - The 'Camouco' Case (Panama v. France – case 5). The Panamanian-registered vessel *Camouco* was arrested by French authorities for fishing for Patagonian toothfish in the EEZ of the French Southern and Antarctic Territories on 28 September 1999 and escorted to Reunion. On 17 January 2000, Panama filed a request to the Tribunal for the prompt release of the vessel and its Master. On 7 February 2000, the Tribunal order the release of the vessel and its crew upon payment of a US$1.2 million bond.
> - The 'Monte Confurco' Case (Seychelles v. France – case 6). The Republic of Seychelles registered *Monte Confurco* was arrested by a French frigate on 8 November 2000 while fishing for toothfish in the EEZ of the Kerguelen Islands in the Southern Indian Ocean. The local district court ruled that the vessel could be released upon posting a bond of 56.4 million French Francs (FF). The Seychelles applied to the Tribunal to have the vessel and Master released and the bond reduced to 9.8 million FF. On 18 December 2000, the Tribunal ordered the prompt release of the vessel and its Master by France upon posting a bond of 18 million FF.
> - Case concerning the Conservation and Sustainable Exploitation of Swordfish Stocks in the South-Eastern Pacific Ocean (Chile/European Union – case 7). The Tribunal

was requested on 20 December 2000 by Chile and the European Union to evaluate whether the European member states were meeting their LOSC obligations with respect to swordfish conservation in the Southeast Pacific and whether the LOSC permits Chile's domestic swordfish conservation measures to be applied on the high seas. On 17 December 2009, the Tribunal discontinued the case at the request of the parties as they had resolved the dispute between them.

- The 'Grand Prince' Case (Belize v. France – case 8). The Belize registered fishing vessel *Grand Prince* was arrested by a French frigate 26 December 2000 while fishing for toothfish in the EEZ of the Kerguelen Islands in the Southern Indian Ocean. On 21 March 2001, Belize applied to the Tribunal to have the vessel and Master released and the bond reduced to 206,149 Euros. On 20 April 2001, the Tribunal ruled that it had no jurisdiction in the matter as it could not be proven that Belize was the flag state of the vessel when the application was made.

- The 'Chaisiri Reefer 2' Case (Panama v. Yemen – case 9). On 3 May 2001, Yemen arrested the fishing vessel *Chaisiri Reefer 2* for allegedly violating Yemeni fisheries laws. Panama requested the Tribunal to order the immediate release of the vessel; however, on 3 July 2001, the case was removed from Tribunal's list of cases as Panama and Yemen reached agreement on the release of the ship and its crew.

- The MOX Plant Case (Ireland v. United Kingdom – case 10). On 25 October 2001, Ireland requested the Tribunal to adjudicate a dispute between Ireland and the United Kingdom over the opening of a mixed oxide fuel (MOX – reprocessed spent nuclear fuel) plant in Sellafield on the Irish Sea. On 3 December 2001, the Tribunal ruled that; a) it had jurisdiction over the dispute; b) provisional measures (an injunction) were not required as the plant would not be operational until 2002; and c) directed both parties to cooperate on exchanging information and monitoring the environmental effects of the MOX plant.

- The 'Volga' Case (Russian Federation v. Australia – case 11). On 7 February 2002, the Australian military arrested the fishing vessel *Volga* on the high seas in the Southern Ocean for allegedly illegally fishing for Patagonian toothfish within the Australian EEZ. Russia requested the Tribunal order the release of the Volga under a $AU 500,000 bond. On 23 December 2002, the Tribunal order the release of the Volga pending provision of an AU$ 1.92 million bond.

- Case concerning Land Reclamation by Singapore in and around the Straits of Johor (Malaysia v. Singapore – case 12). On 5 September 2003, Malaysia requested the Tribunal enact provisional measures against Singapore for land reclamation activities alleged to impinge on Malaysian territory. On 8 October 2003, the Tribunal ruled that provisional measures were not required as Malaysia failed to demonstrate the activities would cause irreversible damage and ordered both parties to cooperate through joint studies and information sharing and ordered Singapore not to conduct its reclamation activities in a manner that might harm the marine environment or affect Malaysian rights.

- The 'Juno Trader' Case (Saint Vincent and the Grenadines v. Guinea-Bissau – case 13). On 26 September 2004, the fishing vessel *Juno Trader* was arrested by Guinea-Bissau for allegedly illegally fishing in its national waters. Saint Vincent and the Grenadines requested that the Tribunal order the prompt release of the vessel and her crew. On 18 December 2004, the Tribunal ordered the release of the vessel under a EUR 300,000 bond.

- The 'Hoshinmaru' Case (Japan v. Russian Federation – case 14). On 1 June 2007, the fishing vessel *Hoshinmaru* was arrested for alleged Russian fisheries violations in

Russia's EEZ. Japan requested that the Tribunal order the release of the vessel under terms and conditions the Tribunal considered reasonable. On 7 August 2007, the Tribunal ordered the release of the *Hoshinmaru* under a 10 million roubles bond.
- The 'Tomimaru' Case (Japan v. Russian Federation – case 15). On 31 October 2006, the fishing vessel *Tomimaru* was arrested for alleged Russian fisheries violations in Russia's EEZ. Japan requested that the Tribunal order the release of the vessel under terms and conditions the Tribunal considered reasonable. On 7 August 2007, the Tribunal ruled that Japan's application was without object (a claim that has become pointless) due to a ruling by Russian courts to confiscate the vessel. The Tribunal noted that confiscation may upset the balance between coastal and flag states.
- Dispute concerning delimitation of the maritime boundary between Bangladesh and Myanmar in the Bay of Bengal (Bangladesh/Myanmar – case 16). The first states to submit a marine boundary dispute to ITLOS. ITLOS issued its judgement in 2012 and ruled that the LOSC Article 15 governed the approach to boundary delineation in this dispute and that equidistance modified by relevant circumstances was the guiding principle for delimiting the territorial sea.
- Responsibilities and obligations of States sponsoring persons and entities with respect to activities in the Area (Request for Advisory Opinion submitted to the Seabed Disputes Chamber – case 17).
- The M/V 'Louisa' Case (Saint Vincent and the Grenadines v. Kingdom of Spain – case 18). On 1 February 2006, Spain arrested the Saint Vincent and the Grenadines registered vessel *Louisa* for allegedly contravening Spanish national marine environmental laws while conducting sonar and caesium magnetic surveys of the seafloor of the Bay of Cadiz to support oil and gas exploration activities. Saint Vincent and the Grenadines requested the Tribunal order the release of the vessel under provisional measures. On 23 December 2010, the Tribunal ruled that there was no real and imminent risk to the rights of the parties and received assurances from Spain that the vessel was being monitored for environmental impacts while being held in Spanish custody.
- The M/V 'Virginia G' Case (Panama v. Guinea-Bissau – case 19). On 21 August 2009, Guinean authorities arrested the oil tanker *Virginia G* while refuelling fishing vessels within Guinea-Bissau's EEZ. The tanker was released after 14 months in Guinean custody, however, Panama claimed that the vessel was damaged during this time and applied to the Tribunal for reparations.
- The M/V 'ARA Libertad' Case (Argentina v. Ghana – case 20). On 4 October 2012, Ghana prevented the military training vessel from departing the port of Tema due to Argentina's default on bonds. Ghana demanded $US 20M for the release of the vessel. On 15 December 2012, the Tribunal ordered the release of the vessel.
- Advisory opinion on the role of the flag state in preventing illegal, unreported and unregulated (IUU) fishing activities (case 21). On 2 April 2015, the Tribunal ruled that flag states are obligated to take necessary measures to ensure vessels flying its flag are not engaged in IUU fishing activities. Furthermore, coastal states may hold flag states liable for vessels flying their flag when engaged in IUU fishing activities in Exclusive Economic Zones.
- The 'Arctic Sunrise' Case (Netherlands v. Russian Federation – case 22). On 19 September 2013, the Russian Federal seized the Greenpeace icebreaker 'Arctic Sunrise' (flying the Netherlands flag) and charged the crew with piracy and hooliganism. The Netherlands applies to the Tribunal on 21 October 2013, for release of the vessel.

> On 22 November 2013, the Tribunal ordered the release of the vessel upon posting a bond of EUR 3.6 million bond.
> - Dispute concerning delimitation of the maritime boundary between Ghana and Cote d'Ivoire (case 23). On 23 September 2017, the Tribunal ruling delineated a boundary between the two states.
> - The 'Enrica Lexie' Incident (Italy v. India – case 24). On 15 February 2012, two Indian fishers were shot and killed by armed guards aboard the Italian oil tanker *Enrica Lexie*. The guards were arrested by India and charged with murder resulting in strained relations between the two states. On 21 July 2015, Italy applied to the Tribunal to direct India to suspend legal actions against the two guards and let them remain in Italy. On 24 August 2015, the Tribunal directed both states to refrain from initiating additional legal proceedings while the Tribunal considered this issue.
> - The M/V 'Norstar' Case (Panama v. Italy – case 25). Between 1994 and 1998, the Norstar delivered fuel to mega-yachts in the Mediterranean outside the territorial seas of Italy, France and Spain. In 1998, the vessel was arrested by Spain in contravention of Italian legislation. On 17 December 2015, Panama applied to the Tribunal for compensation. On 4 November 2016, the Tribunal ruled it had jurisdiction to hear the application.
>
> Source: www.itlos.org/en/cases/

In addition, ITLOS has provided non-binding advice through two advisory proceedings:

1. Request for an advisory opinion submitted by the Sub-Regional Fisheries Commission (SRFC) (Request for Advisory Opinion submitted to the Tribunal); 2013–2015.
2. Responsibilities and obligations of States sponsoring persons and entities with respect to activities in the Area (Request for Advisory Opinion submitted to the Seabed Disputes Chamber); 2010–2011; outlined below.

ITLOS in composed of a number of chambers discussed below.

Seabed Disputes Chamber

The Seabed Disputes Chamber is established under Annex VI, Section 4, Article 35 of the LOSC and governed by Part XI (s. 5) and Part XV of the LOSC. The Seabed Disputes Chamber has jurisdiction within the International Seabed Area (the Area) and is composed of 11 judges selected by the Tribunal every three years. Any mineral disputes in the Area must be settled by the Chamber and parties are not permitted to choose another type of court or tribunal. The Chamber has jurisdiction over disputes between (the LOSC Part XI, section 5, Article 187):

- States over the interpretation of Part XI of the LOSC and associated Annexes.
- States and the Authority concerning the function and actions of the Authority.
- Any type of party with respect to interpretations of contracts or conflicts between parties operating in the Area.
- A state-sponsored contractor and the Authority.
- A party and the Authority for any damage arising out of wrongful acts in the exercise of the Authority's powers and function.

With respect to non-contractual matters, state parties in a dispute may submit cases to a three-member ad hoc chamber convened by the Seabed Disputes Chamber or a special chamber of the ITLOS. With respect to contractual matters, including disputes over work plans, transfer of technology and financial matters, parties will be submitted to binding commercial arbitration.

In 2008, the Authority received applications submitted by Nauru Ocean Resources Inc. (sponsored by the Republic of Nauru) and Tonga Offshore Mining Ltd. (sponsored by the Kingdom of Tonga) for approval of a plan of work for deep seabed mineral exploration. In 2010, Nauru requested that the Authority seek an advisory opinion from the Chamber regarding the extent of the liabilities of a state sponsoring 'seafloor mining in international waters'. Nauru and Tonga were both concerned that the potential liabilities in undertaking seafloor mining could amount to more than the states' GDPs, which effectively excluded developing states from participating in seafloor mining activities.

In 2011, the Chamber produced an Advisory Opinion (case 17 of ITLOS) titled 'Responsibilities and Obligations of States Sponsoring Persons and Entities With Respect to Activities in the Area'. The Opinion determined that states are not liable for damage caused by state-sponsored contractors if the states have domestic laws, regulations and administrative measures to secure compliance by persons under its jurisdiction (Plakokefalos, 2012).

Chamber of summary procedure

The Chamber of Summary Procedure is constituted annually and consists of five members and two alternates. The purpose of this chamber is to quickly resolve issues without extensive formalities or procedures.

Chamber for Fisheries Disputes

The Chamber for Fisheries Disputes, established in 1997, consists of nine members and is established under the Special Chambers provision in Article 15, paragraph of the ITLOS Statute. The Chamber for Fisheries Disputes was established to address disputes concerning the conservation and management of marine living resources.

Chamber for Marine Environment Disputes

The Chamber for Marine Environment Disputes, established in 1997, consists of seven members and is established under the Special Chambers provision in Article 15, paragraph of the ITLOS Statute. The Chamber for Marine Environment Disputes was established to address disputes concerning the protection and preservation of the marine environment.

Chamber for Maritime Delimitation Disputes

The Chamber for Maritime Delimitation Disputes, established in 2007, consists of 11 members and is established under the Special Chambers provision in Article 15, paragraph of the ITLOS Statute.

Special chambers

Article 15, paragraph 2 of the ITLOS Statute enables the establishment of special chambers. The first application of a Special Chamber was in 2000 to address the Case concerning the Conservation and Sustainable Exploitation of Swordfish Stocks in the South-Eastern Pacific Ocean

(Chile/European Community). In 2007, a Special Chamber was established to address maritime boundary disputes.

Commission on the Limits of the Continental Shelf

In accordance with Article 76 of the LOSC the Commission on the Limits of the Continental Shelf (the 'Commission') was established by Annex II of the LOSC. The purpose of the Commission is to make recommendations on the seaward boundaries of the continental shelves for those states whose continental shelves extend beyond their EEZs and to assist states in preparing applications for review by the Commission (Article 3). The Commission is comprised of 21 members with backgrounds in marine mapping/geology elected by state parties to the LOSC for terms of five years with eligibility of re-election (Article 2). States have 10 years after ratifying the LOSC to submit evidence on the extent of their continental shelves (Article 4). As a result, the Commission has for years been backlogged, sorting out the technicalities of various, sometimes competing, submissions.

> **Discussion questions**
>
> - The LOSC is decades old: What significant deep-sea activities did it foresee? Have they all come to pass?
> - What challenges did the Convention not foresee, and how might these challenges be addressed inside and outside of the Convention?
> - How does the LOSC attempt to balance industrial exploitation by developed states with the needs of small island developing states and least developed countries?

Further reading

Bezpalko, I. (2004) 'The deep seabed: Customary law codified', *Natural Resources Journal*, vol 44, pp. 867–887.

Bowen, R. E. (1986) 'The land-locked and geographically disadvantaged states and the law of the sea', *Political Geography Quarterly*, vol 5, no 1, pp. 63–69.

CEPAL (2004) *Major Issues in the Management of Enclosed or Semi-Enclosed Seas, with Particular Reference to the Caribbean Sea*, LC/CAR/L.24, www.eclac.cl/publicaciones/xml/1/20811/L0024.pdf, accessed 13 February 2013.

Churchill, R. (2012) 'The persisting problem of non-compliance with the Law of the Sea Convention: Disorder in the Oceans', *The International Journal of Marine and Coastal Law*, vol 27, pp. 813–820.

Coquia, J. R. (1983) 'Development of the archipelagic doctrine as a recognized principle of international law', *Philippine Law Journal*, vol 58, pp. 143–171.

EIA (2017) *World Oil Transit Chokepoints*, U.S. Energy Information Administration, www.eia.gov/beta/international/regions-topics.php?RegionTopicID=WOTC, accessed 11 May 2019.

Freestone, D. (2012) 'The law of the sea convention at 30: Successes, challenges and new agendas', *The International Journal of Marine and Coastal Law*, vol 27, pp. 675–682.

Freestone, D. (ed.) (2013) *The 1982 Law of the Sea Convention at 30: Successes, Challenges and New Agendas*, Martinus Nijhoff, The Hague.

Institut de Droit International (1891) 'Règlement international des prises maritimes', *ANNUAIRE Hamburg Session*, vol 11, p. 37.

LeGresley, E. (1993) *The Law of the Sea Convention*, publications.gc.ca/Collection-R/LoPBdP/BP/bp322-e.htm, accessed 04 February 2019.

McDorman, T. L. (2008) *Salt Water Neighbors: International Ocean Law Relations between the United States and Canada*, Oxford University Press, New York.

Plakokefalos, I. (2012) 'Seabed disputes chamber of the international tribunal for the law of the sea: Responsibilities and obligations of states sponsoring persons and entities with respect to', *Journal of Environmental Law*, vol 24, no 1, pp. 133–143.

United Nations (1982) *United Nations Convention on the Law of the Sea of 10 December 1982*, www.un.org/depts/los/convention_agreements/texts/unclos/UNCLOS-TOC.htm, accessed 04 January 2019.

United Nations (2004) *United Nations, Committee for Development Policy Report on the Sixth Session (29 March–2 April 2004)*, Economic and Social Council Official Records, Supplement No. 13, www.un.org/en/ga/sixth/72/72_session.shtml, accessed 04 January 2019.

United Nations (2005a) *Convention on the Territorial Sea and the Contiguous Zone*, Treaty Series, vol 516, p. 205, www.un.org/depts/los/convention_agreements/texts/unclos/part2.htm, accessed 04 January 2019.

United Nations (2005b) *Convention on the Continental Shelf*, Treaty Series, vol 499, p. 311, legal.un.org/ilc/texts/instruments/english/conventions/8_1_1958_continental_shelf.pdf, accessed 04 January 2019.

United Nations (2005c) *Convention on the High Seas*, Treaty Series, vol 450, pp. 11 and 82, www.gc.noaa.gov/documents/8_1_1958_high_seas.pdf, accessed 04 January 2019.

United Nations (2005d) *Convention on Fishing and Conservation of the Living Resources of the High Seas*, Treaty Series, vol 559, p. 285, www.gc.noaa.gov/documents/8_1_1958_fishing.pdf, accessed 04 January 2019.

United Nations (2013a) *Official Development Statistics for Small Island Developing States*, unohrlls.org/custom-content/uploads/2013/09/Small-Island-Developing-States-Factsheet-2013-.pdf, accessed 04 January 2019.

United Nations (2018) *Least Developed Country Category: 2018 Country Snapshots*, www.un.org/development/desa/dpad/wp-content/uploads/sites/45/Snapshots2018.pdf, accessed 11 May 2019.

Vasciannie, S. C. (1990) *Land-Locked and Geographically Disadvantaged States in the International Law of the Sea*, Oxford University Press, New York.

Vidas, D. (2000) 'Emerging law of the sea issues in the Antarctic maritime area: A heritage for the new century?', *Ocean Development and International Law*, vol 31, nos 1–2, pp. 197–222.

Chapter 5

An introduction to policy and policy development

Instruments for marine law and policy

Key topics

- 'Policy' can be defined as a plan or course of action to influence and determine decisions, actions and other matters.
- Policy instruments are often broadly classified as regulatory (legal), economic (market-based) or informational (persuasive). Some authors include cooperative (voluntary) instruments as a fourth type.
- Policy analysis encompasses all aspects of how problems are identified, decisions are made, policies are implemented and outputs and outcomes are measured. Policy analysis is the business of informing decision-making.
- Policy analysis may be prospective, where it attempts to forecast the likely outcomes for various policy options and uses existing data, or retrospective, where it is used to assess whether the policy is meeting its goals and objectives.
- The process by which problems are identified, policies are created, outcomes evaluated and policies are revised is termed the 'policy cycle'.
- Unique aspects of the marine environment that affect how policies are formulated include the common property nature of much of the international ocean and that the goods and services the ocean provides to humanity are not normally considered to have economic value.
- At present, no single authority, akin to the nation-state, sets the overall rules on activities on, in and under the world ocean.
- Areas beyond national jurisdiction are managed cooperatively between states using voluntary agreements, some of which (e.g. shipping) have been highly effective. Others (e.g. conservation) have been less effective to date.
- Regulatory approaches to managing the world ocean may be coercive/command and control (where measurable outcomes along with the means to achieve these outcomes are prescribed), performance-based (where measurable outcomes are set), process-based (where processes are developed to minimize risk) or co-regulated (where the regulated are invited to self-regulate).

Introduction

'Law', in the domestic or national context, is broadly understood to be the standards, principles and procedures that must be followed in society and administered through court systems. The definition of 'international law' is less concrete but broadly understood to be the means by which acceptable practices or behavioural norms between states are codified.

Currently, there is no universally agreed-upon definition of 'policy'. (Indeed, in some languages such as German, there is no ready translation.) Broadly, 'policy' can be defined as a plan or course of action to influence and determine decisions, actions and other matters. Deriving from the Greek word *polis*, or 'city', policy is akin to 'citizenship' or belonging to a group in some way. In early Greece, *nomos* was analogous to the term 'law'. However, the concept was broader than law as currently defined and included the customs, principles and need for the *polity* (citizens and their government) to understand *nomos* in order for the laws to be understood. Plato was an early progenitor of the concept of policy and advanced the idea that citizens would better comprehend laws if they understood the motivation underlying their creation.

Policies may be enacted in any situation in which one or more individuals are involved in a common activity or are part of a distinct group or culture. Schools, churches, sports clubs, industry, volunteer groups and governments all establish policies.

Countless decisions are made every day by organizations, be they government or nongovernment. Most of these decisions (e.g. whether to issue a building permit) are not policy decisions but business decisions informed by policies. Policy, therefore, is a way to set strategic direction, intent, objectives and procedures to address problems that are often complex and controversial. Patton and Sawicki (1986) noted that difficult issues require the development of policies and have the following characteristics:

- They are poorly defined.
- Their solutions cannot generally be proven to be correct before application.
- Solutions may not achieve the intended results.
- The best solutions are seldom the least costly and appealing.
- The adequacy of the solution is often difficult to measure against the 'public good'.
- The fairness of solutions is impossible to measure objectively.

Other more routine issues also benefit from development of clear policies, but it is the difficult or 'wicked problems' (Churchman, 1967) that benefit most. Viewed through a socio-economic lens, policy measures are effective when modifying the behaviour of economic agents. These agents may be producers, consumers or suppliers of services. In general terms, public policies aim to modify the course of socio-economic and natural events by (adapted from FAO, 2009):

- Directly supplying goods, services or purchasing power (e.g. the supply of public or merit goods such as ferry services, managing sediment and shorelines, financing the development of maritime industries).
- Promoting/encouraging/supporting (e.g. stimulate the adoption of new technologies for fishing gear modification, promotion of marine export products, supporting sustainable fisheries).
- Imposing/enforcing (e.g. vessel pollution, vessel safety).
- Saving/preserving (e.g. restoration of coastal habitat, preserving coral reefs).
- Preventing (e.g. the spread of invasive species).
- Discouraging (e.g. reducing marine debris).

Policy specific to the actions of government is termed **public policy**. Public policy is made or initiated by government and states the intent of government. It is expressed as laws, regulations, decisions and government actions. Public policy is interpreted and implemented by public and private actors (Birkland, 2011).

There is some debate on the scope of public policy. Certain authors subscribe to the belief that public policy constitutes the sum of all government activities, and others suggest that public policy represents the actions of government and the intentions that determine those actions. Recent discussions on public policy trend towards the latter definition, in which public polices generally refer to changes that are envisioned or in the early stages of implementation rather than changes that are already in place. In this respect, public policy as a 'statement of intent' can be thought to be different than protocols or procedures, which are operational in nature (and ideally are implementing agreed-upon policies).

Public policies have been further subdivided into different types. **Regulatory policies** prescribe specific behaviours or limit types of actions to achieve certain objectives. Regulatory policies are frequently associated with law enforcement, business regulation (e.g. business practices), controlling access to public goods (e.g. air, water) and health and safety (e.g. workplace safety, drug approval). Regulatory policies often have the force of law and provide little to no benefits to those that are regulated.

Distributive or redistributive policies apportion a good or a benefit to a certain segment of the population in order to address one or more public problems. The costs of these benefits are shared across all members of the jurisdiction to which the policy applies. Distributive policies are widely used, especially by politicians in democratic societies. Thus, distributive policy is a tool that can improve a politician's chances for re-election ('pork barrel' politics) as well as push governments into deficit to fund new policy initiatives. Distributive policies may include funding increases for education, subsidies for agriculture, support for research, collection and dissemination of information or the award of infrastructure projects.

Redistributive policies tax one segment of society to provide benefits to another. Redistributive policy is most often used to transfer resources from the 'haves' to the 'have-nots', often in the form of social programs. Common examples of these programs include income stabilization (e.g. unemployment and retirement benefits), welfare, health care, housing and income distribution. However, certain governments use redistributive policies to support policies that aim to reduce the amount of wealth redirected from the richer to poorer segments of a society (e.g. trickle-down economics). The European Union has several international redistribution programs to address the effects of globalization on European nationals including the European Social Fund (ESF), European Agricultural Guidance and Guarantee Fund (EAGGF) and the European Regional Development Funds (ERDF). These programs currently utilize more than 30 per cent of the European Union budget.

One marine example of a redistributive policy, not yet implemented, is the 'Common Heritage of Mankind' principle as applied to The Area beyond national jurisdiction and the sharing of wealth generated mining by its mineral resources (Chapter 4; LOSC, Article 136).

Redistributive policies are approximately 'revenue neutral' in that the overall expenditures from general revenue are unchanged, though the costs of administration may be higher. Because redistributive policies take from some parties in order to give to others, they can be intensely political and divisive.

Constituent policies may establish government structure, document the rules of government conduct (including financial and personnel management) or solidify roles and responsibilities of government organizations. The establishment of new functions of government (e.g. environmental regulatory agencies in the 1970s) or changes in government practices (e.g. adoption of new accounting practices) are good examples of constituent policies.

Symbolic policies (sometimes termed pseudo policies) confer neither economic benefits nor costs on any particular group but instead appeal to collective values, beliefs and perspectives. Examples of symbolic policies include joint statements condemning poor behaviours, such as flags of non-compliance/convenience, or that pressure reluctant states to reduce environmental harms, such as greenhouse gasses and overfishing.

Relationship between law and public policy

The relationship between law and public policy can range from nearly non-existent (e.g. criminal law and crime prevention programs may both address crime but otherwise have no bearing on each other) to inextricably interconnected (e.g. financial laws inform financial policies and vice versa). If public policy is defined as the actions of government, then policies may enable the formulation of laws and, conversely, laws may enable the development and implementation of policies. National marine policies, for example, reflect pre-existing national laws while also leading to the development of new ones.

However, there are clear boundaries between law and policy: While both laws and policies are enacted by governments, legal issues are stated and analysed within the processes and procedures of the legal system. Depending on the type of legal system, legal issues are evaluated solely on the basis of written law, past precedent or both. Law is concerned with the legality of an action and – except in the punishment phase – is blind to extenuating circumstances, ethical and moral consideration and the broader social good. Policy, on the other hand, is concerned with the efficacy of an action and is evaluated using the principles of administration and management. In functioning democracies, policy positions are informed by citizen and stakeholder input, science, social good, value systems and acceptability to citizens.

Other key differences between law and public policy are that public policy is designed to be understood and easily communicated to those whom it affects and those who administer it, who may not have a legal background. It also can be developed and applied much faster than most laws. In a society where technology can rapidly result in unforeseen dilemmas such as bioethics, personal privacy and infectious diseases, policy approaches allow governments to quickly formulate responses to emerging issues that would otherwise go unattended due to the complexity of creating new law. Lastly, policy is a method of testing various approaches to see whether some are more efficacious and could at some point become enshrined in laws.

Developing and analysing policies

Policy analysis is an extensive yet ill-defined topic. It encompasses all aspects of how problems are identified, decisions are made, responses are implemented and outcomes are measured. Simply stated, policy analysis is the business of improving decision-making. Policy analysts are expected to understand the intricacies, complexity and political context of their topic areas. While developing viable alternatives to the status quo, they need to understand and be able to communicate the implications of their solutions. Often, they must provide cost-benefit analyses to their alternatives and then recommend a feasible course of action. As a core function of all governments, policy analysis also applies to any organization that establishes programs intended to resolve social, economic or environmental problems and can be undertaken by any member of the organization, whether internal or external to it.

What separates policy analysis from planning and management is that policy analysis seeks to clarify available (preferred) strategies, intent and objectives to address a particular problem. In contrast, planning and management are typically used to implement a range of already existing policies. For example, a local government may decide that residential population densities are to be increased centrally in order to preserve coastal green space for the public and as a 'future proofing' buffer against increasingly powerful storms ('smart growth'). Planners and managers will subsequently be tasked to implement the policy and may establish processes to change zoning bylaws, (re-)establish coastal wetlands, walks and bike lanes, as well as determining additional incentives and infrastructure necessary to support higher centralized densities.

Participants in, or those affected by, the policy process are often termed 'stakeholders', or sometimes 'actors' or 'agents'. They are generally represented in one of the following three general types:

1. Citizen (public) actors: Represent interests in the environment, health, recreation, gender, local communities, etc. May include scientists and researchers, media and non-governmental organizations.
2. Market actors: Represent the interests tied to the production of public goods and services. May include industry, lobbyists, think-tanks, organized labour, etc.
3. State actors: Represent elected and appointed government officials including aboriginal governments, executive, legislative and judicial administrators. May include government scientists and researchers or public servants whose job it is to fairly consider particular public or private interests.

Approaches to policy analysis

It is outside of the scope of this textbook to provide a thorough treatment of the varied academic perspectives on these delineations. However, we will consider two types of policy analysis in wide usage:

> **Prospective policy analysis**, also termed *ex ante*, *a priori* and *pre hoc* policy analysis, is used prior to or during the development of policies to inform policy development, selection and recommendation to decision-makers. Prospective policy analysis is prescriptive in that it recommends potential solutions. It is also predictive in that it attempts to forecast the likely outcomes for various policy options and uses existing data in both quantitative and qualitative analyses to inform decision-making, for example, responding to likely climate change scenarios.
>
> **Retrospective policy analysis**, also termed *ex post*, *a posteriori* and *post hoc* policy analysis, is applied once a policy has been adopted and implemented. Retrospective policy analysis is evaluative and used to assess whether the policy is meeting its goals and objectives. It often requires data collection and research to describe or interpret whether the policy is meeting specific outputs and outcomes. In certain cases, retrospective analysis will evaluate the present condition of a target (e.g. fish population) of the policy and try to recreate a past historical context to determine whether the policy is working. Most complex policies are difficult to analyse using retrospective analysis since ascribing causal relationships between specific policies and observed outcomes can be exceedingly complicated. Nevertheless, retrospective analysis can reflect the perceptions of success or failure through the eyes of the various stakeholders, including the government itself. In policy, perceptions are equally as important as outcomes and often directly affect one another.

The policy cycle

A number of analysts and governments use what is loosely termed the 'policy cycle' to guide policy development and evaluation. The notion that an ordered approach to decision-making could improve the quality, durability and transparency of government decision was first proposed by Lasswell (1951) and later by May and Wildavsky (1978). The policy cycle is an academic construct designed to be both descriptive and normative. This entails that the policy cycle captures how prudent decisions are made as well as provide a framework to guide government decision-making. It attempts to deliver an archetype that is sensitive to political realities: this realpolitik includes external influences, the need to make decisions quickly and the need to adopt policies palatable to the electorate. In decades past, policies were formulated through a closed, select process where only a few individuals had access to decision-makers. In recent decades, the electorate, industry and non-government organizations have had much more involvement in the development of policy. Driven by the rise of the internet, open government initiatives, participatory democracy and access to information have involved citizens – at least in democratic states – to an extent not contemplated by the early progenitors of the policy cycle.

There is currently no single, agreed-upon model of policy development. However, many interpretations of the policy cycle have common elements that are converging towards a shared understanding of the development and application of policies (Table 5.1, Figure 5.1). All descriptions of the cycle entertain a problem-solving process involving discrete states with multiple actors and institutions. Also, all models also suggest a feedback mechanism where policies are improved or evolve into new policies (Table 5.1).

The policy cycle: problem definition

The policy cycle begins with **problem definition** where an issue comes to the attention of the mandated authority, usually government(s). At this point, the problem is not well defined, the various policy actors (as defined above) may have divergent views on the problem and government may have, or be presented with, a number of incomplete proposals to solve the problem. In this stage, the magnitude and extent of the problem is defined, comparisons to other similar problems are made to see if solutions already exist, data and information are used to further define the problem and estimates of resources required to address the problem are assembled. The problem definition stage is complete when the problem is meaningfully stated, affected parties

Table 5.1 Examples of different interpretations/definitions of the stages of policy development (the 'policy cycle')

Stage in policy cycle	Patton and Sawicki (1986)	Howlett and Ramesh (1995)	Bridgeman and Davis (2000)	(Munger, 2000)	Bardach (2011)
Problem recognition	Verify, define and detail the problem	Agenda setting	Identify issues	Problem formulation	Define the problem
Proposal of a solution	Establish evaluation criteria	Policy formulation	Policy analysis	Selection of criteria	Assemble evidence
Choice of a solution	Identify alternative policies	Decision-making	Policy instruments	Comparison of alternatives	Identify options
Putting the solution into effect	Assess alternative policies	Policy implementation	Consultation	Political and organizational constraints	Select evaluation criteria
Monitoring the results	Display and distinguish among alternatives	Policy evaluation	Coordination	Implementation and evaluation	Project outcomes
	Implement, monitor and evaluate the policy		Decision		Confront trade-offs
			Implementation		Decide
			Evaluation		Tell your story

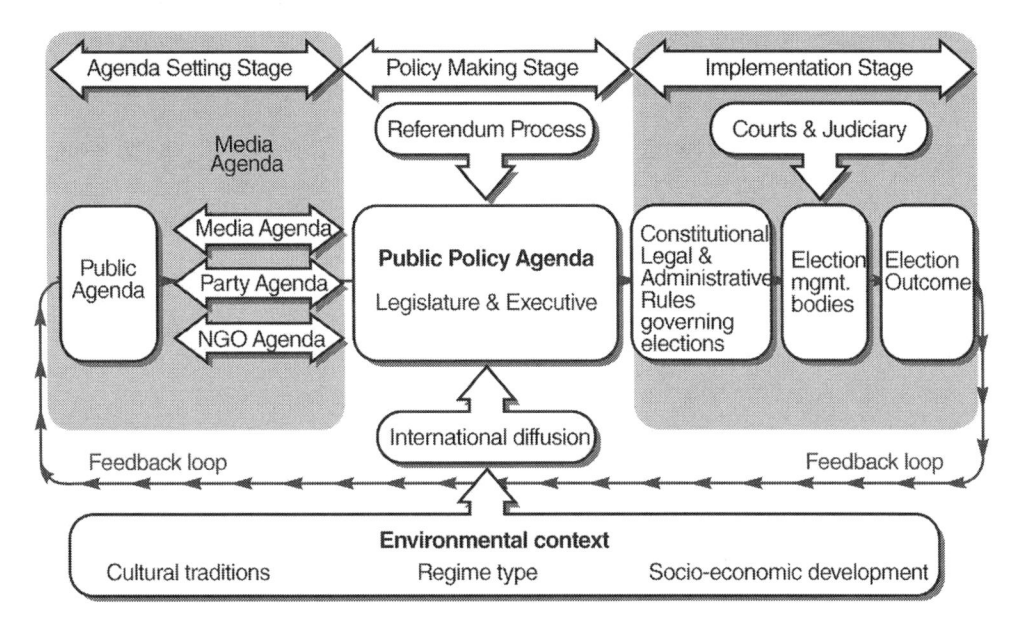

Figure 5.1 The policy cycle and associated actors and institutions

Source: Reproduced with permission from Norris (2011)

are identified and the level of importance of the problem is understood. In defining a problem, key considerations include whether the matter:

- Is wholly or partly within the scope /mandate of the organization.
- Can be effectively addressed ('solved') when historical contexts, cultural milieus and current values are considered.
- Originates as a result of past internal policies (or lack thereof) or is of an external nature (e.g. climate change).
- Originates or is perpetuated as a result of a deficiency of data or information.
- Requires incremental ('tweaking') or transformational change (complete re-thinking) to affect desired outcomes.

The policy cycle: proposal of a solution

The second stage of the policy cycle (recognizing that there are differences in opinion – see Table 5.1) is **proposal of a solution/policy formulation** where policies are articulated. This is usually undertaken by governments or, in the case of marine environments, also by inter-governmental organizations (e.g. the IMO or RFMOs). In this stage, multiple solutions are developed with associated risks, opportunities, benefits and costs. Draft solutions may be developed 'in house' by government(s) or developed through directed stakeholder consultations or open public processes. Solutions normally consider a wide range of options/alternative policies, including the status quo and 'stretch' options. For each option, an assessment of expected effects and impacts as well as the best- and worst-case scenarios should be undertaken in addition to evaluating the likelihood that the option will solve the problem. Many complex policy solutions must consider historical contexts and cultural practices (including Indigenous use), contrasting value systems and past policy decisions, all of which may have helped build trust ('social capital') or distrust. Furthermore, for long-term success, they should anticipate likely changes to the social (e.g. demographics, technology) and natural (e.g. climate change) environments.

Key considerations in developing proposed solutions include:

- Administrative ease (e.g. adapting existing regulations vs. writing new legislation).
- Costs and benefits (and determining who pays and who benefits).
- Effectiveness (i.e. likelihood of meeting government objectives and stakeholder expectations).
- Equity (i.e. whether one or more groups are advantaged or disadvantaged relative to other groups)
- Legality (i.e. whether a solution is likely to withstand a court challenge).
- Political acceptability (i.e. whether parliament and citizens will favourably react to the policy).
- Measurability (i.e. can outputs and outcomes be measured, or proxy indicators developed).

The policy cycle: decision-making

The third stage of the policy cycle is **decision-making** where government(s) adopt a particular solution. In the global context, decision-making often occurs during meetings of parties to international conventions; therefore, the decision becomes synonymous with and referred to as 'convention' or 'agreement' or 'resolution' or simply 'decision' (Chapter 3). Of significant import, decision-makers may not accept a recommended solution and may merge options, change options or ask staff to undertake further analyses and return with revised solutions. This is a normal situation in the formulation of both domestic and international policy. Most solutions are rejected by government decision-makers because they – in reality or perception – are:

- Too expensive.
- Too difficult or complicated to implement.
- Unfair or punitive on certain sectors (e.g. fishing) or states (e.g. less developed states).
- Too much of a divergence from established practices (or sets a new precedent).
- Unlikely to be understood by the public or users.
- Unlikely to solve the problem.
- Could result in unintended consequences.
- Politically risky.

The policy cycle: policy implementation

The fourth stage of the policy cycle can best be described as **policy implementation**, often the most challenging aspect of policy development. Implementing complex policies may require wholesale changes in business processes and practices, human behaviour and infrastructure and technology. Fees and taxes, resource reallocation between users, rules or regulations, and the structure/organization of government(s) can also be significantly altered. Many, if not most, decision-makers underestimate the difficulties in implementing policy and expect new policies to roll out swiftly and unproblematically. Yet, marine policies may take many years to fully implement, especially if the policy mandates changes in infrastructure (e.g. double-hulled tankers) or technology (e.g. turtle excluders for fishing gear).

Key considerations in policy implementation include:

- Confirming responsibilities and mandates.
- Identifying appropriate pilot projects.
- Development and presentation to decision-makers of an implementation plan and communications plan.
- Development of a monitoring plan that evaluates both implementation and effectiveness of the policy.
- Setting future dates when the policy will be reviewed.

The policy cycle: policy evaluation

Lastly, **policy evaluation** is the final stage in the policy cycle. In this stage, the monitoring plan is implemented, and the policy is reviewed at periodic intervals to determine its progress. It should be noted, however, that policy evaluation can be undertaken at any point in the policy cycle and there is much in common between the problem definition and evaluation stages. Theodoulou and Kofinis (2004) define policy evaluation as:

- The assessment of whether a set of activities implemented under a specific policy has achieved a given set of objectives.
- The effort that renders a judgment about program quality.
- Information gathering for the purposes of making decisions about the future of the program.
- The use of scientific methods to gauge the success of its implementation and its outcomes.

Several types of policy evaluations have been proposed, and there is no consensus concerning methods and standards. Evaluation at the various policy cycle stages, however, can be one of two types: **Formative evaluations** are undertaken when policies are either in development or aspects of the policy are already in place. Formative evaluations allow for evaluation and continuous improvement in the first four stages of the policy cycle. In contrast, **summative evaluations** are used for mature policies to assess their effectiveness and to determine if changes in the policy are required.

With respect to the different aspects of policy that could be evaluated, Theodoulou and Kofinis (2004) proposed the following four types of policy evaluation: **Process evaluations** determine how well a policy (or program the policy established) is being administered and focus less on whether the policy is achieving its desired outcome. Process evaluations are generally undertaken by staff directly involved in the program in order to find administrative efficiencies in the delivery of the program. **Outcome evaluations** strive to measure outputs (e.g. number of ships compliant with mandated technologies) to demonstrate the degree of compliance with the policy. In other words, outcome evaluation is used to measure tangible and available results of the policy, often before the policy matures, to provide an indication of whether the policy is working as it should. Outcome evaluation is therefore not used to determine whether the policy has yet achieved its stated goals; rather, it measures tangible progress towards them. **Impact evaluation** evaluates whether the policy is achieving its stated objectives as the decision-makers intended. Impact evaluation is what most consider when thinking about policy appraisal and evaluates whether the policy is working and/or has produced unintended effects. **Cost-benefit analysis** assesses the financial costs of the policy (or program the policy creates) against the societal benefits. In particular, applying cost-benefit analysis to mature programs enables them to be modified. Also, they can be assessed against the economic analysis in the program formulation stage to determine how closely predictive cost-benefit models align with reality.

Regardless of the type of policy evaluation used, there are four drivers for policy evaluation (adapted from Mark et al., 2000):

1. Assessment of merit and worth.
2. Improvement of the program and, by extension, the organization.
3. Oversight and compliance.
4. Development of knowledge.

Policy evaluation can be further separated into process and outcome evaluations. Specific questions asked in process evaluations include whether (adapted from Smith and Larimer, 2009):

- The policy operates according to the relevant rules/laws/obligations.
- The target population is being served by the policy.
- The process matches the goals of the policy.

Three specific questions are often asked in outcome evaluations (adapted from Smith and Larimer, 2009):

1. Is the policy having an impact?
2. How much of an impact is the policy having?
3. If the policy is failing to have an impact, why has it foundered?

Challenges with the policy cycle

While the policy cycle has been in use for several decades and provides a structured approach to addressing complex problems, there remain a number of concerns. The primary concern with the policy cycle is that the model assumes rational behaviour throughout each stage and that governments will follow the model. In reality – and often for very valid reasons – politics, elections, emergencies and events lead to the cycle being circumvented or even marooned. For example, often policies are developed hastily in order to seize political windows of opportunity. Deviations from the structured approach are made in the interests of expediency. In addition, policies are frequently developed by political staff, program managers or third parties (e.g. non-government organizations, industry) who present decision-makers with policy solutions developed outside of the cycle. Lastly, the sequenced, step-wise nature of the cycle rarely coincides with reality. Decision-makers routinely consider multiple, interconnected policies simultaneously such that policy changes are made ad hoc, often with little analysis or empirical evidence. Finally, even in the best-designed processes, mitigating factors frequently arise limiting the validity of a cause-and-effect (causal) evaluation, rendering incomplete, misleading, or erroneous conclusions.

Considerations in selecting public policies

Ideally, public policies are evidence-based, transparent, efficient and accountable and can be effectively implemented. Selecting public policies may be as straightforward as establishing catch limits for a particular fish stock where population trends, life history and efforts expended on catching the stock are known. However, most public policies are complex with direct effects that can only be approximately anticipated and indirect effects that may be completely unpredicted.

As such, policy development requires careful consideration of the development of alternative policies. This section, therefore, outlines a number of considerations when developing public policies. In the real world of politics, interest groups and seizing windows of opportunity, policies are rarely 'bench tested' to the degree necessary to make informed choices. The considerations provided below, however, should guide policy developers towards some of the questions they should ask before bringing alternatives up to decision-makers.

Selecting policies primarily focuses on a number of aspects to inform different alternatives (solutions) that may include:

- The environmental, geographic and socio-economic aspects requiring intervention.
- Available existing policy measures that could be applied to generate the desired changes.
- The expected effects of policy change.
- Who would gain and how much would be gained if the policy measure is implemented.
- Who would lose and how much would be lost if the policy measure is implemented.
- The expected duration of the effects of the policy measure.
- The budgetary implications of the policy measure.
- How the policy measure would be financed.
- Whether the policy measure is legal or consistent with existing law.
- Whether the policy measure requires legislative change or amendment of an international convention.
- Whether those affected by the policy will understand its intent and consequences.
- Whether the policy will be measured to see if it is working.
- Whether the policy could lead to irreversible decisions (painting oneself into a corner).

While policy decisions are not simple, rational discussions, high-level decision criteria can generally be categorized as relating to effectiveness, efficiency or equity of policy tools. Despite strong criteria, however, even the most rational and effective option can still be abandoned in light of political or administrative resistance. As a result, political and administrative support and feasibility are significant criteria for successful policies. Depending on the issue, additional criteria may be relevant. Among these criteria are the scientific or technical effects of an action, the social or cultural acceptability, the legal implications and the uncertainty involved.

One policy tool may not fit all situations. The achievement of deeper structural or economic changes likely requires a multidimensional approach that integrates a number of different policy tools. In considering instrument selection and design issues, it is critical to establish and build on a solid foundation of information. Establishing the review process beforehand is critical to ensuring an approach based on incentives and the use of transparent metrics. Some considerations in the choice and design of policy instruments for environmental management and related examples can be found in Table 5.2.

While there are many considerations in selecting the appropriate policy instrument to manage a particular issue, assessment criteria generally fall under a few broad categories. These include **effectiveness, efficiency, fairness** and **political acceptability**. While different instruments have their strengths and weaknesses, it should be emphasized that the particular circumstances and pre-existing conditions of a management issue will determine the fit and success of a policy tool. This is particularly important in marine environments given the number of states that potentially can be involved in a policy issue. Focussing on the policy objectives, rather than favouring a particular tool, may be the best route to achieving desired outcomes.

Table 5.2 Application of selection criteria for different categories of policy instruments

Criteria	Environmental Regulations	Using markets	Creating Markets	Informational Instruments
Effectiveness (immediacy & certainty)	High	Moderate	Moderate – Low	Low
Efficiency (financial costs)	Generally high administrative and enforcement costs	Low implementation, but business compliance and revenue collection costs	Set-up and allocation costs complex and involved	High costs for research and advertising, low cost for transparency mechanisms
Fairness (of costs and benefits)	Resistance depends on regulatory target, positive public image for addressing industrial violations and 'polluter pays' principle	Low resistance as long as costs are proportionally distributed, and ability to pay is equivalent	Equity considerations for access limitations to public goods (e.g. parks), local impacts may still be high, under regional trading scheme	Low resistance, positive associations with transparency increasing accountability, can provide short-term advantage to those who don't participate
Political acceptability	Government resistance due to needs for enforcement and administration; Industry push back on costs and limitations to technological innovation	Often favoured, but depend on precedent for charges, risk public reprisal for 'just another tax'; economic disincentives often met with resistance where industry based on profiting from externalities	Theoretically appealing, but complicated to implement and requires considerable inter-governmental cooperation to apply trading schemes across borders	Often viewed as a soft policy option, but low cost and positive messaging and accountability held in high regard

Source: Adapted from Sterner and Coria (2012)

Effectiveness

Considering the potential efficacy of public policies at the development stage will inform alternatives and is a key component of the solutions when presented to decision-makers. Effectiveness is not simply a question of whether the policy will achieve the desired outcome but considers when policies will become effective, the degree of risk to those affected by the policy, certainty that the policy will achieve its intended objectives, buy-in from those affected by the policy and what constitutes 'success'. For example, contrasting effectiveness between regulatory instruments and market-based approaches will require consideration of whether the immediate results but higher costs to industry using regulatory instruments is more desirable than delayed results and economic flexibility using market-based instruments (MBIs).

Analyses of effectiveness borrow measurement and evaluation techniques from other disciplines and industry sectors. Also, they may utilize evidence-based techniques and consider direct and indirect responses on other policies and programs. As discussed previously, effectiveness is often separated into institutional and impact effectiveness. **Institutional effectiveness** (output effectiveness) is defined as the having effective policies/programs in place, and

impact effectiveness (outcome effectiveness) is whether the policies/programs are meeting their objectives. For example, Kenya has sufficient policies to address the management of marine oil pollution (institutional effectiveness), but whether these policies avert oil pollution (impact effectiveness) is unknown (Ohowa, 2009).

Vital to evaluating policy effectiveness is the identification of performance criteria (indicators) to support both the evaluation of alternative solutions prior to committing to a policy and to evaluate the efficacy of the policy once implemented. Performance criteria should be selected that can be routinely monitored and that correlate with one or more goals or objectives of the policy (Mazur, 2010). Table 5.3 outlines the primary types of performance criteria for policies, and Table 5.4 illustrates key considerations for the selection of performance criteria.

Table 5.3 Main types of performance criteria for policies and marine examples

Type of criteria	Example
Benchmarks	Comparison with a documented best-case performance related to the same variable within another entity or jurisdiction.
	Example: Number of fisheries in a jurisdiction that are certified by the Marine Stewardship Council.
Thresholds	The value of a key variable that will elicit a fundamental and irreversible change in the behaviour of a system. The policy is evaluated based on its role in making the system move towards or away from the threshold in any given period.
	Example: Maximum sustainable yield of a fishery.
Principles	A broadly defined and often formally accepted rule. Principles should have relevant performance measures.
	Example: Transparency shall presume fisheries data will be publicly available unless explicitly determined to be confidential (using transparent criteria and processes).
Standards	Minimum requirements for properties, procedures, or environmental quality.
	Example: Bilge water quality standards.
Policy-specific targets	Determined in a political and/or technical process taking past performance and desirable outcomes into account.
	Example: Setting targets for offshore hydrocarbon production.

Source: Modified from Pintér et al. (2000)

Table 5.4 Key considerations for the selection of performance criteria

Clarity/Transparency	Can be understood by laypersons
Policy relevance	Relates to the policy objective(s)
Analytical relevance	Accurately measures the problem
Responsiveness	Responds to changes in management
Time horizon	When results can be expected
Accountability	Identifies responsibility
Robustness	Cannot be manipulated/massaged
Measurability	Can be measured over the life of the policy

Source: Modified from Jesinghaus (1999)

The European Environment Agency (2005) developed the following perspectives on policy evaluation:

- Governance can make or break the success of a policy: Policies will be ineffective unless suitable governance exists.
- By tackling problems at the source, economic instruments can be fruitful ingredients in the policy mix: Effective policies utilize economic incentives to discourage the production of externalities (e.g. pollution) that are costly to address.
- In assessing a policy's goal-achievement, it is important to distinguish between different types of goals: Policy evaluation must recognize that multiple goals and objectives exist and that some goals will be met while others will be unattained.
- Data limitations are demanding but not insurmountable: Policy evaluation is frequently limited by a lack of data. Nonetheless, sufficient data can often be obtained to understand the policy.
- Effectiveness evaluations are complex and require multidisciplinary efforts.
- Effectiveness evaluations contribute to capacity building and shared policy learning.

Efficiency

Financial costs, as well as other non-market goods and services that influence the overall level of social well-being, are another consideration in selecting a policy. Strategies that achieve their goals at a minimum cost maximize limited resources for addressing other problems. Cost dimensions can be considered as they apply to society, government agencies, and targeted individuals and organizations. The financial spectrum of instruments ranges from low-cost voluntary programs, to regulatory options with high administrative costs. The limitations of defining environmental and associated health costs in economic terms have made efficiency arguments – that is, weighing regulatory expenses against overall costs to the economy – controversial (see Box 5.1). Often, these negative costs have largely

---

Box 5.1 The Weitzman Theory

In 1974, Martin Weitzman explored whether it would be better to diminish certain forms of pollution by setting emission standards (direct regulation of quantifies) or by charging the appropriate pollution taxes (indirect control by prices). He demonstrated that where the value of marginal benefits and costs associated with abatement are uncertain but where economic costs are likely to increase faster than environmental benefits, a smaller overall welfare loss is likely to result from a pricing mechanism than from a strict quantity constraint. Termed the 'efficiency rule', Weitzman recognized that if there is a risk that policy-makers are unable to determine the optimal pollution level, it is best to regulate a quantity target (even with an increase in abatement costs). However, if environmental impacts are escalating but the increased compliance costs are of greater concern, then a price approach is a more efficient choice. As a result, the most efficient policy for managing hazardous pollutant discharges or safeguarding threatened species or habitats would likely be a form of quota or access restriction. Meanwhile, tax assessments would be an appropriate instrument for dealing with a relatively benign pollutant or slowly accumulating environmental threat.

Four decades later, most economists continue to support Weitzman's conclusions.

Source: Weitzman (1974)

been externalized or ignored, and critiques of prescriptive regulatory policies have focussed on the real financial costs of development, administration, enforcement and potential limitations to business innovation. Externalities may not only give rise to an inefficient use of the resource, but they can result in costs being imposed on others that are widely perceived as unjust.

Economic or market-based instruments (MBIs) are often positively associated with lower financial costs. However, there are still efficiency considerations for selecting the appropriate MBI, be it quantity, price or market enhancement policy. Allocation of permits or licenses has been applied to diverse types of resources, ranging from fisheries to the control of water pollution. Implementing a quantity-based trading instrument requires substantial upfront development costs to negotiate the contracts of exchange and to absorb legal, brokerage and insurance fees. Trading systems can also involve significant time and effort on the part of governments – with due consideration of the potential efficiency and equity implications involved. In particular, it has been a complex and protracted exercise for policy-makers looking to establish a domestic permit trading system for greenhouse gas emissions. For a price or market-enhancement approach, trading costs are avoided, but compliance costs still apply, as do costs associated with revenue collection or subsidy allocation by government. Efficiency implications are also associated with evasion under a price-based approach. This is opposed to a trading system, where the tax or charge is commoditized and converted into a tradable 'asset' that can be exchanged.

Fairness

Fairness can be a significant issue in selecting a policy tool. Public policies can have disparate effects on different people that vary by location, ethnicity, income or occupation. The distributional consequences of a policy typically indicate a great deal about who supports and who opposes a policy. Although 'polluter pays' is at first glance a 'fair' principle, in practice it can sometimes lead to counter-intuitive results (e.g. providing a 'license to pollute'; or the 'cost of doing business'); furthermore, coercive strategies are often difficult to implement or assess (e.g. is there anything to pay for an oil spill at sea that disperses on its own?). Finally, when a small group or sector bears the cost of a policy, they have significant motivation to organize and resist implementation while the majority that benefits are often widely distributed and without a coordinated voice. There will be trade-offs between costs and benefits with any policy; government bureaucracy, however, may not be the most responsive to changing conditions affecting the cost-benefit relationship. How policies proportionally affect vulnerable or affluent constituencies can also influence the public's perception of a policy's social equality. Voluntary instruments typically have less resistance from targeted sectors and are relatively easy to implement but may not be viewed seriously and consequently suffer from 'free riders'. Larger environmental concerns, such as widespread pollution, are typically best mitigated by economic or regulatory approaches. Market-based instruments have been controversial in the allocation of the wealth associated with these resources. Although MBIs typically do not privatize resources as conventional wisdom suggests, to some degree they do privatize access to (and use of) those resources. And since access rights can be very valuable, these rights may represent a substantial amount of wealth. While the ability to reclaim and protect common resources can motivate sustainable behaviour, the ethical issues raised by its distribution among competing claimants remain a sizable challenge.

Political considerations

The political environment, as expressed through the desires of elected and non-elected politicians, can distort or limit the available public policy options. As a result, even successful public policy can be cut short with changes in government, and failures are a frequent occurrence.

Government mandates can have a major impact on instrument choice, as can the perceptions of public acceptability. As policy trends often follow political agendas, the economics of voluntary or non-regulatory approaches and streamlining or deregulating existing policies have frequently been favoured. Despite evidence that they could be effective, many economic disincentives are seen as difficult to implement due to constituent push back. Additional regulation is often met with resistance due to increased political and administrative burdens for implementation and enforcement.

Specific political considerations in the development of policies include:

- Ensuring that key influencers do not affect outcomes that favour certain groups over others.
- Ensuring that the instrument does not create equity issues and benefits 'free riders'.
- Ensuring that policies are understood and accepted by the public. Certain instruments, including taxes, subsidies and trading schemes, can be controversial. So too are more complex combinations of policy instruments. In contrast, the public generally understands simpler concepts such as 'polluter pays' (although they may be more difficult to implement).
- Ensuring bureaucratic capacity to implement a policy.
- Maintaining relationships with sub-national levels of government. For example, regional governments often prefer MBIs over state control in instituting pollution standards.
- Maintaining relationships with other states. Cross-border policies often face additional political and diplomatic considerations. In addition, international agreements often restrict the use of certain policy instruments.
- Maintaining relationships with the media and other 'influencers' in order to accurately and effectively convey the intent of the policy.

Other evaluation criteria

Effectiveness, efficiency, equity and political considerations presented earlier in this chapter offer some general criteria for evaluating policies. Yet, there are various other assessment processes that advance detailed lists of evaluation criteria for policy selection. Some additional criteria and associated examples for selecting environmental policies can be found in Table 5.5.

Methods and tools for selecting and analysing policies

Once the criteria have been analysed for each alternative, a decision still has to be made. Policy decisions are often described as being logical, ordered and merit based. While there are a number of analytical tools that are helpful for synthesizing relevant information and determining the most appropriate and strategic policy approach, there are many paths in the selection process and many opportunities for judgement to play a role.

In the idealized 'rational model', policy development is based on a hierarchical normative process (Fischer et al., 2006). It often begins by clearly identifying the issue that needs to be addressed and stating the objectives of the policy. Consideration of the historical background and the severity of the problem are diagnosed at this juncture. Depending on priorities, alternative options and public expectations, government chooses whether to address the problem. If government action is required, existing policies should be assessed to avoid overlap. This involves directing policies at the appropriate level of government (national or regional) and engaging staff regarding the operational responsibilities for administration and enforcement. Consultation with the public and stakeholders is advisable since it provides a sounding board for

Table 5.5 Criteria and considerations in the selection and design of policy instruments

Ownership/rights	Policy instruments need to be sensitive to potential compensation if rights are affected.
Diffuse/point source problem	Certain policy instruments (e.g. tradeable permits or load-based licensing systems) require knowledge of who is polluting in order to be effective. Otherwise, other instruments (e.g. ambient standards) may be more appropriate.
Single-issue multiple benefits	Instances where addressing single issues (e.g. water quality) result in multiple benefits are well served by combinations of instruments or flexible instruments.
Available information	Market-based instruments generally require reliable data on sustainable yields and an understanding of issues such as compliance costs. Regulation and financial incentives may be preferable in information-poor environments.
Proportional cost of tool	Taxes, fines and permit prices must be sufficiently large such that they change behaviour and incent change.
Intended outcome	The instrument should drive behaviours towards clear, science-based, quality or quantity standards.
Efficiency gains	Instruments should attempt to realize efficiency gains, as entitlements tend to move to producers with the highest marginal returns.
Ongoing incentives	Economic instruments, such as permit systems, which provide incentives for self-management by industry, are typically preferred to those that are dependent on funding for administration and enforcement.
Timing	Instruments that can be quickly enacted are preferable to those that may take some time to implement if an environmental threat is imminent.
Flexibility	Instruments may need to be adaptable to respond to new knowledge produced by scientific research and monitoring. Irreversible decision-making should be avoided.
Equity aspects	Instruments should be reviewed for any differential impacts on those affected by the policy. Considerations should include charges for 'public rights'; differential treatment of similar entities; flat charges or levies which act regressively, impacting those least able to pay; and prohibitively large charges or subsidies for industry restructuring.
Transaction costs	For MBIs, consideration of the effects of transaction costs on efficiency may be required.
Community acceptance	The instrument must be perceived as legitimate if it is to be effective. Some instruments, including taxes and levies may not garner public support if the public believes their purpose is to benefit certain groups or are tax grab.
Transparency	Transparency (sharing of information, and open process) improves social acceptance of the instruments and also improves policy outcomes by minimizing the risk of fraud, corruption and mismanagement of public funds, and supports a level playing field for business.
Administrative feasibility and costs	The instrument selected ideally should be the most cost-effective to administer. Considerations in determining cost-effectiveness should include enforcement and monitoring costs as well as the capacity of the government to administer the program.
Dependability or certainty	There should be certainty that the instrument will drive the desired change even when knowledge about likely responses is uncertain.
Precaution	Instruments should reflect the precautionary approach where a lack of scientific information cannot be used to justify inaction or weaken the policy. Indeed, a lack of information should justify greater caution, until more information becomes available.

Source: Adapted from Robinson and Ryan (2002)

them to propose alternatives, identify impacts and ensure their concerns are addressed; in short, it builds trust. Policy selection may also have legal implications and needs, such as identifying the constitutional and statutory basis for the policy as well as clearly defining the legal entities to which it applies. Alternative examples and case studies of responses from other jurisdictions with similar issues can provide additional insight on the best course of action. Cost-benefit analysis or another policy selection tool may be used to support the final decision. Finally, establishing a clear scope and communications strategy to ensure the objectives are clear, consistent, understandable and accessible will help ensure the success of a policy during implementation and beyond.

Although an efficient process of a rational decision is an appealing means of selecting the best policy option, policy changes rarely emerge from an orderly cost-benefit analysis alone. Critics argue that merit-based criteria and bureaucratic process do not adequately explain reality; government decision-making does not always follow a strictly logical progression, as some issues become 'hot' overnight, resulting in large changes rather than measured improvements. As such, a number of alternative policy decision models have been tabled. These are discussed in Table 5.6.

In reality, policy selection processes often involve decisions made with a high degree of uncertainty and expediency and based on ill-informed political instincts. An alternative to the rational selection model likens the policy stream to a 'primeval soup' where the primary players – experts, researchers, policy analysts and politicians – create a messy pool where ideas are floated, exchanged and combined. The method for decision-making involves convoluted discussions and debate that results in particular ideas rising or sinking in popularity. While many ideas are proposed, the ones that survive meet some criteria or mandate of the contemporary policy environment. The most basic criteria for successful policies in this decision-making model are those that are available now, technically feasible, politically acceptable and flexible to future uncertainties.

Ultimately, policy decisions are in the hands of the decision maker. There are a number of technical methods, however, for synthesizing policy options, which aid in selecting the optimal policy. It may be necessary to conduct more than one form of analysis to assess whether the preferred policy option depends on the decision method chosen. These methods are outlined in the following section.

Table 5.6 Common models/methods/approaches/theories used to describe how public policy decisions are made

Model	Description
Rationalist	Decisions are made with the best information for the best interests of society
Bounded rationalist	Complete rationality cannot be assumed in policy making
Incrementalism	Policy is slowly improved over time
Group theory	Groups compete to influence decisions
Elite theory	Elites influence government in their own self-interest
Public choice theory	Governments operate in their own self-interest
Normative	Decisions are made based on value judgements rather than empirical information
Valuative	Policies are assessed to determine their value

Source: Adapted from Fischer et al. (2006)

Dominance

Under the dominance rule, policy options, when compared to alternatives that are not superior in any factor, are eliminated. If a policy option does not meet or exceed set criteria, then logically it should be eliminated from further consideration. Nonetheless, there should be consensus on the ranking of decision criteria, for rarely is one option completely dominated by another. Typically, an objective method cannot be used to find the optimal policy choice. In fact, if one alternative clearly dominates the others on every decision criterion, it may be suspected that the criteria were inadequate or purposefully designed to make the dominant alternative the obvious choice. Dominance can be useful for eliminating substandard alternatives and reducing the number of options requiring further consideration, or it may be instructive in highlighting deficiencies in the decision criteria (Loomis and Helfland, 2001).

Criteria ranking

Following on dominance comparisons, criteria ranking prioritizes one or more criteria. Recognizing that some criteria are more important than others does not mean that other criteria are irrelevant; however, they should receive less weight in determining the appropriate policy. Some criteria may be deemed 'critical' or 'deal breakers', while additional criteria may be used in a secondary decision process or as a tiebreaker between comparable alternatives. Criteria ranking requires very clear priorities and is generally a less flexible approach (Loomis and Helfland, 2001). Furthermore, simple additive approaches, while intuitive, may be mathematically incorrect and produce misleading results. Common mistakes include adding together completely different criteria (statistically orthogonal – where the square root of the sum of squares should be used).

Cost-benefit analysis

Using a cost-benefit criterion, economic efficiency becomes the guiding benchmark. The best policy should maximize the net benefits to society. Typically, this is without regard to who gains or loses. As the gains outweigh the losses, then it is possible for the losses to be compensated. Weitzman (1974) emphasized the importance of including natural capital (the environment) in accounting for economic growth (see Box 5.1). However, using environmental resources can cause negative environmental externalities – essentially, intangible costs such as the depletion of natural stocks or degradation of the environment. In order to fully analyse the cost benefit of a management policy that involves natural capital, intangible variables must be assigned values or 'shadow prices'. As an example, in the cost-benefit analysis of permitting an offshore wind farm operation, the lost intangible value associated with the scenic views must be priced and factored in as a cost. Cost-benefit analyses have been criticized as disguising subjective value assessments as economic equations, yet they have proven useful in establishing practical estimates for environmental damages and abatement costs. In addition, Schulz and Schulz (1991) propose that cost-benefit estimates can assist in:

- Making the economic dimension of environmental degradation clearer.
- Making the environmental debate more objective.
- Directing scarce financial resources to the most urgent environmental issues.
- Making polluters aware of the costs of their actions.
- Further developing statistical measures of welfare.

In recent decades, many studies have estimated the value of a wide variety of ecosystem services. From an economic perspective, the value of ecosystem services is infinite for humanity would not exist without them. As one might suspect, the cost-benefit approach has a few structural limitations stemming from the general problem of putting a value on a public good. Measurement of the tangible benefits associated with any policy involves assessing the rate of change in ecosystem services from existing levels and the benefits from the service in terms of damage avoided. Although estimating the value of ecosystem services can help illuminate the extent of potential environmental damage, estimating the terms is inherently uncertain. Calculating the monetary values for damages due to degradation of coastal water quality, for example, requires multiple uncertainties in estimating the damages to shoreline property values, fishery losses, health and recreation, etc., over both the short and long term. Cost-benefit analysis also has challenges when considering the problems of (adapted from Hanley, 1992):

- Differences between citizen and consumer values.
- Detailing the complexity of ecosystems.
- Irreversibility and uniqueness.
- Intergenerational equity and discounting.

While acknowledging the many conceptual and empirical problems inherent in producing such estimates, the economic valuation process has proven to be an instructional exercise with multiple benefits including:

- Making the range of potential values of ecosystem services more apparent.
- Establishing a first approximation of the relative magnitude of global ecosystem services.
- Setting up a framework for their further analysis.
- Flagging those areas most in need of additional research.
- Stimulating additional research and debate.

Costanza et al. (1987) presented one of the first international syntheses of ecosystem valuation. They estimated values for ecosystem services per unit area by biome, multiplied this value by the total area of each biome and then summed over all services and biomes. Most of the problems and uncertainties encountered in their work indicated that estimates typically represent a minimum value that would probably increase with additional effort in studying and valuing a broader range of ecosystem services, with the incorporation of more realistic representations of ecosystem dynamics and interdependence and as ecosystem services become more stressed and 'scarce' in the future.

Willingness to pay for environmental quality – and thus the preferred trade-off between environmental and other goods – depends upon social preference and the need to assay the preferences of future generations. In practice, these issues can only be made explicit through the political process. Assessment of benefits can nevertheless make these processes more informed and transparent.

National studies are another form of cost-benefit analysis. These weigh the consequences of additional taxes and regulation with gains in market efficiency and increases in net revenue in order to create a more comprehensive policy picture. National studies often employ independent third-party analysis to evaluate management issues. In marine environs, such an analysis might study the potential impacts and policy options for limiting coastal water pollution to target levels. Comprehensive environmental policy plans provide target values for emission, waste and resource extraction up to a future date and estimate the costs and macroeconomic effects of several different policy options. In-depth analysis would consider the significance of legislation, regulation, charges and taxes and increased investment. Scenarios typically compare the measures required and impacts of achieving stringent targets, stabilizing emissions, or keeping the status quo (i.e. taking no action) (Loomis and Helfland, 2001).

Not all environmental harms can be costed (e.g. what price the extinction of a small worm, a songbird, or the loss of a picturesque view?); therefore, a cost-benefit analysis should be seen as an instructive but limited assessment tool. Invariably, other non-monetary considerations will also factor, as outlined below.

Multi-criteria decision analysis

It has long been recognized that policy decisions are informed by socio-economic, environmental, cultural, political and ethical considerations. Accordingly, there has been considerable effort to develop tools to integrate disparate data sets so that options can be developed and ranked, either qualitatively or quantitatively. Multi-criteria decision analysis (MCDA) methods are a systematic approach to combine different data inputs with stakeholder input and benefit/cost analyses to rank policy alternatives (Huang et al., 2011). MCDA quantifies public feedback and non-monetary factors to enable the comparison of policy options. For example, someone shopping for a new car will have certain preferences (e.g. price, reliability, style, fuel economy) that are more important. Some of these approaches can be quantified (e.g. fuel economy) while others (e.g. style) are subjective. Through the quantification of subjectives, MCDA approaches attempt to find the best alternative.

These techniques have been widely applied to many fields – reducing contaminants entering aquatic ecosystems, optimizing water and coastal resources, sustainable energy and transport policy, to name a few. Furthermore, they can be linked with adaptive management approaches where uncertainty is recognized in natural resource management. The three primary MCDA approaches include multi-attribute utility theory (MAUT), outranking and the analytic hierarchy process (AHP). These approaches build on criteria ranking (see above) by assigning weights to each variable and summing the weighted criteria for each option. Each approach produces an overall score for each option based on values for alternatives multiplied by weights; however, there are some differences between the approaches, which are discussed below.

Multi-attribute utility approach

Like other MCDA approaches, MAUT assigns weights to each variable and sums the weighted criteria for each option. However, MAUT includes a utility function that scales the importance or impact of each attribute from 0 to 1 with 0 representing the worst preference and 1 the best (Loomis and Helfland, 2001). While the absolute numbers for weights are not important, the relative values are, highlighting the subjective component of MAUT. Identifying and assigning weights can be controversial, and results are sensitive to differences in opinion about the relative importance of each criterion. Sensitivity analysis, where various uncertain parameters are changed to see if the results change, can assist in this process. If the order of the policy options changed with only minor adjustments to the weighting, then certainty in assigning the weights is more important. MAUT is mathematically hobbled by correlated variables and treating these aspects the same as others that are unrelated (orthogonal). In addition, MAUT methods are challenged by both uncertain or fuzzy information and by information expressed in quantities other than a ration or interval scale.

Outranking approaches

Outranking approaches are based on 'voting theory' where the alternative a is deemed better than alternative b if the number of votes (or criteria) indicating that alternative a is better than alternative b is larger than the number of votes indicating the opposite. A single, correct alternative is not developed using outranking approaches; rather, the output is the degree of dominance of one alternative over another. Outranking approaches accept incomplete value information and are suitable in situations where there are a finite number of discrete alternatives from which to choose and where the input information is descriptive in nature. Criticism of outranking approaches focuses on the complexity of algorithms that are not understood by decision-makers and that the method does not always consider that over-performance on one criterion may not make up for under-performance on another. Popular outranking approaches include PROMETHEE (Preference Ranking Organizational Method for Enrichment Evaluation) and ELECRE (Elimination and Choice Expressing Reality).

Analytic hierarchy process

The AHP uses pairwise comparisons to solicit whether criterion a is preferential over criterion b. These criteria are then weighted according to importance and summed. Similar to outranking approaches, AHP does not identify a 'correct' alternative but one that reflects the preferences of the participants in the process. The approach is suited to situations where the inputs are composed of expert knowledge and subjective preferences. In addition, qualitative criteria can be applied in the evaluation. Weaknesses of AHP include algorithms that can yield illogical or mathematically unsound results, and the methodology may not reflect the participant's true preferences.

Screening

Depending on the situation, there can be a set of criteria that present an obvious requirement for any policy. Legal obligations or threshold values present some examples of where a screening approach may assist in determining the preferred policy approach. Screening ensures that selected policies acknowledge and meet the key requirements, although in some uncertain situations achieving targets or meeting thresholds may require additional analysis. If no, or only a few, options exist that meet the screening criteria, more alternatives may be required. Also, screens do not provide methods to decide among multiple options that pass the screen.

Cost-effectiveness analysis

Cost-effectiveness analysis (also known as least cost analysis) is used to identify the most expedient financial course of action for achieving an objective. Once options for achieving the target are identified, the avenue with the lowest present value of costs is selected as the most cost-effective choice. Rather than attempting to identify and value the benefits, it is implicitly assumed that they outweigh the cost. Cost-effectiveness analysis is appropriate in situations where valid and reliable estimation of the benefits of alternative options is not feasible. For example, cost-effectiveness analysis is suited to conditions where clear and defensible environmental goals exist that can be measured in terms of biophysical units such as minimum water quality standards. It can also be used to identify the most effective option for a fixed amount of funding that has been allocated to achieve a policy objective. One drawback of cost-effectiveness analysis is that it usually does not identify the incremental benefits of specific additional actions – only those meeting the minimum standard. It also cannot gauge the willingness of society to pay for improvements in environmental quality; both these are important considerations in many decision contexts. Nevertheless, cost-benefit analysis is, where practicable, the preferred tool for decision support.

Counterfactual analysis

Counterfactual analysis is a type of **impact analysis**. Simply put, it attempts to determine what would happen in the absence of a policy. In doing so, it attributes cause and effect between interventions and outcomes. This is often necessary in policy analysis because a simple comparison of outcomes before and after a policy's implementation does not demonstrate the effect of the policy. Counterfactual analysis can exploit experimental design (e.g. comparing outcomes against control or comparison groups not affected by a policy) or hypothetical prediction (e.g. interviews with experts to determine outcomes without policy intervention).

Counterfactual analysis follows a logical process with the following steps (Figure 5.2). Construction of a base scenario, which is a description of the current system without policy intervention (WoP) (i.e. a state that is assumed to represent the situation if the policy measure is not implemented). This will be the reference scenario, also called benchmark or baseline, for the impact analysis of policies. The reference scenario is then described using some indicators, chosen on the basis of the type of policy that one would analyse. If, for example, the policy measure aims at protecting marine mammals and endangered marine species, one would use indicators like the cetacean stranding rate, population growth rate, their habitat requirements and their distribution. These become reference indicators for the reference scenario. After building and describing the reference scenario, the analysis then focuses on the construction of one scenario that integrates the expected impacts of the policy option. This is the scenario with policy (WiP). If more than one policy option has to be analysed, the analyst can build different scenarios 'with' policy. The scenarios WiP are usually built as a modification of the WoP scenario.

Quantitative versus qualitative analysis

Policy analysis employs a range of qualitative and quantitative research methods to analyse and compare existing policies and possible modifications to them. It is not possible to neatly decouple qualitative and quantitative analysis, as many approaches have mixed natures: however, qualitative approaches for policy impact analysis essentially make use of non-numerical information. Qualitative research often gains a general sense of phenomena and forms theories that can be tested using quantitative research. This research includes value judgements of key informants, non-structured interviews of stakeholders, focus groups, panel discussions etc. These approaches typically compare qualitative outcomes relating to public perception and can be very detailed and context-specific in order to feed the decision-making process. Many of these methods derive from ethnography, a discipline that uses fieldwork to provide descriptive studies of human societies.

In contrast, quantitative approaches are rooted in empirical methods of investigation. These focus on unbiased measurement of numerical data, usually gathered by means of surveys based on structured questionnaires with close-ended questions, or the data can be garnered from other statistical sources (e.g. national census, national accounts, custom data, international databases). Researchers treat information by means of mathematical methods in order to derive expected responses to variables affected by policy interventions. Quantitative approaches may deal with physical data (e.g. physical quantities of inputs and outputs, number of tonnes landed

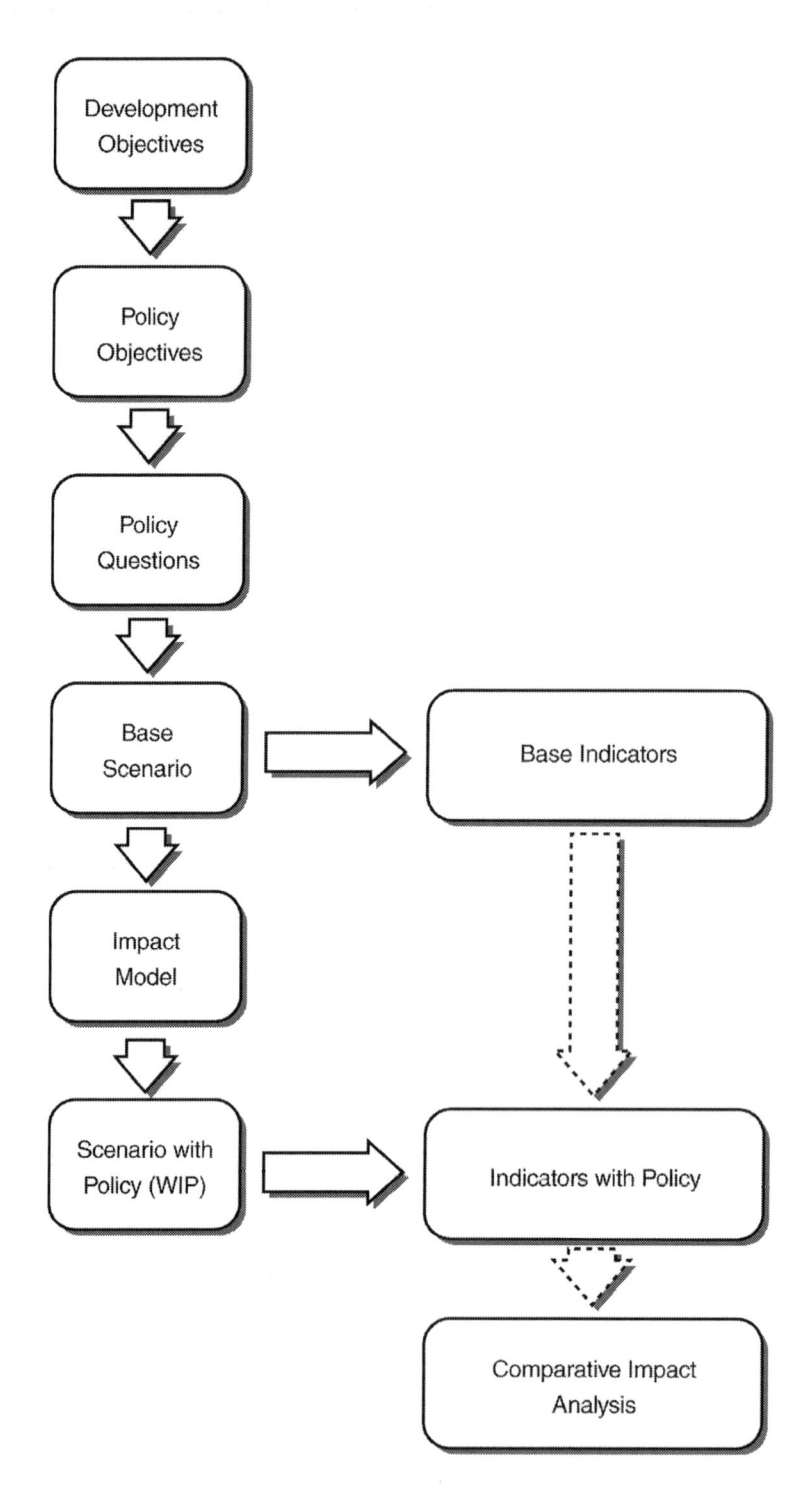

Figure 5.2 Flow chart of counterfactual analysis process

Source: Adapted from Bellù and Pansini (2009)

by a fishery, number of oil spill incidents) and provide decision-makers with information on policy impacts in physical terms.

A subset of quantitative analysis, monetary evaluations are often executed using currency as a common unit of measure. Physical quantities are converted into monetary values by means of prices expressed in monetary units. In these cases, decision-makers are provided with summary information expressed in monetary terms or in percentage variations of monetary variables (e.g. income or expenditure of various groups, GDP growth rates, value added variations, budgetary implications) (Costanza et al., 1987). Although money is the most common currency, others can be used; in the case of fisheries, these might be jobs (fishing and processing), vessels affected, communities affected, or total landings (maximizing protein). The use of different currencies will produce different results; hence, using more than one is to be recommended, so as to remind everyone of the trade-offs involved.

The use of economic-environmental indicators

Indicators can bridge qualitative and quantitative methods and link economic and environmental developments. This bridge facilitates the integration of environmental and economic policies (see Box 5.2). Economic-environmental indicators are necessary for monitoring the discrepancy between actual economic developments and targets to meet environmental objectives and, once calculated, to inform policy adjustments. Recording market reductions in output due to environmental degradation, spending on abatement, waste disposal or environmental charges may give a more accurate comparison of economic costs associated with environmental policies than survey-based results (Dahl, 2000).

Available information on the state of the environment may be translated into measures of flows of environmental services and estimates of damage caused by degradation of environmental resources. The use of technology, such as satellite monitoring, can be used as part of a comprehensive approach linking physical stocks of resources to national balance sheets and resource use to national accounts. Given that opportunity costs can be calculated by using targeted

---

Box 5.2   Constructing a green GDP

Including the costs of environmental degradation to better reflect measurements of national welfare, such as gross domestic product (GDP), has involved numerous alternative 'green' calculations. While there has been significant support for the green GDP concept, critics assert that indicators, such as the GDP, require a standard judgement of environmental quality. For many reasons, however, these (necessarily arbitrary) indicators cannot capture all the dimensions of national situations. Some critics also argue that the current GDP measure is aimed at monitoring factual market activity rather than general welfare levels and as such is not the appropriate type of indicator for including measures on the state of the environment. Consequently, the treatment of environmental spending and degradation will probably not be explicitly included in GDP calculations. Despite imperfections, knowledge of environmental policy objectives makes it possible to calculate the cost of the measures needed to achieve such objectives and hence to construct a 'greener' GDP.

Source: Navrud and Pruckner (1997)

abatement costs, these prices can be connected to information on actual pollution or resource stocks in physical units. Still, many practical difficulties regarding both the actual measurement of physical stocks and their conversion into monetary values exist.

Role of science in policy making

The role of science in policy and politics is a complex topic that is explored in a growing body of scholarly literature (Andresen et al., 2000; Pielke, 2008). Effective environmental policy considers a wide range of inputs but depends on science for a number of functions. Scientific knowledge is employed at many stages in the policy development process – to identify emerging problems, contribute to policy formation, evaluate impacts and assess the state of natural resources and provide a rational, quantitative framework for resolving conflicts. In an ideal and healthy democracy, policy-makers and citizens need accurate, relevant and unbiased science to decide how best to proceed on issues. One of the fundamental roles that science plays in policy development is providing 'raw material' to inform systematic policy decisions (Margules and Pressey, 2000).

Scientists and science can and often do play useful roles in policy and democratic politics, but they can also confuse, muddle, or otherwise impede sound decision-making. Science is essential for effective leadership; however, there is a well-documented need to improve the flow of information in both directions between scientists and decision-makers. Scientific information for policy is required to be accurate, relevant, credible and politically legitimate, yet political interference, competing interpretations, fluctuations in funding, existing legal and economic imperatives, the pace of new information and even an absence of basic research can challenge the role of science in the policy process.

The successful and effective use of science advice in decision-making has been characterized by decisions where:

- There has been full consideration of the best available science.
- The scientific knowledge used is sound (systematic, peer reviewed, independent, free from bias or bias is revealed).
- The full diversity of scientific thought from relevant disciplines is considered.
- Risks and uncertainties are explicitly considered.
- There is an acknowledgement that the 'best available' science does not encompass all possible knowledge.
- Decisions are transparent and open – all pertinent information, data, assumptions, values and interests are identified and accessible.
- A rationale for major decisions is provided/available to all interested parties.
- Adaptive management principles have been incorporated and the decision is subject to ongoing review.
- Formal institutional processes exist to ensure accountability.

Scientific advice is often modified by social, economic and political considerations (Margules and Pressey, 2000). Recognizing this, and that there are major inconsistencies in how many scientists treat 'facts' versus 'values', it is important to state explicitly that there are choices and options outside the linear model available for how scientists can participate in policy and politics. On one end of the spectrum is the pure scientist, who plays a role that is as independent and isolated from policy and politics as possible. For a pure scientist, research is published in academic journals, and no effort is made to help resolve society's current policy challenges. At the other end of the spectrum is the issue advocate, a scientist who uses his or her science and scientific credentials to advocate for policies.

The need for marine policy

The unique structure and function of oceans means that their management is necessarily different from terrestrial environments (see Chapter 2). To more effectively discuss the efficacy of the various legal and policy approaches to marine management, these unique characteristics must be considered.

Common property characteristics

As discussed in Chapter 3, much of the marine environment is a common property resource (open access) and, as such, is freely available to all. Past experience with common property resources (e.g. air and water pollution) has repeatedly demonstrated that resources are often poorly managed (degraded, depleted) when owned by no one (or no community) and individuals are left to act for their own singular benefit, as was suggested by Hardin (1968). Unfettered individual access to marine resources, when combined with modern technology, can lead to a rapid depletion of resources due to 'gold rush' behaviour as well as low profits stemming from overcapitalization and excessive costs. In turn, these low profits can occasionally lead to dangerous working conditions and poor product quality (Deacon, 2009).

Most discussion around the perils of open marine access relate to the rapid depletion of fisheries. This depletion is a result of fishers having few economic incentives to consider the long-term sustainability of fish stocks. However, other marine sectors – including transportation, energy development and mining – also lack incentives to minimize environmental harm and maximize long-term opportunities.

Unpriced positive externalities

The marine environment provides a range of social and environmental benefits that are enjoyed by humanity. These public benefits, termed **ecological goods and services**, are widely valued and necessary to support social and economic activities. Yet they remain unpriced (meaning they have no direct economic value). Examples of positive externalities include biodiversity and ecological values (e.g. nutrient cycling), aesthetics and recreational amenities, health, cultural attachments and water and air quality.

Marine environments provide many important services that are not normally considered to have economic value; the oceans assimilate human wastes from terrestrial and freshwater environments and, through fixing atmospheric CO_2, buffer the effects of climate change. In addition, the oceans contain innumerable raw materials that could provide new sources of food, fibre, medicines and technologies. Maintaining functioning marine environments, then, can be viewed as 'insurance' against current and future negative human impacts on terrestrial and marine environments. Meanwhile, forward-thinking marine policies can convert unpriced externalities into social and environmental benefits. For example, the strategic disposal of retired ships and offshore platforms can create artificial reefs that benefit fish populations and produce opportunities in fishing or dive tourism.

Pollution and other negative externalities

The value and productivity of marine resources can also be threatened by pollution, ocean warming and acidification (as a result of climate change), habitat loss, invasive species and other factors that impose costs that are not reflected in market processes. Just as a market framework can undervalue a resource when positive externalities are present (because the full set of beneficiaries are unrepresented), the presence of negative externalities can lead the market to over-estimate the value of (and hence over-produce) a good or activity.

Characteristics of marine policy

The relationship between public policy and law within a state's coastal waters (territorial waters and EEZs) is similar to the domestic relationship between law and policy within a state's sovereign lands. The primary difference is that international law intrudes into a state's coastal waters to a greater extent than a state's terrestrial areas. Specific examples include the right of innocent passage and the requirement to accommodate the needs of other states (e.g. profit-sharing from the extended continental shelf).

The relationship between public policy and law in ABNJ is very different. Through international conventions and agreements, the high seas and deep seabeds are managed as a commons. As such, international law in ABNJ is largely cooperative, operates outside the court system (with exceptions in the case of disputes) and is not generally punitive (see Chapter 3).

Areas beyond national jurisdiction (ABNJ) are governed by an amalgam of cooperative regulatory policies. While the theoretically voluntary nature of the regulatory policies in ABNJ would suggest that they may be less effective than equivalent domestic regulation, this has not been the case in practice. The voluntary nature of ABNJ agreements has resulted in some of the most innovative and cooperative conventions conceived. While many conventions, including those regulating pollution and dumping, have rates of compliance exceeding similar domestic laws, other conventions, including those on environmental protection, have proven to be more aspirational in nature with little in the way of enforcement 'teeth' to promote compliance. (An exception to this general rule is CITES – Convention on the International Trade in Endangered Species of Wild Flora and Fauna.)

Given that international organizations have limited budgets and few to no powers of taxation, distributive and redistributive policies are usually limited to appeals to states that are largely symbolic in the hope of ensuring rights of passage, access to developing nations to fish and limits on pollution. (A notable exception being the International Seabed Authority.) As ocean governance evolves, constituent policy is frequently used in the establishment and improvement of governance mechanisms.

Introduction to marine policy instruments

At present, no single authority, akin to the nation-state, sets the overall rules for activities on, in and under the world ocean. As discussed in earlier chapters, the legal powers of governments over marine resources vary from sovereign rights in nearshore areas to 'economic' rights in offshore areas. While perhaps an over-simplification, the primary purpose of marine management is to sustain economic, cultural and environmental productivity over time. This is the essence of sustainability. It is achieved through policies that correct for inefficiencies in consumption and negative externalities. Depending on the circumstances, a number of types of policies – termed 'policy instruments' – are applied to this end.

Policy instruments are often broadly classified as **regulatory** (legal), **economic** (market-based), or **informational** (persuasive). Some authors include **cooperative** (voluntary) instruments as a fourth type. Commonly categorized as 'carrots, sticks, or sermons', policy instruments have evolved over time. 'First generation' regulatory instruments addressed immediate concerns (e.g. social failure, tragedy of the commons) that required government intervention. 'Second generation' instruments use markets to correct for human behavior or limit externalities or in instances where government regulation has failed. 'Third generation' instruments are 'suasive', meaning that the regulated enact self-regulation in their own best interests (Table 5.7).

These policy instruments are not mutually exclusive. The differences between standards and taxes, for example, may not be readily apparent to those affected by the policies. The World Bank (1997, Table 5.8) suggests four categories of environmental policies: using markets, creating markets, regulations or informational instruments (including public engagement). The approaches presented here emphasize the importance of individual conditions. Some examples of policy instrument applications for the marine environment can be found in Table 5.9.

Table 5.7 Types of policy instruments

Regulatory instruments	Economic instruments	Cooperative instruments	Informational instruments
Command and control	Environmental taxes/charges	Voluntary agreements/ codes of conduct	Environmental information
Performance-based	Tradeable permits	Roundtables	Education
Process-based	Subsidies	Mediation	Awareness (e.g. eco-labels)
Co-regulation	Financial funding	Certification	

← High ——————— Degree of state intervention ——————— Low →

Source: Adapted from Böcher (2012)

Table 5.8 Taxonomy of policy instruments

Regulations	Using markets	Creating Markets	Informational Instruments
Standards	Liability	Property rights	Information supply
Bans	Subsidy reduction	Tradable permits and rights	Research
Permits and quotas	Environmental charges and taxes	International offset systems	Education and public outreach
Zoning/plans	User charges		Transparency mechanisms
	Deposit-refund systems		Product certification
	Targeted subsidies		Performance measures
	Insurance		Collaborative governance
	Grants and loans		Direct government

Source: Adapted from World Bank (1997)

Table 5.9 Sample applications of policy instruments for marine management

Policy Instrument	
Direct Provision	Subsidies and subsidy reduction
	Deposit-refund schemes
Detailed regulation	Creation of property rights
	Common property resources
	Legal mechanisms, liability
Flexible regulation	Voluntary agreements
Tradable quotas or rights	Information provision, labelling
Taxes, fees, or charges	International treaties
Example Marine Applications	Industrial pollution fees
Provision of marine parks	Fisheries
Wastewater treatment	
Establishing shipping lanes	Hazardous waste management
Spatial zoning and planning	Private marine parks,
Regulation of fishing (by dates, equipment, location)	Exclusive fishery rights
Bans on trade to protect biodiversity (e.g. seahorses)	Community-based management
Bans on dumping	Liability bonds for oil and gas transport and
Low sulphur fuel standards	mining, hazardous waste management
Water quality discharge standards	Standards for marine ecotourism operators
Ballast water treatment	Sharing of transboundary fish stocks
Individually tradable fishing quotas	Marine Stewardship Council certification for
Emissions permits	sustainable fisheries
Fishing licenses	International treaty on Law of the Sea
Park fees	

Source: Adapted from Sterner and Coria (2012)

Regulatory instruments

The objectives of regulations are manifold. Such instruments exist to minimize damage to natural resources, ensure equitable access to valuable resources and monitor the use of these resources. They also encourage or restrict certain behaviours, ensure competition and establish positive social outcomes. In other words, regulation balances rational individual behaviour with the broader societal good to maintain both the economic and environmental integrity of the 'commons'.

While regulation broadly refers to the process by which activities are administered, regulation as a legal construct establishes obligations that require permits or prohibit particular activities, and so forth, in order to implement existing legislation. Regulations control risks and reduce harms. For example, domestic law may allow for the establishment of pollution limits. Regulation then sets specific limits and may mandate how the limits are to be met (e.g. type of technology). It is therefore a legal tool. When considered in the domestic context, regulatory tools are a traditional form of policy. States (or sub-national governments) act as a trustee, controlling access and use of environmental resources. Regulations can include standards, bans, permits or quotas. They can also limit the temporal or spatial extent of an activity (through zoning).

Regulation is also used in international law. International conventions on pollutants, greenhouse gasses, ocean dumping and endangered species are based on regulatory approaches. These approaches are popular because they promise predictability and effectiveness. In addition, both the regulator and the regulated understand what is required of them. Regulation can be particularly effective in waters in ABNJ and states' EEZs (see Box 5.3) since point source discharges, including ships, aircraft and platforms, are known and can be monitored. In contrast, regulation is less effective in territorial waters. These areas are subject to considerable non-point sources of pollution (i.e. from multiple sources including land sources), making it difficult to ascribe discharges to specific culprits.

Box 5.3   Regulating fisheries discards

The unintentional catch (by-catch) of non-target species that are thrown back to the sea are known as discards. Currently, approximately 20 per cent of the total global fisheries catch amounting to approximately seven million tons is discarded, with shrimp and demersal (bottom) trawl fisheries responsible for about half of these discards. Discards contribute to both economic inefficiencies and ecosystem impacts. As such, they are recognized as a major fisheries management problem that can be improved through better regulatory policies. A number of factors contribute to discards including:

- Biological (spatially and temporally coincident fish populations, similarities in size and habitat use).
- Technological (fishing gear unable to distinguish between species).
- Behavioural (customs or practices that contribute to by-catch).
- Economic (inability or unwillingness to adopt selective fishing gear).
- Institutional (lack of regulations prohibiting by-catch).

The primary approaches to regulating fisheries discards are via the following types of regulatory instruments:

Input controls

- Mandating selective fishing gear to avoid by-catch.
- Spatial and temporal fishery closures.
- Limiting the number and size of vessels.
- Bycatch rules.

Output controls

- Catch limits to cap the proportion and total amount of by-catch.
- Transferrable bycatch quota which can be bought and sold.
- Mandatory landings to improve information on by-catch and incent fishers and processors to utilize by-catch.

> **Monitoring and reporting**
> - Catch log recording and control.
> - Landing reports and control.
> - Inspections at sea.
> - Monitoring of fishing grounds.
>
> Non-regulatory tools may include economic incentives to adopt certain gear types and education to change fishers' and consumers' behaviour.
>
> Source: Modified from Johnsen and Eliasen (2011)

While all regulatory approaches are coercive in nature, they can take a number of forms:

Coercive or **command and control regulation** sets a specified, measurable level of activity (e.g. pollution maximum), and the 'control' dictates how the level of activity is to be achieved (e.g. technology). Command and control approaches evolved from engineering policies and were first widely applied in the United States in the early 1970s through the *Clean Air Act* and *Clean Water Act*.

Performance-based regulation specifies the outcomes ('output standards') to be achieved but not the means by which these outcomes are to be attained. Performance-based regulation can benefit the regulator since regulations framed as objectives/outcomes are relatively easy to draft. A central benefit to the regulated party includes the ability to choose the most efficient processes and technologies to meet the regulation. On the downside, difficulties with performance-based regulation include additional costs for the regulator to verify compliance. In addition, performance-based regulations may be difficult to develop if the desired outcomes are not apparent. Most states that use performance-based regulation have adopted 'safe harbour' ('deemed to comply') provisions that limit liability and incent the regulated to develop innovative practices.

Process-based regulation is an incentive-based approach that requires the regulated to develop processes that minimize risks. The premise behind this approach is that businesses have an economic incentive to avoid regulation and, consequently, will develop lower cost solutions to mitigating risks than would be achieved through command-and-control or performance-based regulation. The most frequent marine use of process-based regulation is managing seafood safety. Australia, Canada and the United States allow seafood producers/processors to develop management plans that identify and address risks at key points of the production cycle.

Co-regulation is a partnership between the regulator and regulated that moves the burden of identifying and mitigating risks onto the regulated. Under this model, codes of practice are jointly developed that are enforced by industry or industry associations rather than states or international organizations. The benefit of co-regulation to the regulator is reduced costs (e.g. monitoring, enforcement), voluntary participation, and greater compliance. Benefits to the regulated include lower costs than traditional regulation, industry-wide compliance and demonstration of responsibility to consumers. A challenge with co-regulation, however, is that industry self-regulation may result in anti-competitive practices such as exclusion of new entrants into an industry.

(OECD, 2002)

Regulations are enforced through legal tools that include penalties, fines, liability rules and performance bonds. They focus on the punitive consequences attached to violations of conditional access or misuse. Regulations or other market-based instruments must be backed up with some threat; in the case of non-compliance, this is typically financial. When a law is broken, fines are imposed to remove the advantages gained from non-compliance. While penalties for conviction of an offence range from fines to imprisonment, in practice, legal maximum penalties are rarely imposed. As a result, violators often assess the risks of being caught and calculate the potential fines as part of the cost of doing business.

However, when the offence is serious (e.g. affecting public safety), fines or other sanctions are not considered sufficient, and punishments under the criminal law system may be applied to offending individuals. Liability for damages and risks related to individuals seeking compensation also influences environmental responsibility. Liability can be partial, if the cause of the damage is due to negligence or lack of acceptable precautions, or it can be full, where injured parties have the right to compensation regardless of any precautions taken. Although both forms of liability should induce necessary environmental precautions, the costs for litigation often prevent affected individuals from pursuing compensation. Partial liability can also reduce a company's precautionary measures. Since reporting on environmental performance acknowledges that products or practices may be harmful, the defence of ignorance is removed.

Strict or full liability makes sense where, as with oil spills, precautions can best be taken by one of the parties (and where large numbers of third-party victims make bargains between the parties difficult or impossible) (Box 5.4). Strict liability gives more rights to claimants but has

---

Box 5.4 Liability and marine oil spills in the United States

In the case of marine oil spills in the United States, there are a wide range of laws with complex implications for liability. The *Oil Pollution Act* of 1990 (OPA 90) was enacted largely in response to the major spill in Prince William Sound, Alaska, in 1989 that had resulted from the grounding of the *Exxon Valdez*. The OPA 90 makes parties responsible for oil spills strictly liable for damages caused. It 'channels' liability for oil spills by specifying exactly who is to be treated as the responsible party. In the case of spills from vessels, the owner/operator of the vessel is the responsible party. For offshore facilities like the Deepwater Horizon (responsible for the world's second largest marine oil spill in 2010 – Chapter 9), the holder of the drilling permit (e.g. BP) is the responsible party.

The OPA 90 also includes liability caps, traditional in maritime law, that include some exceptions (e.g. for gross negligence) and vary depending on the type of spill and damage caused. Prior to OPA 90, liability was limited in most cases to the value of the vessel. However, the OPA 90 did not pre-empt criminal or state law, so the availability of alternative legal regimes (civil or criminal) and jurisdictions (state and federal) add significant complexity and uncertainty to the damages phase of litigation. As a result of the *Exxon Valdez* disaster, the federal government sought recovery of damages to Prince William Sound by filing criminal charges under the *Migratory Bird Treaty Act* (MBTA) for causing the death of protected birds and the *Refuse Act*, for dumping of 'refuse'. Since the MBTA is a strict liability statute, the prosecution does not have to show that the defendants intended to harm wildlife or prove that the defendants knew their actions would lead to an oil spill to find liability. Violations of these laws carried penalties and required restitution in the form of compensation for damages to the natural environment. As a result of civil and criminal

> charges, Exxon pled guilty and reached a settlement involving significant payments to the federal government for clean up and damages. Claimants in a separate class action lawsuit (mainly fishermen) were initially awarded $US 5 billion, but Exxon spent 20 years appealing the award. Finally, in 2008, the Supreme Court reduced assessed damages to $US 500 million. It remains to be seen whether private suits for economic damages due to the Deepwater Horizon spill will be subject to the OPA 90 limits, currently capped at $US 75 million.
>
> Source: Faure and Wang (2006, 2008)

been criticized for escalating demand on the court system and preventing economic activity that is exposed to such risks. However, even strict liability may not be a sufficiently forceful instrument; in some cases, there may be a long chain of causality, considerable distance in space or time between negligent action and realized environmental cost. Also, a confusing array of parties may be involved, some of which might no longer exist. To be effective in such cases, liability must be joint and retroactive so that all parties involved face the consequences (Sterner and Coria, 2012).

Regulatory standards

Prescriptive **regulatory standards** are the most commonly used instruments in natural resource management, likely due to their simplicity in principle and typically short timelines for application. Regulatory standards can take one of three forms: **Ambient standards** place limits on the total concentration of a pollutant (e.g. contaminants, noise) in a geographic area. **Emission standards** place limits on individual sources (e.g. ships, platforms). **Design standards** specify a particular type of pollution control technology or production process (e.g. oil spill mitigation requirements for commercial vessels).

Design standards are, however, widely criticized for their lack of incentives to develop alternative solutions that arrive at the same desired result (e.g. life cycle or 'cradle to grave' approaches). Yet, depending on the situation, there are a number of reasons to consider design standards, including situations where (after Sterner and Coria, 2012):

- Technical and ecological information is complex.
- Only a few technologies exist, and one is superior.
- Monitoring costs are high, or monitoring pollution is difficult.

An example of a specific mandated technology is the regulation for the prevention of pollution by oil addressed in Annex I of the International Convention for the Prevention of Pollution from Ships (MARPOL). The Convention not only includes requirements for technical standards in construction (e.g. double hulls for oil and chemical tankers), and equipment (e.g. oil/water separators) but also inspection (including enhanced surveys on existing tankers), record keeping (e.g. oil record books, cargo record books) and in-port procedures (e.g. receipts for the discharge of wastes to shore, reception facilities, tank cleaning in port).

Regulatory approaches may restrict the location or timing of an activity (e.g. marine protected areas and no-take zones) or ban a particular product or process (e.g. fishing gear restrictions). Marine examples include bans on dynamite fishing and the use of cyanide fishing for the live fish trade, as well as proposed bans on fishing gear like the 'hulbot-hulbot' in the Philippines (a form of seining that involves throwing a large rock tied to a net into the sea and dragging it underwater) (White et al., 2002). Because of the destructive impact to coral reefs – a critical habitat and spawning ground of the fishery resource – these bans are considered reasonable.

Success of regulatory approaches in the marine environment

A major criticism of regulatory policy is that to be effective, monitoring and enforcement are required indefinitely. Situations where monitoring is problematic – including non-point source pollution where many emissions are small, dispersed and mobile – may necessitate regulating preventative design standards. In the case of hazardous wastes, it may not be the use of the product, but its disposal that is the chief concern.

In addition, regulatory approaches are often challenging to operate in developing economies. Institutions may be ineffectual, underfunded or corrupt; monitoring and enforcing regulations can be near impossible. The immediate pressure of building the economy or feeding the population can override investment in guaranteeing the sustainability of even the most valuable resources (Sterner and Coria, 2012).

The following examples illustrate the variety of marine activities that have been regulated both in territorial waters and ABNJ.

A review of the regulation of sea cucumber fisheries in the territorial waters of 62 states concluded that depleted and overexploited fisheries were on average managed by 2.6 regulatory measures (Purcell et al., 2013). In contrast, better managed fisheries were regulated by an average of 4.7 regulatory measures. The most successful regulatory measures included fleet (vessel) controls, limited entry controls and harvest closures (see Chapter 8). However, enforcement capacity and poverty were also strongly correlated to the sustainability of sea cucumber fisheries worldwide.

Albert et al. (2013) reviewed domestic (including sub-national) and international regulatory approaches to the management of ballast water, a source of biological invasive species (see Chapter 8). They concluded that some jurisdictions (e.g. IMO) establish regulatory standards that emphasize international consensus, practicability and cost of treatment in exchange for a higher risk of invasions. Meanwhile, other jurisdictions (e.g. United States) have questioned whether existing ballast water standards sufficiently protect against invasive species. Troublingly, the study concludes that without a better scientific understanding of what standards are required to prevent or minimize invasions, regulations are, by and large, merely guesswork.

Tributyltin (TBT) had been a popular anti-fouling compound used in marine paints since the 1960s (Gipperth, 2009). TBT compounds are hazardous to marine life and may enter the human food chain. The domestic and international response to regulating organotin compounds (including TBT) is twofold: First, many states have established ambient standards for maximum concentrations in marine waters and sediments. Second, states began to ban or regulate organotin compounds in the 1980s. These domestic efforts were followed by the 2001 IMO anti-fouling Convention that entered into force in 2008 (see Chapter 9). Parties to the Convention are required to prohibit and/or restrict the use of harmful anti-fouling systems on ships flying their flag. So too are ships that operate under their authority and all ships that enter a port, shipyard or offshore terminal of a Party.

While banning organotin compounds and establishing ambient standards are well intentioned, they have created both anticipated and unanticipated consequences. The anti-fouling Convention does not regulate alternatives to organotin compounds. As a result, a number of new chemicals have appeared on the market that appear to have similar deleterious effects as TBT. Shippers, in an effort to control costs, may register their vessels in a flag state that is not a party to the Convention or that lacks domestic organotin regulations (the 'race to the bottom'). Moreover, states with stores of organotin compounds that they can no longer use may dispose of them in an unsafe manner. Bans on these substances may also inadvertently increase species invasions (Gipperth, 2009).

Economic instruments

Critics of regulatory approaches have proposed economic instruments as a complementary or alternative approach to achieving public policy objectives. Rather than trying to prescribe behaviour through regulation, market-based approaches rely on consumers responding to market-driven price signals to change behaviours and achieve policy goals. Expanding on the 'polluter pays' principle, economic instruments incorporate the 'user/beneficiary pays' principle along with the principle of full cost recovery. If a policy changes the cost of a good, service, activity, input, or output, then it is a **market-based instrument** (MBI).

There are many definitions for MBIs depending on the context in which they are applied. Most commonly, MBIs are policies that encourage specific behaviours through market signals and incentives rather than through explicit directives regarding mechanisms, levels or methods. While regulations can also affect the market, they allow little flexibility in the means by which goals are achieved. Furthermore, regulations do not impact cost or price directly. Thus, MBIs may be able to achieve policy goals more cost-effectively than regulatory mechanisms. Nevertheless, MBIs and regulatory instruments are not mutually exclusive, as any policy still requires appropriate legislative or regulatory backing.

MBIs can also be employed to provide diverse social benefits. These may include payments to Indigenous peoples for protecting goods and services, developing alternative income sources in rural/remote areas, or fostering income diversification opportunities (Greiner, 2013). Usually, however, MBIs are normally associated with addressing environmental or public health issues including pollution (e.g. waste management, greenhouse gasses), resource use (e.g. fisheries) or protecting environmental goods and services (e.g. pollination, waste assimilation).

The earliest applications of MBIs were effluent charges (polluter pays) instituted in France (1968), Netherlands (1969) and Germany (1981). One of the most well-known applications of MBIs is fuel taxes that, through increasing the cost of driving, result in more energy efficient cars. In Europe, for example, 40–60 per cent of the cost of fuel is taxes. As a result, European cars produce two to three times less CO_2 emissions than equivalent vehicles in the United States.

MBIs are generally classified into the following categories:

- **Price-based instruments**: Taxes, charges or subsidies that establish (or help to establish) a specific regulated price for resources or outputs that are deemed to have a social or environmental cost associated with them. Advantageously, they can work as a positive or negative incentive. For example, a price may be set for the creation of a unit of pollution.
- **Quantity-based instruments**: Also termed 'rights-based' instruments, these establish quantitative limits on resource use (or environmental goods and services). Advantages are that rights and obligations are often tradable. For example, permit trading systems and offset schemes create a 'right' to pollute that can be transferred or sold.
- **Market-enhancement instruments**: Also known as 'market-friction' approaches, these include a range of policy actions and approaches from subsidies to establishing property rights. Their purpose is to correct existing market imperfections or 'frictions' that lead to poor resource use. Eco-labelling programs, for example, make it easier for consumers to select goods and services that align with their environmental preferences.

Ultimately, MBIs are designed to ensure that resources are used more efficiently. However, they also offer an opportunity to generate revenue. Making the markets work more efficiently involves establishing prices that address market failures, be they externalized costs or under-priced resources. The proceeds of taxes or levies can provide further environmental subsidies, reduce broader income taxes or be channelled into constructive programs such as monitoring and enforcement.

Use of MBIs requires the following conditions to be met (after Stavins, 2003):

- A pre-existing market or capacity to create or award property rights.
- A measurable target resource that is influenced by cost and profit considerations.
- A target resource that, if MBIs are applied, can be used to attain public policy goals.
- A homogenous metric (indicator) that can be used as the focus of the instrument. The metric should correspond to and drive behavioural changes that are consistent with the resource management objectives.

Importantly, the effectiveness of MBIs will generally be a function of how well each of the above conditions is met. The focus on quantitative metrics means that MBIs typically have already identified measurable goals, such as catch quotas, kilos of discharge, or production dollars. MBIs with potential application for marine resource management include charges and taxes, subsidies and tax concessions, property rights and market support and tradable permits and quotas (Greiner et al., 2000). A summary of MBIs with application to marine management issues can be found in Table 5.10.

Table 5.10 Summary of advantages and disadvantages of market-based instruments' application in the marine environment

Instrument Type	Advantages	Disadvantages	Relevance
Emission and effluent charges/ taxes	Low transaction and compliance costs for firms or individuals. Promote technological innovation. Create long-term incentives. Raise revenue. Create flexibility for polluters. Useful when damage per unit of pollution varies little with quantity of pollution.	Setting charge/tax at the right level. Require monitoring.	Discharge from point sources. When monitoring is viable at a reasonable cost. When pollution abatement is feasible.
Product/user charges	Reduce use of harmful products (for product charges). Raise revenue. Create flexibility for users. Simple to administer.	Setting charge at the right level. Require monitoring. Often weak link to pollution.	Products used in large quantities. Where not feasible to monitor pollution from individual sources. For products whose demand or output is sensitive to price changes. Effective when sources are numerous and damage per unit of pollution varies little with pollution quantity.

(Continued)

Table 5.10 (Continued)

Instrument Type	Advantages	Disadvantages	Relevance
Deposit-refund systems	Reduces volume of waste/pollution. Encourage safe disposal, reuse and recycling. Create flexibility for users.	Transaction costs may be high. Markets for recycled products may not be well developed.	Most effective if applied to products with an existing distribution system and where reuse and recycling is technically and economically feasible. Where problems are related to waste disposal.
Tradable permits	Allocation of resources to highest valued use. Reduced information needs for regulators. Certainty regarding pollution or resource use levels. Compliance cost can be reduced. Flexibility for polluters. Create long-term incentives.	Establishing an efficient market. Setting overall level and initial allocation of permits. Transaction costs involved in trading.	Where environmental impact is independent of pollution source. When environmental impact does not correlate with time of production. Where there are enough sources to establish a market. Effective when damage per unit of pollution varies with amount of pollution.
Subsidies and tax concessions	Encourage actions to overcome environmental problems.	Externalities are not internalized by polluter. May reward poor environmental performers. May pay those who would undertake activity even without subsidy. Cost to budget. May stimulate too much activity.	Where other instruments do not work or are too expensive.
Property rights/market support	Enable goods/services to be identified, which can then have value attached, which in turn allows trading to occur. Improves efficiency of market.	May not be feasible where environmental benefits exist, for which payment cannot be extracted. Jurisdictional issues can arise – e.g. international 'commons' such as the ocean or atmosphere offer limited opportunity to enforce property rights on the full set of users.	Where ownership of 'environment' is uncertain or non-existent. Where significant information asymmetry is present. Where environmental or social values are unpriced.

Source: Adapted from UNEP (2004)

Price-based instruments: taxes and charges

A charge or tax can be envisioned as a 'price' to be paid on the use of the environment. This 'price' can be levied on producers or consumers and is a very effective means of changing behaviour. Taxes that are levied to internalize social or environmental costs – such as pollution – on private third parties, causing the externality are known as **Pigouvian taxes**. The objective of Pigouvian taxes is to drive down pollution output by ensuring the economic benefits of further production exceed the environmental costs associated with it. User charges are a related method, but rather than taxing activity levels or particular behaviours, they involve a fixed charge for the use of environmental services. These charges are typically assessed as an access or license fee. Product charges or taxes could be applied to goods that have an environmental impact anywhere along the line from production to consumption. They can be assessed on a product, such as the use of petroleum, or on a product characteristic, such as the carbon content in petroleum (Sterner and Coria, 2012).

Taxes and charges are suited to situations where the individuals responsible for the problem can be readily identified. However, the costs of monitoring and enforcement may still be high because of the need to ensure that all individuals responsible for the problem are subjected to the tax. Where a tax or charge is issued, the party responsible for the damage to the environment bears all of the costs of changing resource use or management practices.

More complex price-based instruments include integrated strategies, such as ecological fiscal reform (EFR) or ecological tax reform (ETR). These approaches adjust existing taxes to phase out products and to discourage certain practices. Alternatively, new ecological taxes provide incentives to reduce environmental impacts and potentially 'recycle' the revenue from the new taxes. Revenue can be recycled in numerous ways including reducing existing tax rates; providing new credits or subsidy programs; providing refunds to taxpayers to encourage positive behaviours; or supporting the development and upgrades to green technologies. While determining the level of taxes necessary to drive behavioural change can be hard to predict, complementary tax reduction enhances political acceptability. It does so by ensuring there is no increase in the overall tax burden.

Sterner and Coria (2012) suggest that taxes and charges can be appropriate in situations where:

- The risk of 'excessive economic costs' dominate concerns about potential costs associated with insufficient environmental action.
- The action or commodity to be taxed is the direct cause of the negative externality being targeted or is closely connected to it.
- The reasons for the tax/charge are understood by affected parties and there is an understanding of how revenues will be used.
- There is clarity over the rate of tax/charge necessary to change behaviour(s) such that externalities are reduced or eliminated.
- Unfair competitive advantages are avoided.
- There is a desire for improved cost recovery or expanded revenue.

Quantity-based instruments: tradable permits/rights

Tradable permits allocate property rights over a resource that was previously common property. This property could be in the form of air or marine fisheries. The purpose of permits is to mitigate risks and impacts. Permit systems broadly work in two ways: first, 'cap and trade' systems establish an allowable overall level of pollution or extraction and create a limited number of permits that allow an activity. Markets then establish economic values for the permits. Users that stay below their allotted level may sell their surplus permits to others or apply them to offset other practices in their business. The second system is 'baseline and credit'. These systems do not limit the absolute amount of pollution or extraction. Instead, a baseline allowance is established for each user (e.g. facility, land owner), and users may sell unused credits to others. Both systems operate on the premise that rewarding those who can reduce their externalities at the lowest costs will maximize the economic value extracted from all permits. These two systems can also include 'banking' and 'borrowing' (intertemporal trading) schemes. This incents polluters to voluntarily reduce emissions through retrofitting or closing down less efficient sources. The credits from these actions can be applied to future projects, which, in turn, accelerate emissions reductions. In addition, banking can dampen price fluctuations. Over time, either system may ramp down their numbers to reduce overall pollution.

Tradable permits evolved from regulatory permitting approaches that were deemed to be more efficient if permits were allowed to be bought and sold. Tradable permits are normally used once regulations have fully allocated either the maximum acceptable pollution or the resources to be exploited. Perhaps the most successful permit scheme is the United States cap and trade system for sulphur dioxide (SO_2). In 2001, a mere six years after its initiation, this system reduced SO_2 emissions by 40 per cent and reduced compliance costs by 43–55 per cent (Burtraw and Palmer, 2003).

Numerous fisheries are managed today using tradable permits (Individual transferrable quotas – see Chapters 8 and 12), and the IMO is currently investigating whether tradable permits should be applied to shipping in order to reduce CO_2 emissions (which are to be reduced by 50% of 2008 levels by 2050). The growth in permit trading is fostered by technological advances that make it economically feasible to monitor use and changes in environmental quality. This information informs decisions on economic trade-offs and alternative resource use options.

Stavins (2003) suggests that trading systems are likely to be feasible and appropriate for situations in which:

- There is a willingness to grant private property or exclusive access rights to a common property resource.
- The traded commodity is relatively homogeneous (e.g. CO_2, a fish stock).
- Resource or environmental conservation is a principal objective.

Tradable permits have been shown to be effective in theory (through economic modelling) and in practice. Among the benefits of this system are certainty that environmental or resource use thresholds will not be exceeded, economic growth will continue without compromising environmental quality or resource yields and finding the most cost-effective methods to reduce production of externalities. Disadvantages to permit trading schemes are centred on both the costs of establishing initial allocations and operational transactions. Transaction costs include obtaining market-relevant information, negotiating trades and compliance and verification costs. A summary of considerations in the design of permit trading schemes is provided in Box 5.5.

Box 5.5 Considerations in establishing permit trading systems

Defining what commodity is to be traded: Most permit trading systems specify physical quantities such as tonnes of a pollutant (e.g. SO_2, CO_2) or harvested resources, such as fish. If these quantities are difficult to measure and enforce, however, then it may be more feasible to measure transactions such as fuel purchases instead of emissions.

Defining what sectors trading will be applied to: Permit trading (even for a single pollutant) can be applied to certain sectors (e.g. coal power generation) or all sectors (e.g. all industrial facilities that emit SO_2). Schemes that cover all sectors are believed to be more economically efficient as abatement costs are spread over a much larger part of the economy.

Defining where a commodity is to be traded: In the case of pollutants, their impacts and abatement costs (mitigating actions) may vary depending on geographic location. As such, allocations and permit prices may need to be adjusted to account for these differences.

Determining what other measures may be required to achieve objectives. Permit trading alone may be insufficient to achieve anticipated reductions in pollution or extractive use. Other, complimentary measures such as regulatory schemes, subsidies and tax credits (to foster innovation and technological change) may be required for permitting systems to be successful.

Determining who is allowed to hold a permit: Permits may be assigned directly to 'downstream' users (e.g. polluters, fishers), 'upstream' users (e.g. producers or processors of fuels) or both ('hybrid system').

Determining how permits are allocated: Permits can be granted either to existing users (grandfathering) or through a competitive process (auctions). Grandfathering is straightforward as existing resource users are granted permits that may be adjusted for the output of each user. Furthermore, when users exit the market their allocations can be reallocated to new entrants. The disadvantages of grandfathering are that the practice creates instant windfalls for users, especially historical polluters, that receive free allocations and that the opportunity to raise revenue (through auctions) is lost. In contrast, auctions set a true market price for emissions/resource yields and do not limit new entrants into the market. Auctions also provide an opportunity to raise substantial revenue and can be structured to give local/Indigenous participants an opportunity to receive allocations.

Setting the rules for participation: Rules can be established on who is allowed to participate (e.g. third-party brokers, small users), whether participation is voluntary or mandatory, the minimum size of the trades or whether government approval of trades is required.

Establishing compliance and penalty regimes: Compliance regimes set dates by which users must report on their actual allocations, something often not straightforward. Intense trading often occurs prior to these dates as users in deficit attempt to purchase credits from users in surplus. Penalty regimes set the rules (e.g. financial penalties) for non-compliant users.

Source: Modified from OECD (2004, 2008)

Market-enhancement instruments

Substantial gains can be made in environmental protection simply by reducing existing frictions (inefficiencies) in market activity. As discussed previously, market-enhancement instruments are the means by which existing market imperfections or 'frictions' are corrected. Most instruments operate in two ways: First, they improve the information available to consumers, investors or producers. This enables them to make more informed (efficient) decisions about the environmental or social consequences of their purchases, their investments or their production. Second, they reduce transaction costs, which lead to improved environmental and social outcomes. Some examples of market-friction reductions include the following.

Subsidies and tax concessions

A subsidy is a payment for certain activities that the government wishes to promote. A tax concession reduces tax assessments on those engaged in such activities. The objective of subsidy and tax reduction measures is to provide incentives to reduce costs and expand activities that address environmental and social problems. There are many similarities between subsidies and tax concessions, and, from an economic perspective, they are generally considered as 'negative taxes' in contrast to taxes and charges.

There are several types of subsidies: **Production subsidies** are government payments to producers either to increase the production of a good or service (e.g. fishing, offshore wind energy) or decrease some externality (abatement) of production (e.g. sewage treatment). **Export subsidies** are government payments to producers to gain international market access. **Employment subsidies** are government payments to reduce unemployment, train workers or encourage the growth of certain fields. **In-kind** subsidies are non-monetary support provided by governments. The provision of infrastructure, services to assist exporters, preferential access to natural resources or limitations on liability exemplify in-kind subsidies. **Procurement subsidies** are government purchases (e.g. ship building) to support domestic producers or certain sectors. **Market price support** is when governments establish minimum prices or enact tariffs and trade barriers to protect certain producers.

All of these subsidies may be time-limited to encourage growth in a certain part of the economy, or they may become a permanent part of the economy. Subsidies may be appropriate in cases where a socially or environmentally beneficial activity can be reliably targeted. Producing a benefit is generally not equivalent to ceasing to generate a cost. Offshore windfarms, for example, may receive subsidies or tax concessions since 'green energy' is necessary to meet greenhouse gas reduction targets. In this instance, the subsidy is either a one-time payment to help launch the operation or a deficiency payment to compensate for the difference between the target price and the actual market price of the product or service. However, subsidies frequently have unintended effects and often become 'perverse subsidies'. For example, many energy exporting states heavily subsidize domestic energy use (which nullifies conservation efforts), and many agricultural states subsidize agricultural industries (which can result in wasteful practices and over-production).

Subsidies have a long history in marine fishing. Currently, the world's fishing fleet receives annual subsidies of $US30–40 billion. The consequence has been extreme excess fishing capacity of 40–60 per cent or, stated differently, up to 2.6 million unnecessary fishing vessels (Sumaila et al., 2016). These subsidies lower retail prices (and thus increase consumer demand) and create a 'race to fish'. Unsustainable catches and often poor safety practices for fishers are the sad result.

Tax concessions are similar to subsidies in that both approaches attempt to grow targeted aspects of the economy. Concessions can be conceived as a subsidy to a producer's tax burden. Concessions can take diverse forms: **tax exemptions**, where no tax is paid; **tax credits**, where reduced tax is paid; or **tax deferrals**, where tax payments are delayed. Tax concessions are criticized because they are often less transparent than subsidies and, once enacted, are difficult to repeal.

Property rights and market support

The peril of open access (advocated by Grotius, Chapter 3) is a frequent theme found throughout this text. Individuals, corporations and states have few incentives to produce, regulate, manage or monitor the production of marine-derived goods and services efficiently in ocean areas which are shared with others. Low profits due to overcapitalization and excessive costs, dangerous working conditions due to low profit margins and poor product quality further illustrate this peril (Deacon, 2009). While most discussion around the perils of open access relates to the rapid depletion of fisheries, other sectors, including transportation, energy development and mining, also lack incentives to minimize environmental harm and maximize long-term opportunity.

From an international perspective, the global ocean commons is gradually being enclosed. The proliferation of declarations of EEZs and extended continental shelves over the past several decades has resulted in 95 per cent of the world's fisheries and nearly all of the world's discovered offshore hydrocarbon resources resting within state governments. The establishment of EEZs has been a significant step towards economic rationality and enclosing the commons. Nonetheless, many of the most important global fisheries, including tuna, occur primarily outside of EEZs.

Marine property rights, therefore, rest primarily with the state, which may then confer ownership onto sub-regional governments, Indigenous groups or private parties. However, states may allow open access to certain resources within their jurisdictions or they may lack the ability to control and regulate access (e.g. failed states). Yet, there are some situations where it may be more beneficial for resources to remain in the commons. Situations where state control is weak and citizens are poor (unable to invest in powerful technologies) may benefit from retaining a common property system (Sterner and Coria, 2012).

An emerging trend in the creation and trade of property rights is assigning rights and value to environmental services such as carbon sequestration and biodiversity protection. The development of markets for these types of services in the marine environment is challenging for two reasons. First, the diverse nature of the ecosystem services produced in different geographic areas gives rise to measurement and accounting difficulties. Second, the relationship between actions and environmental outcomes is poorly understood. This understanding is complicated by the time lags involved and the unpredictability of variables such as ocean currents and temperature. Almost inevitably, problems of who should be legally responsible for underwriting the environmental improvements are generated (Davis and Gartside, 2001).

Informational instruments

Information significantly affects the policymaking environment, so much so that it can also be considered a policy instrument in and of itself. Information provides the foundation for all instruments, as generally every policy requires data on technology and ecology. There are a broad range of explicit educational and information-based policy tools. These can act as stand-alone instruments or supplement and reinforce the effectiveness of other mechanisms. Within the context of the marine policy environment, the main informational categories include supply, research, education and public outreach campaigns, transparency mechanisms and product certification.

Information supply

Information supply is fundamentally important in changing attitudes and behaviours. Studies have shown that environmentally inappropriate behaviour often arises not from selfishness motivations but rather from ignorance among user groups. There is a significant relationship between engagement in environmentally responsible behaviours and knowledge of environmental issues and conservation activities. This is consistent with surveys and community direct consultations with resource users, which identify research-based advice as a major need for improving management. Recognizing the central role of information supply can assist the establishment of effective environmental policies.

Where informed resource users have self-interest in protecting biodiversity, appropriate information programs can occasion substantial benefits. For example, enlightened self-interest may be sufficient to encourage traditional fishers to voluntarily discontinue the use of a net type which damages fish stocks and which would ultimately impact their livelihood. In this instance, the fishers would be informed that there are exclusionary devices which reduce by-catch and the effort and time spent sorting their catch would be minimized. Although the precautionary principle has come a long way in influencing policy decisions made with little baseline scientific or economic understanding, setting serious limitations before any scientific connections have been established is usually seen as overly restrictive. Lack of information, however, is a common source of failure in a range of policies. Providing information is a powerful means by which governments can facilitate market operation and improve decision-making by their participants, especially when supported by the user communities themselves. Nevertheless, as with all cooperative voluntary arrangements, the problem of 'free-riders' who continue to use wasteful or harmful methods cannot be completely solved.

Research

When it comes to providing information on the impacts and effects of different environmental activities, research is required for an evidence-based understanding. Funding scientific and economic research aimed at understanding the dynamics and interrelationships in ecological and economic systems can assist in determining values and risks for natural resource management purposes and reflect these in public policy. Some of the benefits of research are private or, at least, can be recovered efficiently through the sale of goods and services, while other benefits are of a 'public good' nature. Research is usually most cost-effective when it is conducted by scientists or specialist organizations in association with resource users who have local expertise, on behalf of policy-makers (Greiner et al., 2000).

Education and public outreach

Advertising and granting awards to those who adopt best management practices potentially provide powerful incentives to shift public perception concerning environmental problems (Box 5.6). Well-defined advertising campaigns can simultaneously provide information and influence behaviour. There is also evidence that prizes and awards can play an important role in raising awareness of environmental issues and changing attitudes. In cost-benefit terms, prizes and awards are usually very inexpensive to create and maintain, yet they yield meaningful results. If they are developed at multiple levels, awards can engage the interest of substantial sections of the community, often securing free publicity and raising consciousness and influence motivation at minimal cost. However, it must be acknowledged that such campaigns have not always been successful; much is contingent on the tactics employed (McKenzie-Mohr and Smith, 1999).

Environmental education and awareness campaigns, like voluntary initiatives, are ultimately considered a soft policy option. Focussing on encouraging change rather than requiring action or providing broad financial support to do so, they typically function best as part of a larger strategic approach. Educational initiatives play an important role in motivating participation and sustaining behavioural changes in individuals, households and industry. Ambitious education and awareness have also been proven to effectively reduce the need for economic incentives and regulatory enforcement.

Box 5.6 World Wildlife Fund's (WWF) International Smart Gear Competition

One example of a marine conservation award is the World Wildlife Fund's (WWF) International Smart Gear Competition. First held in 2005, the competition brings together the fishing industry, research institutes, universities and government to inspire and reward practical, innovative fishing gear designs that reduce by-catch – the accidental catch and related deaths of sea turtles, birds, marine mammals, cetaceans and non-target fish species. WWF believes the Smart Gear competition will help catalyze a wide-ranging, multidisciplinary response by encouraging creative thinkers everywhere to share their ideas for modified fishing gears and procedures that increase selectivity for target fish species and reduce by-catch. The competition is open to eligible entrants from any background, and entrants have included gear technologists, fishermen, engineers, chemists and inventors.

WWF offers more than $US 50,000 in prize money to attract innovative ideas that may prove to be a valuable solution to some of the most pressing bycatch problems in fisheries around the globe. Financial support for the competition is provided by a number of government departments including National Oceanic and Atmospheric Administration (NOAA), Canadian Department of Fisheries and Oceans (DFO), as well as support from a number of foundations and corporations.

An international panel made up of gear technologists, fisheries experts, representatives of the seafood industry, fishermen, scientists, researchers and conservationists judge the entries.

Winning both the Grand Prize ($US 50,000) and the Special Tuna Prize ($US 7,500) in 2011 was an entry from the captain of a Japanese tuna vessel. The design, called the 'Yamazaki Double-Weight Branchline', works by increasing the sinking rate of fishing gear, making it more difficult for seabirds to chase baited hooks. Used in conjunction with other devices, it can reduce seabird mortality by almost 90 per cent.

Runners up, each receiving $US 10,000, include the 'SeaQualizer', designed to reduce fish mortality in the recreational fishing industry, and the 'Turtle Lights for Gillnets', aimed at reducing the capture of marine turtles in gillnets. This device reduced turtle interactions by up to 60 per cent without affecting target catch rates or catch value.

Following the award, WWF works with the inventors to develop their ideas into usable products that enable smarter fishing.

Source: www.smartgear.org

Transparency mechanisms: corporate and state of the environment reporting

Governments seeking to advance sustainable development are increasingly employing strategies that encourage, support, mandate or directly demonstrate more environmentally sound business practices. Disclosure is designed to support market-based incentives (such as product differentiation), public awareness and pressures, and government enforcement. Transparency mechanisms can be advanced on at least two different levels: disclosure of government inspection and factory audit reports, or disclosure of corporate environmental responsibility (CER) practices, including both the positive and negative impacts of business operations on a range of environmental issues (Box 5.7). Requiring that industry track and transparently disseminate information on the quality

> **Box 5.7 Biodiversity audits**
>
> Building on the success of applying transparency mechanisms to industry, an emerging motivational strategy is the biodiversity audit. This innovative approach involves a small audit team containing scientific and other expertise as well as including respected members of the local community. Their role is to assess the biodiversity and significance of a specific area. Audits are conducted periodically and results published locally to raise awareness of biodiversity issues and to enable longitudinal comparisons of the biodiversity characteristics. If audits are coupled with a system of prizes and awards, their capacity to raise consciousness and influence attitudes could be further advanced in a manner that is cost-effective and efficacious.
>
> Source: Hill (2005)

of their environmental practices can encourage action through accountability (Wahba, 2008). Australia, China, Denmark, the European Union, France, Germany, India, Norway, Spain, Sweden and the United States are among the countries that have developed governmental policy initiatives to promote sustainability reporting or environmental, social and governance (ESG) disclosure. Stock exchanges in Brazil, China, Malaysia, Pakistan, Singapore and South Africa are also playing a pivotal role in requiring or recommending listed companies to disclose sustainability performance information. Recent research from the auditing firm, KPMG, shows that, in 2011, 95 per cent of the Fortune Global 250 companies voluntarily provide information on their sustainability policies and performance – an increase from 50 per cent in 2008. Studies have also found that there are some significant correlations between corporate management's perceived importance of environmental factors and environmental reporting practices, suggesting environmental disclosure influences corporate practices (Marshall, 2003). While the impact of transparency is hard to separate from the effects of other regulatory policies, some facilities have responded to rankings or public reporting obligations. Ultimately, increasing community and facility awareness of emissions – be it through internet access to pollutant registries, inventories, right-to-know publications or reports – has made it easier to evaluate sources of environmental contaminants and analyse policy instrument effectiveness.

Product certification

Also known as eco-labels, environmental product certification comprises a number of labelling systems for food and consumer products. The roots of eco-labelling can be found in the growing global concern for environmental protection on the part of the public, businesses and governments. A certified 'eco-label' identifies a product that meets specified environmental performance criteria (standards). Eco-labels are often voluntary, but disclosing the environmental impact of some products can also be mandated by law. Usually directed at consumers, but also suppliers, the intent of certification and labelling is to make it easy to take environmental concerns into account when shopping. Some labels quantify pollution or energy consumption by way of index scores or units of measurement; others simply assert compliance with an established standard or certification (Salzhauer, 1991).

The International Organization for Standardization (ISO) has developed a series of standards and guidelines for use in various programs, including eco-labelling. Type I environmental labelling, or the ISO 14024 standard, relates to a wide range of environmental issues associated with a product or service. The certification criteria are developed by a third party and are based on life cycle considerations (i.e. assessment of product from raw material extraction to end of life). Type I certification is limited to environmental leaders in a given product or service category.

Currently, the Ecolabel Index (www.ecolabelindex.com) tracks 463 eco-labels in 199 states that span 25 product sectors. Marine eco-labelling can be separated into four broad types:

1   Seafood product labels: These have independent certification based on agreed-upon standards, third-party certification, chain of custody and certification of individual stocks. In most instances, the products are labelled to inform consumer choice. Examples include Marine Stewardship Council (discussed in Box 5.8), Naturland, and Friend of the Sea.
2   Consumer awareness initiatives: These labels recommend to consumers what types of fisheries products to buy based on overviews of current information on overall stock status and fishing methods. Individual products are not labelled, but restaurants can partner to display the relevant labels. Examples include SeaChoice, Seafood Watch and FishWise.
3   Eco-labelled shipping: These labels are relatively new, and currently the German Blue Angel program is the most renowned. The label's launch coincides with publication by the European Commission of a new EU strategy on air pollution from shipping that aims to reduce SO_2 and particulate emissions in the short term and NOx and other problems in the long term (ED 21/11/02).
4   Tourism eco-labels: These initiatives are aimed at strengthening sustainability practices through tourism services. With Canada's Clean Marine Green Leaf Eco-Rating Program, operators follow environmentally sound marina practices. Galapagos Quality is a seal of quality given to tourism businesses in Galapagos that have voluntarily committed to meeting environmental standards and requirements.

Box 5.8   Product certification example: Marine Stewardship Council

The Marine Stewardship Council's (MSC) fishery certification program and seafood eco-label recognize and reward sustainable fishing. MSC is a global organization working with fisheries, seafood companies, scientists, conservation groups and the public to promote the best environmental choice in seafood. MSC certification assures consumers worldwide that seafood products bearing the MSC eco-label come from fisheries that have passed scientific, independent assessment for environmental and management practices. In order to become certified, fisheries must meet the MSC environmental standard for sustainable fishing.

The MSC standard has three overarching principles that every fishery must prove that it meets:

Principle 1: Sustainable fish stocks: The fishing activity must be at a level that is sustainable for the fish population. Any certified fishery must operate so that fishing can continue indefinitely and is not overexploiting resources.
Principle 2: Minimizing environmental impact: Fishing operations should be managed to maintain the structure, productivity, function and diversity of the ecosystem on which the fishery depends.

> **Principle 3: Effective management:** The fishery must meet all local, national and international laws and must have a management system in place to respond to changing circumstances and to maintain sustainability.
>
> **What does this mean in practice?**
>
> Under the MSC program, every fishery is measured against these principles. The actions that different fisheries take to show they meet the three principles vary in each case, taking into account the unique circumstances of the fishery. For example, a fishery could be asked to show that:
>
> **Principle 1**: It has reliable data on the age and gender patterns of fish populations to prevent too many young fish being caught and that other factors that affect the health of the stock – such as illegal fishing – have been considered.
> **Principle 2**: Measures are in place to limit by-catch (living creatures caught unintentionally). This could mean changing how fish trimmings are discarded so that seabirds are not drawn towards hazardous fishing gear.
> **Principle 3**: Vessel owners have signed a code of conduct, shared GPS data or undertaken research to ensure their fishery is well managed. Effective management also ensures that all vessels will, for example, change their fishing gear or, when required, respect closed zones.
>
> Eligibility for certification is assessed and awarded by third-party certifiers. When a fishery meets the MSC standard for sustainable fishing its certificate is valid for five years. During this period, the performance of the fishery will be reviewed at least once a year to ensure that it continues to meet the MSC standard. After five years, the fishery must be reassessed in full if it wishes certification. Once a fishery has been certified, all companies in the supply chain – from boat to plate – that want to sell seafood from the fishery with the blue MSC eco-label must be certified as meeting the MSC chain of custody standard.
>
> Despite their intentions, product certification still raises controversy surrounding implementation. Fees upon successful certification provide a perverse incentive to the certifier to certify as many actors as possible. MSC programs have been criticized for their failure to demonstrate positive environmental impacts in many fisheries as well as certifying fisheries that may not be sustainable (Jacquet et al., 2010).

Performance measures

Performance measures can be part of an environmental reporting or certification process or an independent mechanism that assesses whether the goals and objectives of existing policies have been reached. Measures identify strategic outcomes, track progress and determine where adjustments should be made. They are valuable policy tools for improving accountability, informing decision-makers, demonstrating program effectiveness and increasing political support and financial resources. Performance measures can be qualitative or quantitative assessments of outputs, outcomes or processes. However, identifying the right suite of measures and appropriate targets can be challenging, and managing complex outcomes over time requires expertise and financial support. In addition, achieving targets can be influenced by factors outside the policy process, making high-level outcomes difficult to assign and account for internally (Mazur, 2010).

Voluntary instruments

One of the most effective ways to build motivation is to pursue structures that build community and industry ownership. Voluntary initiatives are an alternative option to more expensive legislative policy tools or programs. Voluntary instruments and mechanisms rely neither on coercion nor on the continuation of substantial financial incentives; rather, they rely on voluntarism and self-regulation. Among the mechanisms are self-regulation by industry, projects supported by non-government organizations, community groups or stakeholders and direct contracts with stakeholders (e.g. covenants on resource use). Voluntary instruments may also involve challenging business, industry, or government to improve their environmental standard or practices; in return, they garner public acknowledgement or avoid future regulation. While the costs of development and administration for voluntary initiatives are lower, the lack of regulatory consequences has made their effectiveness highly controversial.

Voluntary initiatives are particularly vulnerable to 'free riders', where those who choose not to participate avoid the costs of improvements incurred by competitors. In addition, not all resource users will make rational choices. Policy must also take account of the minority who may be irrational, incompetent or inflexible, underlining the need for a regulatory safety net. The lack of reporting or monitoring also makes it difficult to assess whether and where progress has been made. Proponents of voluntary programs suggest they promote an ethic of resource custodianship and a cooperative model for achieving environmental goals. However, there are concerns that working closely with industry lowers the bar and prevents setting more ambitious targets. As a result, government proposals for voluntary initiatives in response to environmental concerns are often viewed as falling short of necessary or serious action.

Whether voluntarism and self-regulation are dependable in delivering targeted environmental outcomes depends on the extent of the gap between the public interest in environmental protection and the private interests of resource users. Where society's interest in protecting resources and the resource user's interest substantially coincide, self-regulation may be a cost-effective and appropriately non-interventionist strategy. Generally, the smaller the gap between interests, the more dependable voluntary mechanisms will be.

Unfortunately, in many circumstances, there is a considerable gulf between the public interest in resource conservation and the private interests of individual consumers. For example, whereas many fishermen may identify with the conservation objectives of voluntary protection of nursery grounds for juvenile fish, their self-interest in protecting adult spawning resources is not so apparent. More commonly, there is a perceived tension between maximizing the use of resources and protecting stocks. While the preservation of spawning grounds may arguably provide long-term benefits to users (increasing the productivity of the stock), these benefits are less immediate and tangible than the increase in short-term catch yields that fishing the rich spawning grounds promises. For economically marginalized resource users, evidence suggests that short-term production pay-offs are often perceived to outweigh the longer-term benefits of conservation, presenting a major limitation of self-regulation and other motivational-based approaches to resource protection.

The application of voluntary instruments may be most effective at a small scale or in situations where parties already participate in self-regulation and more coercive policy is not necessary. Because of the risks and lack of dependability, relying on voluntary instruments as an exclusive means for environmental protection can work only if resource users appreciate the value of protection and comprehend how their own self-interest is served in protecting it. Informational and educational incentives are extremely important in this respect. However, in virtually all cases, voluntary programs need to be supported by other mechanisms like price, property rights and regulation. Nevertheless, voluntarism still has an important role, particularly where environmental threats require active community participation. The challenge is to build a stewardship ethic, normalize the necessary conservation practices and harness altruistic behaviour.

Policy-makers also need to be cognizant of the traps of using financial incentives to jumpstart voluntary programs. If payments for conservation are subsequently withdrawn, then the benefits of those conservation efforts may be lost. Moreover, removal of the financial support may convert a carefully cultivated custodial ethic into one that is antagonistic towards the costs of conservation. One way to address the risk of reactive behaviour is to emphasize that any subsidies are only transitional in nature.

Although it is important to recognize and design policy that acknowledges self-interest, it is not a full explanation for an individual's behaviour. In contrast, altruism, community and respect for broad conservation objectives can also drive behaviour, and mechanisms that seek to support and harness such motivations can be integrated into a policy approach. If, for example, limited resources were put into encouraging and supporting resource users willing in principle to enter voluntary agreements, this might provide a cost-effective mechanism, even though it is only applicable to a small minority. If it were coupled with appropriate tax concessions, then many more users might be encouraged to participate. Successful programs may also promote awareness of resource conservation and community programs, thereby contributing to motivational change – a factor acknowledged in the Convention on Biological Diversity.

To summarize, except in those specific circumstances where the public interest in protecting biodiversity conservation and private interest substantially coincide, and where biodiversity loss is reversible, voluntary mechanisms are most successful when employed as a supplement to other instruments to avert irreversible losses. These mechanisms, in most cases, need to be supported by policies that ensure dependability and recognize the need for precautionary measures (Boardman, 2009).

Discussion questions

- Describe a public policy that you believe to be successful and why.
- Explain the differences between a law, a regulation and a policy.
- Explain how fisheries regulators have more policy instruments at their disposal than parks departments.
- What type of policy instrument(s) are likely to be most successful in addressing marine pollution issues between states?

Further reading

Albert, R. J., Lishman, J. M. and Saxena, J. (2013) 'Ballast water regulations and the move towards concentration-based numeric discharge limits', *Ecological Applications*, vol 23, no 2, pp. 289–300.

Andresen, S., Skodvin, T., Underdal, A. and Wettestad, J. (2000) *Science and Politics in International Environmental Regimes: Between Integrity and Involvement*, Manchester University Press, New York.

Bardach, E. (2011) *A Practical Guide for Policy Analysis: The Eightfold Path to More Effective Problem Solving*, CQ Press, Washington, DC.

Bellù, L. G. and Pansini, R. V. (2009) *Quantitative Socio-Economic Policy Impact Analysis: A Methodological Introduction*, FAO, Rome.

Birkland, T. (2011) *An Introduction to the Policy Process: Theories, Concepts, and Models of Public Policy Making*, M.E. Sharpe, Armonk, NY.

Boardman, R. (2009) *Canadian Environmental Policy and Politics: Prospects for Leadership and Innovation*, Oxford University Press, Don Mills, ON.

Böcher, M. (2012) 'A theoretical framework for explaining the choice of instruments in environmental policy', *Forest Policy and Economics*, vol 16, pp. 14–22.

Bridgeman, P. and Davis, G. (2000) *Australian Policy Handbook*, Allen & Unwin, Sydney.

Burtraw, D. and Palmer, K. (2003) *The Paparazzi Take a Look at a Living Legend: The SO2 Cap-and-Trade Program for Power Plants in the United States*, Resources for the Future, Washington, DC.

Churchman, C. W. (1967) 'Wicked Problems', *Management Science*, vol 14, no 4, pp. 141–146.

Costanza, R., d'Arge, R., de Groot, R., Farber, S., Grasso, M., Hannon, B., Limburg, K., Naeem, S., O'Neill, R. V., Paruelo, J., Raskin, R. G., Sutton, P. and van den Belt, M. (1987) 'The value of the world's ecosystem services and natural capital', *Nature*, vol 387, pp. 253–260.

Dahl, A. L. (2000) 'Using indicators to measure sustainability: Recent methodological and conceptual developments', *Marine and Freshwater Research*, vol 51, pp. 427–433.

Davis, D. and Gartside, D. F. (2001) 'Challenges for economic policy in sustainable management of marine natural resources', *Ecological Economics*, vol 36, no 2, pp. 223–236.

Deacon, R. T. (2009) *Creating Marine Assets: Property Rights in Ocean Fisheries*, PERC Policy Series, No 43, Bozeman, Montana.

EEA (2005) *Policy Effectiveness Evaluation: The Effectiveness of Urban Wastewater Treatment and Packaging Waste Management Systems*, www.eea.europa.eu/publications/brochure_2006_0305_111039, accessed 04 January 2019.

FAO (2009) *Quantitative Socio-Economic Policy Impact Analysis: A Methodological Introduction*, FAO, Rome.

Faure, M. and Wang, H. (2006) 'Economic analysis of compensation for oil pollution damage', *Journal of Maritime Law and Commerce*, vol 37, pp. 179–217.

Faure, M. and Wang, H. (2008) 'Financial caps for oil pollution damage: A historical mistake?', *Marine Policy*, vol 32, no 4, pp. 592–606.

Fischer, F., Miller, G. J. and Sidney, M. S. (2006) *Handbook of Public Policy Analysis: Theory, Methods, and Politics*, Marcel Dekker, New York.

Gipperth, L. (2009) 'The legal design of the international and European Union ban on tributyltin antifouling paint: Direct and indirect effects', *Journal of Environmental Management*, vol 90, pp. 86–95.

Greiner, R. (2013) 'Social dimensions of market-based instruments: Introduction', *Land Use Policy*, vol 31, pp. 1–3.

Greiner, R., Young, M. D., McDonald, A. D. and Brooks, M. (2000) 'Incentive instruments for the sustainable use of marine resources', *Ocean Coastal Management*, vol 43, no 1, pp. 29–50.

Hanley, N. (1992) 'Are there environmental limits to cost benefit analysis?', *Environmental and Resource Economics*, vol 2, no 1, pp. 33–59.

Hardin, G. (1968) 'The tragedy of the commons', *Science*, vol 162, no 3859, pp. 1243–1248.

Hill, D. A. (2005) *Handbook of Biodiversity Methods: Survey, Evaluation and Monitoring*, Cambridge University Press, Cambridge.

Howlett, M. and Ramesh, M. (1995) *Studying Public Policy: Policy Cycles and Policy Subsystems*, Oxford University Press, Toronto.

Huang, I. B., Keisler, J. and Linkov, I. (2011) 'Multi-criteria decision analysis in environmental sciences: Ten years of applications and trends', *Science of the Total Environment*, vol 409, pp. 3578–3594.

Jacquet, J., Pauly, D., Ainley, D., Dayton, P., Holt, S. and Jackson, J. (2010) 'Seafood stewardship in crisis', *Nature*, vol 467, pp. 28–29.

Jesinghaus, J. (1999) *Indicators for Decision-Making*, European Commission, Joint Research Centre, Brussels.

Johnsen, P. and Eliasen, S. Q. (2011) 'Solving complex fisheries management problems: What the EU can learn from the Nordic experiences of reduction of discards', *Marine Policy*, vol 35, no 2, pp. 130–139.

Lasswell, H. (1951) 'The policy orientation', in D. Lerner and H. Lasswell (eds.), *The Policy Sciences*, Stanford University Press, Stanford, CA.

Loomis, J. and Helfland, G. (2001) *Environmental Policy Analysis for Decision Making*, Kluwer Academic Publishers, Dordrecht, The Netherlands.

Margules, C. R. and Pressey, R. (2000) 'Systematic conservation planning', *Nature*, vol 405, pp. 243–253.

Mark, M., Henry, G. and Julnes, G. (2000) *Evaluation: An Integrated Framework for Understanding, Guiding, and Improving Public and Non Profit Policies and Programs*, Jossey-Baso Inc., San Francisco, CA.

Marshall, R. S. (2003) 'Corporate environmental reporting: What's in a metric?', *Business Strategy and the Environment*, vol 12, no 2, pp. 87–106.

May, J. V. and Wildavsky, A. (1978) *The Policy Cycle*, Sage, London.

Mazur, E. (2010) 'Outcome performance measures of environmental compliance assurance', *OECD Environment Working Papers*, vol 18, no 18.

McKenzie-Mohr, D. and Smith, W. (1999) *Fostering Sustainable Behaviour: An Introduction to Community-Based Social Marketing*, New Society Publishers, Gabriola Island, British Columbia.

Munger, M. (2000) *Policy Analysis*, W.W. Norton and Co, New York.

Navrud, S. and Pruckner, G. J. (1997) 'Environmental valuation: To use or not to use? A comparative study of the United States and Europe', *Environmental and Resource Economics*, vol 10, pp. 1–26.

Norris, P. (2011) 'Cultural explanations of electoral reform: A policy cycle model', *West European Politics*, vol 34, no 3, pp. 531–550.

OECD (2002) *Regulatory Policies in OECD Countries: From Interventionism to Regulatory Governance*, Annex 2, OECD, Paris.

OECD (2004) 'Political economy of tradable permits: Competitiveness, co-operation and market power', in F. J. Convery, L. Dunne, L. Redmond and L. B. Ryan (eds.), *Greenhouse Gas Emissions Trading and Project Based Mechanisms*, OECD, Paris.

OECD (2008) *Environmentally Related Taxes and Tradeable Permit Systems in Practice*, OECD, Paris.

Ohowa, B. O. (2009) 'Evaluation of the effectiveness of the regulatory regime in the management of oil pollution in Kenya', *Ocean and Coastal Management*, vol 52, no 1, pp. 17–21.

Patton, C. V. and Sawicki, D. S. (1986) *Basic Methods of Policy Analysis and Planning*, Prentice-Hall, Englewood Cliffs, NJ.

Pielke, R. A. (2008) *The Honest Broker: Making Sense of Science in Policy and Politics*, Cambridge University Press, Cambridge.

Pintér, L., Zahedi, K. and Cressman, D. (2000) *Capacity Building for Integrated Environmental Assessment and Reporting: Training Manual*, International Institute for Sustainable Development for the United Nations Environment Programme, Winnipeg.

Purcell, S. W., Mercier, A., Conand, C., Hamel, J.-F., Toral-Granda, M. V., Lovatelli, A. and Uthicke, S. (2013) 'Sea cucumber fisheries: Global analysis of stocks, management measures and drivers of overfishing', *Fish and Fisheries*, vol 14, pp. 34–59.

Robinson, J. and Ryan, S. (2002) *A Review of Economic Instruments for Environmental Management in Queensland*, citeseerx.ist.psu.edu/viewdoc/download?doi=10.1.1.194.3099&rep=rep1&type=pdf, accessed 04 January 2019.

Salzhauer, A. L. (1991) 'Obstacles and opportunities for a consumer ecolabel', *Environment*, vol 33, pp. 10–37.

Schulz, W. and Schulz, E. (1991) 'Germany', in J. P. Barde and D. W. Pearce (eds.), *Valuing the Environment: Six Case Studies*, London: Earthscan.

Smith, K. and Larimer, C. (2009) *The Public Policy Theory Primer*, Westview Press, Boulder, CO.

Stavins, R. N. (2003) 'Experience with market-based environmental policy instruments', in K. Mäler and J. R. Vincent (eds.), *Handbook of Environmental Economics*, Elsevier, Amsterdam.

Sterner, T. and Coria, J. (2012) *Policy Instruments for Environmental and Natural Resource Management*, RFF Press, New York.

Sumaila, U. R., Lam, V., Le Manach, F., Swartz, W. and Pauly, D. (2016) 'Global fisheries subsidies: An updated estimate', *Marine Policy*, vol 69, pp. 189–193.

Theodoulou, S. Z. and Kofinis, C. (2004) *The Art of the Game: Understanding: American Public Policy Making*, Thomson/Wadsworth, Belmont, CA.

UNEP (2004) *Economic Instruments in Biodiversity-Related Multilateral Environmental Agreements*, UNEP, Nairobi, Kenya.

Wahba, H. (2008) 'Does the market value corporate environmental responsibility? An empirical examination', *Corporate Social Responsibility and Environmental Management*, vol 15, no 2, pp. 89–99.

Weitzman, M. (1974) 'Prices vs. Quantities', *Review of Economic Studies*, vol 41, no 4, pp. 477–491.

White, A. T., Courtney, C. A. and Salamanca, A. (2002) 'Experience with marine protected area planning and management in the Philippines', *Coastal Management*, vol 30, no 1, pp. 1–26.

World Bank (1997) *World Development Report 1997: The State in a Changing World*, Oxford University Press, Oxford.

Chapter 6

Marine environmental protection policy

International efforts to address marine pollution and protect marine biodiversity

Key topics

- Threats to biodiversity can be broadly categorized as a result of overharvesting, pollution, habitat loss, introduced species, ocean acidification and global climate change.
- The first international marine treaties were established to address the exploitation of marine mammals and fish stocks.
- The Law of the Sea Convention (LOSC) affirms that states have the rights to exploit their living marine resources through to their exclusive economic zone (about 200 nautical miles offshore). It also requires states to protect and preserve the marine environment.
- The LOSC is an overarching treaty covering all human activities on the ocean. It requires that states prevent, reduce and control pollution and interference with the ecological balance of the marine environment; protect and preserve rare or fragile ecosystems as well as the habitat of depleted, threatened or endangered species and other forms of marine life; and prevent the spread of alien (introduced) species.
- Other binding conventions that contribute to marine environmental protection fall under the umbrella of LOSC and include the 'London Convention' (on dumping), 'Ballast Water Convention', the International Convention for the Prevention of Pollution from Ships ('MARPOL'), the 'Basel Convention' (hazardous waste movement and disposal), the World Heritage Convention, and the Convention on Biological Diversity.
- There are several non-binding agreements that also contribute to marine environmental protection, including the 'Global Program of Action' under UN Environment, 'Plans of Action' under the Food and Agricultural Organization of the UN, and UN General Assembly resolutions.
- Ond hundred forty-three states participate in 18 Regional Seas programmes that address the degradation of coastal and marine areas.

Introduction

For centuries, the oceans were envisaged as immutable and immune to human activities. Fish were plentiful, and the capacity for the oceans to absorb human wastes was believed to be unlimited. Hugo Grotius's 17th Century legal principle of 'mare liberum' (free seas – Chapter 3) argued in favour of unfettered access to the ocean and its resources by all states – a worldview that continues to underpin the actions of many states and their nationals to this day. The unregulated environment created by the free seas doctrine predictably led to countervailing claims by coastal states, which continue to extend ever-further seaward, evidenced by growing territorial seas, followed by Exclusive Economic Zones (EEZs) and expanding outer continental shelf claims that continue through to present.

Chapter 1 explored the human impacts to the world ocean from overharvesting, pollution, habitat loss and degradation, introduced species and global climate change. It is not the intention of this chapter to revisit these topics; instead, this chapter will present the primary international mechanisms that govern the protection and preservation of the oceans. Important regional conventions and examples of bilateral and multilateral agreements are also reviewed.

History of international marine environmental law and policy

The first marine environmental treaties were established to prevent the exploitation of fish stocks. Early treaties between Great Britain and France (1839) and the North Sea states (1882, Belgium, Denmark, France, Germany, Netherlands, United Kingdom) set the stage for negotiations between states to address marine environmental issues (Table 6.1). Treaties ensued on fur seals in the North Pacific (1911), whaling (1946) and fishing gear restrictions (1946). The first convention to specifically mention the conservation of living resources was the first Law of the Sea Convention in 1958 (Chapter 4). Other conventions followed in order to arrest and clean up oil pollution (1971), prohibit dumping (1972), protect natural heritage (1972) and address regional marine environmental issues (1974 – Baltic Sea).

Marine environmental law is part of the broader subject of international law, which has been formed through International Court of Justice (ICJ) decisions, treaties and accepted common practice ('customary international law', see Chapter 3). It was not until the 1972 UN Conference on the Human Environment (UNCHE) in Stockholm, Sweden, however, that several key principles in international environmental law were formally recognized through a global consensus. Attended by representatives from 113 states, the Conference produced the 'Stockholm Declaration' in which 26 principles were ensconced. The declaration recognized the impacts of humans on the natural environment, some of the causes of these impacts (e.g. under development, over population, lack of education) and the need for global cooperation to address these issues. While all the principles are broadly applicable to marine environments, Principle 7 specifically obliges states to prevent marine pollution. In addition, the UNCHE produced an Action Plan with 109 Resolutions and one Declaration. Perhaps the most significant achievement of the UNCHE was the establishment of the United Nations Environment Program (UNEP). The UNCHE subsequently became the template for UN environment conferences in 1992, 2002, and 2012.

The 1987 World Commission on Environment and Development (WCED) was the next major advancement in international environmental cooperation. Known as the 'Brundtland Commission', its charge was to unite countries to jointly pursue sustainable development. The key output of the Commission was to define sustainable development as 'development that meets the needs of the present without compromising the ability of future generations to meet their own needs' (WCED, 1987). In addition, the report recommended that 12 per cent of each state should be protected. The Commission's report, published as *Our Common Future* (1987) contained a chapter outlining the threats to marine areas and opportunities for improved oceans management. Specifically, the Commission proposed measures to:

- Strengthen capacity for national action, especially in developing countries.
- Improve fisheries management.
- Reinforce cooperation in semi-enclosed and regional seas.

- Strengthen control of ocean disposal of hazardous and nuclear wastes.
- Advance the Law of the Sea.

While the WCED's work did not establish new international law, many of the concepts in *Our Common Future* were incorporated into the 1992 UN Conference on Environment and Development (see below). It is through this avenue that the WCED became an important contributor to current international environmental law.

The following section outlines the key agreements, conventions and treaties established to manage marine biodiversity.

Conventions that pertain to marine environmental protection

The LOSC

Part XII of the LOSC (see Chapter 4) addresses the protection and preservation of the marine environment. It restates earlier marine protection commitments from the 1958 Convention for the High Seas (Articles 116–120). Promoting the 'optimum utilization' of living resources, it echoes the Convention's call for catch limits based on best available science and ensuring that fisheries do not negatively affect other species (Article 61). The LOSC affirms that states have the rights to capitalize on their living marine resources in their sovereign waters but must do so in a manner that protects and preserves the marine environment (Articles 192–193). Specifically, the LOSC provides direction on three specific marine protection topics:

First, states must 'prevent, reduce and control pollution from any source, and interference with the ecological balance of the marine environment' and ensure that their activities do not cause damage by pollution in other jurisdictions (Articles 145 & 194). The LOSC addresses four sources of pollutants that states are obliged to address:

1. Toxic, harmful or noxious substances introduced into the marine environment by land-based sources, the atmosphere or dumping.
2. Pollution from vessels.
3. Pollution from installations related to offshore hydrocarbon development or seabed mining.
4. Pollution from other installations. (Article 194)

States may not reduce pollution in their jurisdictions by transporting pollution to other jurisdictions (Article 195).

Second, states must protect and preserve rare or fragile ecosystems as well as protect and preserve rare or fragile ecosystems as well as the habitat of depleted, threatened or endangered species and other forms of marine life. (Article 194).

Third, states must prevent the spread of alien (introduced) species (Article 196).

The LOSC also delineates a global cooperation regimen to ensure that states cooperate to protect the marine environment. In particular, states must cooperate to develop science-based rules and standards to maintain environmental quality (Articles 197, 201), notify other states that are or may be affected by pollution (Article 198), jointly develop plans to respond to environmental emergencies (Article 199), undertake joint scientific research (Article 200) and cooperate on monitoring and environmental assessments (Articles 204–206).

In addition, more developed states are obligated to assist less developed states to improve their scientific and research capacity related to the protection of the marine environment (including ensuring less developed states receive priority funding for these programs). An obligation also exists to apprise less developed states of the effects of their activities on marine environmental quality (Articles 202–203).

The majority of Part XII, however, addresses the requirements for states to develop national (domestic) legislation to contribute to the aims of Part XII. Specific LOSC direction to states includes the requirements to adopt and enforce laws and cooperate across jurisdictions to prevent or minimize pollution. Land-based sources (Articles 207 and 213), seabed activities in sovereign areas (Articles 208 and 214), activities in the Area (seafloor beyond national jurisdictions; Articles 209 and 215), dumping (Articles 210 and 216), vessels (Article 211) and the atmosphere (Article 212) are specific pollution-related issues mentioned in Park XII.

Additionally, the LOSC articulates the requirements for enforcement by flag states. States must ensure that vessels flying their flag comply with all necessary international and national rules and regulations established to protect the marine environment (see Chapter 9). Flag states are also required to ensure that vessels flying their flag carry the necessary documentation to demonstrate compliance and to undertake inspections and investigations, with other states if necessary, to guarantee compliance (Article 217).

Port states are entitled to undertake investigations of ships operating voluntarily within their territorial waters or EEZs where evidence suggests that a vessel operating on the high seas was in contravention of international rules with respect to Part XII of the LOSC (Article 218). Port states may also prohibit a vessel from operating in their national waters if the vessel is believed to be unseaworthy and a risk to the marine environment (Article 219). Likewise, coastal states may inspect, institute proceedings or detain a vessel transiting its territorial waters or EEZs where evidence suggests that the vessel was in contravention of international rules with respect to Part XII of the LOSC (Article 120).

Port and coastal states may exercise their rights under Part XII under the following conditions:

- Flag states and their vessels have an opportunity to present evidence during investigations or hearings (Article 223).
- Enforcement against foreign vessels may only be undertaken by warships, military aircraft or other clearly marked government vessels (Article 224), and enforcement actions must not subject ships and their crews to risk (Article 225).
- Detained foreign vessels must be released without delay and, if repairs are needed, be permitted to transit to the nearest repair facility (Article 226).
- Foreign vessels are not to be discriminated against (Article 227).
- Proceedings against foreign vessels with alleged violations outside of the port or coastal state's territorial sea are suspended if the flag state initiates proceedings. The exception is where major damage was caused in the port or coastal state's EEZ (Article 228).
- Penalties may not be imposed on any vessel beyond three years after the incident was alleged to have occurred (Article 228).
- Only monetary penalties can be levied against foreign vessels determined to have violated national or international laws within a coastal or flag state's territorial sea (Article 230).
- Flag states will be immediately notified of all measures taken against vessels flying their flag (Article 231).
- Warships or other state vessels are immune from enforcement obligations but are expected to meet the spirit and intent of Part XII (Article 236).

While the LOSC addresses pollution from land-, sea-, and air-based sources, the Convention is noticeably silent on other aspects of protection of the marine environment. Habitat loss, climate change, marine genetic resources and overharvesting are ignored, but this is unsurprising given that these issues were either unknown or only emerging during final (third) round of negotiations from 1973–1982.

The World Heritage Convention

The 1972 World Heritage Convention, which predates LOSC, was the first international vehicle for protecting places of 'outstanding universal value', taking into consideration both cultural and natural heritage (WHC, 1972). Increasingly applied in national waters, the designated sites can cover vast areas (e.g. Great Barrier Reef [Australia], Papahānaumokuākea [USA] and Phoenix Islands [Kiribati]), as well as more typically smaller nearshore sites. To maintain World Heritage status, States must demonstrate that sites have operational management plans that are protecting the identified cultural and natural values. There is growing interest in considering how the WHC's marine coverage could be expanded further offshore (UNESCO, 2011, Recommendation 5).

The Convention on Biological Diversity

The Convention on Biological Diversity (CBD) was signed in June 1992 at the United Nations Conference on Environment and Development (the Rio 'Earth Summit' or UNCED) and entered into force in December 1993 (UN, 1992). The CBD was the first global agreement specifically intended to conserve and sustainably use biodiversity with its attendant genetic components. Biodiversity in the CBD is defined as 'variability among living organisms from all sources including, inter alia, terrestrial, marine and other aquatic ecosystems and the ecological complexes of which they are part; this includes diversity within species, between species and of ecosystems'.

The CDB also recognizes the role(s) of Indigenous cultures, women and poverty reduction in biodiversity conservation. The CDB commits more developed states to provide funding to less developed states in order to further the aims of the Convention.

While the text of the Convention does not specifically address marine environments, it does consider the marine environment as a component of biodiversity (Article 2) and that the CBD is to be applied to the marine environment consistent with the laws of the sea (Article 22.2). The Parties to the Convention established 'Marine and Coastal Biodiversity' as one of its thematic programs and marine and coastal biodiversity was an early priority for the COP. At the second COP in 1995, Decision II/10 (CBD, 1995) committed signatories to address marine biodiversity issues and provided some overall guidance on the architecture of this program. In addition, COP II produced a global consensus on the importance of marine and coastal biodiversity. Known as the 'Jakarta Mandate on Marine and Coastal Biological Diversity', the mandate reaffirms that 'there is a critical need for the COP to address the conservation and sustainable use of marine and coastal biological diversity and urge Parties to initiate immediate action to implement the decisions adopted on this issue' (CBD, 2000a).

The Jakarta Mandate further articulates the following six implementation principles:

1. Apply the ecosystem approach.
2. Apply the precautionary principle.
3. Recognize the importance of science.
4. Utilize a roster of experts.
5. Involve local and Indigenous communities and consider traditional knowledge in decision-making.
6. Implement programs at the national, regional and global levels.

At the COP 4 meeting in Bratislava, Slovakia, in 1998, the Parties adopted Decision IV/5 on the conservation and sustainable use of marine and coastal biological diversity, including a Multiyear Programme of Work on Marine and Coastal Biological Diversity. Coral reefs and coral reef bleaching induced by climate change were particular focuses of this COP (CBD, 1998).

At the COP 8 meeting in Curitiba, Brazil, in 2006, the Parties adopted Decisions VIII/21, 22 and 23. These stressed the value of researching and protecting seabed habitats, the import of integrated marine and coastal area management and the recognition that protected areas are an essential tool for the conservation of marine biodiversity (CBD, 2006). The meeting's delegates set a target of 10 per cent for the protection of the world's ecological regions, including coastal and marine areas.

At the COP 9 meeting in Bonn, Germany, in 2008, the Parties adopted Decisions IX/20 to compile and synthesize scientific information on the effects of ocean fertilization (to sequester CO_2 and address climate change) on marine biodiversity (CBD, 2008). The Parties also settled on criteria for the delineation of ecologically and biologically significant marine areas and criteria for establishing networks of marine protected areas (see Chapter 12).

A number of other agreements have been framed under the Convention, including:

- The Cartagena Protocol on Biosafety (adopted 2000, entered into force 2003) that aims to ensure the safe handling, transport and use of organisms that have been modified using modern biotechnology. It takes measures against the adverse effects on biological diversity and risks to human health (CBD, 2000b).
- The Nagoya Protocol on Access and Benefit Sharing (adopted 2010, entered into force 2014) that governs the access to and benefits derived from genetic resources. Specifically, it attempts to address the needs of those searching for genetic resources (bioprospecting) with those who live on or govern the lands and coastal areas that contain these genetic resources (CBD, 2010).

The World Summit on Sustainable Development and the protection of the marine environment

The WSSD was held 26 August–4 September 2002. Termed the 'Earth Summit 2002' or 'Rio+10', the purpose of the Summit was to focus on global problems (e.g. poverty, conflict, disease, discrimination) inhibiting sustainable development that could be addressed through multilateral cooperation. The meeting produced the Johannesburg Declaration on Sustainable Development, which proclaimed an aspirational, non-binding suite of resolutions (framed as sustainable development outcomes), furthering the intent of the UNCHE and UNCED. The Declaration was executed through a Plan of Implementation of the World Summit on Sustainable Development.

The Plan of Implementation formulated a number of marine-related objectives. While reiterating many binding actions from previous international agreements, it also introduced a number of new commitments and implementation dates. Specifically, states agreed to restore depleted fish stocks by 2015 (s.31(a)), establish a global network of representative marine protected areas by 2012 (s. 32(c), see Chapter 12), and establish process to report on the state of the marine environment by 2004 (s. 36(b)).

Rio + 10 was followed by Rio + 20 in 2012. Conference attendees adopted a document titled *The Future We Want*, which re-iterates commitments from previous conferences but updated to reflect technological changes (e.g. the importance of connections to the internet) and the state of the global economy (e.g. the need to address youth unemployment). Sections 158–177 address marine issues and, consistent with the previous sections, reiterate past promises. Unlike previous

WSSD declarations, however, few new targets were made, though it does commit to achieving 'significant reductions in marine debris' by 2025. The document also expresses concern for the possible unintended consequences of ocean fertilization – at the time seen by some as a possible way to increase carbon uptake by the ocean to stem the effects of increasing CO_2 levels in the atmosphere.

Threat-specific conventions and the protection of the marine environment

Binding conventions

A number of international agreements exist to protect oceans from specific human impacts. The most significant conventions are listed below.

The convention on the prevention of marine pollution by dumping of wastes and other matter (1972)

After MARPOL (Chapter 9), the Convention on the Prevention of Marine Pollution by Dumping of Wastes and Other Matter (the 'London Convention' or 'London Dumping Convention') is perhaps the next strongest international law related to marine environmental protection. Adopted in 1972 and entered into force in 1975, the Convention is administered by the International Maritime Organization (IMO). It prohibits the deliberate dumping of different types of hazardous materials from ships, aircraft and platforms, including ships, aircraft and platforms themselves (IMO, 1972). The Convention does not apply to discharges from land-based sources (e.g. outfalls), waste generated as a result of the normal operation of ships, intentional placement of materials for purposes other than dumping (e.g. seawalls) and materials produced as a by-product of fish processing or seabed mining. Articles in the Convention may be contravened if the safety of humans or structures is compromised by unforeseen events (e.g. storms). The Convention contains 22 Articles and 3 Annexes and has been ratified by 87 states. It applies to all territorial waters and high seas but not internal waters.

In 1996, the Protocol to the Convention on the Prevention of Marine Pollution by Dumping of Wastes and Other Matter 1972 ('London Protocol'; together with the London Convention called the LC/LP) was added as a separate agreement to the Convention, later revised and entered into force in 2006. It is intended to eventually replace the 1972 London Dumping Convention and currently has 51 parties. The London Protocol introduced a new approach to marine waste management and introduced the concepts of the 'precautionary principle' and the principle of 'polluter pays' and stressed the need to avoid transferring wastes from one type of receiving environment (e.g. the ocean) to another (e.g. the land). While the London Convention prescribed the substances and compounds (e.g. organohalogen compounds) that were prohibited from dumping, the Protocol – recognizing the pace in development of new compounds that may negatively impact marine and human health – prohibited the dumping of all materials with the exception of those listed in its Annex I. Thus, the Protocol reverses the onus of proof and was a key development in the evolution of marine governance.

The London Protocol includes three Annexes (which in no way relate to the Annexes of the London Convention) as follows:

1   Annex I specifies the wastes and other matter that may be dumped. The Protocol bans the dumping of all waste except for dredged material, sewage sludge, fish waste from processing,

man-made structures, inert or inorganic geological material, bulky items composed of inert materials that will not interfere with navigation or fishing and CO_2 for sub-seabed sequestration in geological features (i.e. not on the deep-sea floor). A permit is required for any dumping activity under the Protocol.

2   Annex II outlines the assessments that must be undertaken prior to consideration for dumping. Specifically, the purpose of Annex II is to prevent waste from being produced (e.g. product life cycle management), consider alternatives to ocean dumping once waste has been produced (e.g. recycling) and ensure that the material considered for ocean dumping is free from or contains limited hazardous materials (e.g. treatment). Annex II describes the processes and options to be considered prior to issuing a dumping permit.

3   Annex III outlines the arbitral procedure to be used between contracting parties.

Permits are granted by the states that have ratified the Convention. States are required to establish an appropriate authority to issue permits as well as record the location, nature and quantities of the dumping of permitted materials that must be provided to the IMO. Permits are to be issued by the port state where the material is loaded onto ships or the flag state if material is loaded onto a ship from a port state not party to the Convention.

Incineration of industrial wastes and sewage sludge is prohibited. Other wastes may be incinerated at sea under special permit. Dumping of low-level radioactive wastes are prohibited.

Signatories to the Convention agree to cooperate with the IMO and other international organizations (e.g. UNEP) to assist member states with training, to provide the necessary equipment that complies with the Convention and to facilitate the management of wastes to avoid their disposal at sea.

International convention for the control and management of ships' ballast water and sediment

Ballast is a broad term for any liquid or solid carried by ships to increase draft, change trim, regulate stability or maintain structural integrity of the ship. Modern steel ships, including cruise ships, tankers and bulk cargo carriers, use water for ballast that is pumped in or out of separate compartments depending on ship's cargo and anticipated weather and sea state conditions. Most modern ships are built with ballast water tanks that hold between 25–30 per cent of their dead weight tonnage. Upwards of 100,000 tonnes of ballast water can be contained in the largest bulk carriers (Branch and Roberts, 2014).

The environmental, economic and public health implications of the use of water as ballast are significant. While the world ocean are part of a single, large water body, marine biological communities are constrained to certain geographic areas. This separation is caused by land masses, currents, unfavourable temperature regimes, or biological control (competition and predation). This geographic isolation has resulted in the evolution of countless unique marine species over the past several million years. Historically, some of these species moved and flourished in other parts of the ocean either through their own locomotion or if they were advected by currents, storms, attached to debris or as hosts on a species that were able to move long distances (Davidson and Simkanin, 2012).

Modern marine shipping, however, provides another means (termed a 'vector') for the transport of species into marine areas they would otherwise have slight chance of reaching. Ships may inadvertently move species between areas when species attach to a ship's hull, are captured when ballast water is pumped into ballast tanks or reside in a ship's sea chests (piping and chambers used to bring water on board a ship for ballast and other purposes such as cooling). The speed of modern ships combined with the cleanliness of modern ballast water tanks increases the likelihood of an organism surviving its voyage in ballast tanks. Furthermore, shipping now moves over 80 per cent of the world's commodities and approximately 3 to 5 billion tonnes of ballast water internationally each year. This provides ample opportunities for the movement of species across the oceans. It is estimated that shipping is responsible for at least one third of documented marine invasions and that at any given moment at least 7,000 different species are being carried in ships' ballast tanks around the globe (Davidson and Simkanin, 2012).

While modern ships have grates and mesh on water intakes to prevent larger species from entering ballast tanks, these preventative measures are incapable of excluding viruses, bacteria, microbes and smaller zooplankton. In addition, most marine organisms (even sessile species attached to the ocean floor) have planktonic life cycles where eggs, larvae and juveniles are available to enter ballast tanks and, once released at the ship's destination, may settle into their sessile adult phase (Chapter 2). The environmental and economic impact of introduced species from shipping is large; perhaps the most well-known invasions are the incursion of the North American Comb Jelly (*Mnemiopsis leidyi*) into the Black and Azov Seas and the introduction of the Eurasian Zebra Mussel (*Dreissena polymorpha*) into the Laurentian (North America) Great Lakes. *Mnemiopsis* was introduced into the Black Sea in the 1980s and has reached densities of 1kg of biomass per cubic metre (400 animals per cubic metre) and contributed to the collapse of commercial fisheries in the Black Sea. The Zebra Mussel was accidentally introduced into the Great Lakes in 1998 and now infects approximately 40 per cent of internal waterways in the United States and Canada, requiring up to $US 100 million annually on control measures (globallast.imo.org).

Currently, the IMO is most concerned with the following invasives:

- Cholera (*Vibrio cholera*) (various strains).
- Cladoceran Water Flea (*Cercopagis pengoi*).
- Mitten Crab (*Eriocheir sinensis*).
- Toxic algae (red/brown/green tides) (various species).
- Round Goby (*Neogobius melanostomus*).
- North American Comb Jelly (*Mnemiopsis leidyi*).
- North Pacific Seastar (*Asterias amurensis*).
- Asian Kelp (*Undaria pinnatifida*).
- European Green Crab (*Carcinus maenas*).

While initial IMO guidelines on the control and management of ballast water were published in 1991 and subsequently revised in 1997, the International Convention for the Control and Management of Ships' Ballast Water and Sediment was adopted 13 February 2004 (IMO, 2004). The Conference was attended by 75 states, 1 IMO associate member, 2 governmental organizations and 18 non-government organizations. The Convention entered into force 8 September 2017, 12 months after the Convention was signed by at least 30 countries representing at least 35 per cent of the world's fleet tonnage. To date, the Convention has been signed by 66 countries representing 75 per cent of the world's fleet tonnage.

The purpose of the Convention is to prevent, minimize and ultimately eliminate the transfer of harmful aquatic organisms and pathogens through the control and management of ships' ballast water and sediments (Article 2.1). Key Articles of the Convention include:

- Article 3: Outlines the application of the Convention and instances where the Convention does not apply.
- Article 4: Requires parties to the Convention to adhere to the standards and requirements outlined in the Annex. It also requires parties to cultivate national policies, strategies and programs for ballast water management.

- Article 5: Requires parties to establish sediment reception facilities where the cleaning and repair of ballast water tanks occurs.
- Article 6: Requires parties to cooperate on scientific and technical research and monitoring.
- Article 7: Requires parties to survey ships flying their flag or operating under their authority to ensure compliance with regulations in the Annex.
- Article 8: Violations of the Convention are the responsibility of the flag state to investigate and pursue.
- Article 9: Port states may inspect visiting ships, but inspections are limited to the verification of a valid Certificate, the inspection of the Ballast Water record book and the examination of the ship's ballast water.
- Article 10: Port states may warn, detain or exclude a ship that is detected to have violated the Convention.

The Annex contains the regulations for the control of ships' ballast water and sediments. Key regulations include:

- Regulation B-1: Requirement for each ship to have a Ballast Water Management plan. The plan includes safety procedures for ballast water management, how the ship will meet the regulations as outlined in the Annex and detailed procedures for the disposal of sediments at sea and on shore.
- Regulation B-2: Requirement for each ship to have on board a Ballast Water record book that retains records for a period of two years.
- Regulation B-3: Establishes a tiered implementation timetable – based on the construction date of the ship – for the implementation of regulations.
- Regulation B-4: Outlines standards for ballast water exchange, stating that whenever possible ballast water be exchanged at least 200 nmi from shore in depths at least 200m.
- Regulation C-1: Parties, individually or jointly, may develop measures that exceed the regulations in the Annex.
- Regulation C-2: Permits port states to identify areas where uptake of ballast water should not occur.
- Regulation D-1: Requires ships exchanging ballast water to meet a 95 per cent volumetric efficiency in the exchange.
- Regulation D-2: Requires ships to discharge less than 10 viable organisms per cubic metre greater than or equal to 50 micrometers in minimum dimension and less than 10 viable organisms per millilitre less than 50 micrometers in minimum dimension and greater than or equal to 10 micrometers in minimum dimension. Also sets out the indicator microbes to be used (cholera, *Escherichia coli*, Intestinal Enterococci) as a human health standard.
- Regulation D-3: Active substances used to treat ballast water must be approved by the IMO.
- Regulation E-1: Sets out a survey schedule for ships 400 gross tonnes or larger to which the Convention applies.
- Regulation E-5: Ballast Water Certificates are valid up to five years.

Basel Convention on the Control of Transboundary Movements of Hazardous Wastes and their Disposal

The Basel Convention on the Control of Transboundary Movements of Hazardous Wastes and their Disposal (the 'Basel Convention') was adopted 22 March 1989 and came into force in 5 May 1992. There are currently 187 parties to the Convention (UNEP, 1989).

The Convention resulted from the growing public awareness in the 1970s and 1980s of the dumping of toxic wastes from more developed to less developed nations that often lacked the capacity to effectively store and process these wastes. Africa, in particular, was a destination for much of these wastes. Hazardous waste exports received public attention in the 1980s with 'toxic trader' ships that moved wastes to Africa, Eastern Europe and other regions. Perhaps the most famous of these was the *Khian Sea*, which in 1986 was loaded with approximately 12 million kg of municipal incinerator and waste ash containing numerous toxic compounds. Over a period of 18 months, the ship was renamed twice (*Felicia*, *Pelicano*), refused entry into at least 11 countries and dumped approximately 1800 tonnes of ash on a Haitian beach. The ship was eventually found empty in Singapore (Jaffe, 1995).

The purpose of the Convention is to 'protect human health and the environment against the adverse effects of hazardous wastes'. This is achieved primarily through controls on the transboundary movement of hazardous wastes. Objectives of the Convention are to:

- Minimize the production of hazardous wastes and hazardous recyclable materials.
- Ensure hazardous materials are disposed in an environmentally responsible manner, preferably close to the source of production.

The Convention was amended at the third meeting of the Conference of the Parties on 22 September 1995 (UNEP, 1995a). Termed the 'Ban Amendment', it prohibits the export of all hazardous wastes covered by the Convention from more developed countries (Organization for Economic Cooperation and Development, European Community, Liechtenstein) to less developed countries. Ninety-five states are signatories to the Ban Amendment, but not all states have ratified the Amendment and therefore it is not yet in force. However, many nations subscribe to the principles of the Ban Amendment and the European Union fully complies with the Amendment under its Waste Shipment Regulation.

The Basel Convention is viewed as highly successful. Specific achievements are the development of technical guidelines for specific waste issues worldwide. Even though they are non-binding from an international legal perspective, they are now accepted as national policy in most states.

Non-binding agreements

Global Programme of Action for the protection of the marine environment from land-based activities

The Global Programme of Action (GPA) for the protection of the Marine Environment from Land-based Activities is a UNEP-led program adopted 3 November 1995 by 108 states. Termed the 'Washington Declaration', the GPA is not legally binding but rather an implementation framework to protect the health, productivity and biodiversity of coastal and marine environments from land-based sources of pollution (UNEP, 1995b; Osborn and Datta, 2006). The GPA is concerned with nine specific impacts on coastal and marine environments: sewage, persistent organic pollutants, radioactive substances, heavy metals, oils (hydrocarbons), nutrients, sediment mobilization, litter and the physical alteration/destruction of habitats. The Programme does not identify specific policy or legislative tools that states should adopt, nor does it develop global 'one size fits all' solutions to various land-based threats. Instead, the GPA is designed to assist states in their efforts to protect and restore coastal and marine environments. Five primary goals underpin the Programme:

1 Identify the sources and impacts of land-based sources of marine pollution.
2 Identify priority problems for which action is needed.
3 Set management objectives for these problem areas.
4 Develop strategies to achieve these objectives.
5 Evaluate the impacts of these strategies.

Specifically, the GPA encourages states and regions to develop National Programmes of Action (NPAs) to identify priorities, devise action plans and monitor performance. NPAs are specifically tailored to individual states or regions and may utilize regulatory, market-based, voluntary, or other means to reduce impacts on marine environments. To date, 72 countries are developing or have completed their NPAs.

The GPA was reviewed in 2001, the outcome of which is the Montreal Declaration on the Protection of the Marine Environment from Land-Based Activities. The Declaration focusses attention on municipal wastewater, ecosystem approaches to coastal and ocean governance and financing of the GPA. The second review of the Programme was completed in 2006. The review recommended that the NPAs include valuations of ecosystem goods and services, integrate freshwater and coastal management, address wastewater issues and link GPA activities with the amelioration of poverty. Nutrients, wastewater and marine litter were the principal subjects at the third review in 2012. The fourth review was held in Indonesia in 2016, calling for acceleration of implementation. Noteworthy here is the recent increase in international commitments and initiatives to address marine plastic pollution, which are independent of the GPA (e.g. the Third UN Environmental Assembly Resolution 8; the G7 Plastics Charter [2018], and the Commonwealth Clean Oceans Alliance [2018]).

Important marine regional agreements

The proliferation of international law in the 20th Century demonstrated that states could, without difficulty, enter into agreements with one another on various issues. While most states have many bilateral agreements with neighbouring jurisdictions, the concept of states within a defined geographic area subscribing to an action plan to address one or more issues was rarely seen in areas outside of the European Union until the 1970s.

UNEP Regional Seas Programme

The UNEP Regional Seas Programme was initiated in 1974 to address degradation of coastal and marine areas. The Programme is administered by UNEP's Oceans and Coastal Areas Programme Activity Centre (OCA/PAC) in Nairobi, Kenya. The objectives of the Programme are to:

- Control marine pollution and to protect and manage aquatic resources through the advancement of regional conventions, guidelines and action plans.
- Understand and monitor pollution and its impacts on human health and marine resources.
- Cooperate on the sustainable development of coastal and marine resources.
- Support developing countries in the protection, development and management of marine and coastal resources.

Source: www.unep.org/regionalseas/default.asp

The Programme considers both the socio-economic and environmental aspects within a geographic area and is therefore one of the earliest attempts at region-based sustainable development through international cooperation.

Early Regional Seas agreements focussed almost exclusively on marine pollution and dumping as well as scientific research and large-scale monitoring. While entrained seas (e.g. Mediterranean) were found to contain significant levels of various types of pollution, the open oceans were generally found to be much cleaner; by the 1990s, pollution levels in most marine areas were stable. This suggested that the Programme might benefit from a revised focus. As a result, UNEP revised its Regional Seas policies in 1983 to focus on Integrated Coastal Area Management (see Chapter 12) with a specific emphasis on land-based sources of pollution and the human activities responsible for degradation of coastal environments. The revised Programme also focussed on pollution control measures, environmental impact assessment, training, public awareness and collaboration between jurisdictions (Akiwumi and Melvasalo, 1998). Linked into the Regional Seas Programme is the Global Programme of Action for the protection of the Marine Environment from Land-based Activities (discussed previously). Its purpose is to reduce, control and/or eliminate marine degradation from land-based sources.

Regional Seas Programmes have been initiated by states either through multilateral negotiations and treaties (e.g. CCAMLR, OSPAR and HELCOM, in Antarctica, northeast Atlantic and Baltic) or through a formal request to UNEP. In the latter case, national and technical experts are tasked to develop an action plan which is enabled (in most cases) through a regional convention with associated protocols specific to particular regional issues. These agreements are termed Regional Seas Conventions and Action Plans (RSCAPs). Many of the conventions have multiple protocols addressing different threats, and, in many cases, the development of a Regional Seas Programme begins with the simultaneous adoption of a regional convention and subsequent protocols. The action plans are financed by participating states, although states can request that UNEP administer the funding under the Financial Rules and Regulations of the United Nations. The UNEP Environment Fund also provides seed funding to support start-up costs (Verlaan and Khan, 1996).

Currently, 143 countries participate in 18 Regional Seas Programmes (or associated partner programs) that represent all the populated marine coasts of the world (Figure 6.1). The Programme is unique in the UN in that it is the only UN body with regional, autonomous structures. The five partner programs consist of the Antarctic, Arctic, Baltic Sea, Caspian Sea and North East Atlantic Regions, which are not under the auspices of UNEP. Several Regional Seas Programmes merit further discussion.

UNEP Regional Seas Programme: Mediterranean

The Mediterranean Sea covers 2.5 million square km, averaging a depth of 1500m and bounded by 46,000km of coastline. With only two connections to the global oceans at the Strait of Gibraltar (14km wide) and the Suez Canal (less than1km wide), turnover of the Sea is likely greater than a century. Although the Mediterranean comprises only 0.82 per cent of the surface area and 0.32 per cent of the volume of the world ocean, the Sea is estimated to contain at least 7 per cent of all known marine species. Moreover, approximately 30 per cent of the area's fauna species are endemic to the Mediterranean (UNEP, 2012). Meanwhile, 460 million people reside in the Mediterranean's watersheds and this population is expected to grow to 520 million by 2025.

In 1974, the UNEP Governing Council endorsed the Mediterranean Sea as a high-priority area for the protection of living resources and the prevention of pollution. The Mediterranean Action Plan (MAP) was in place a year later. It called for integrated planning and development, the coordination of pollution research and monitoring, the elaboration of robust financing mechanisms and the establishment of a convention and related protocols (Vallega, 2002).

Marine environmental protection policy   159

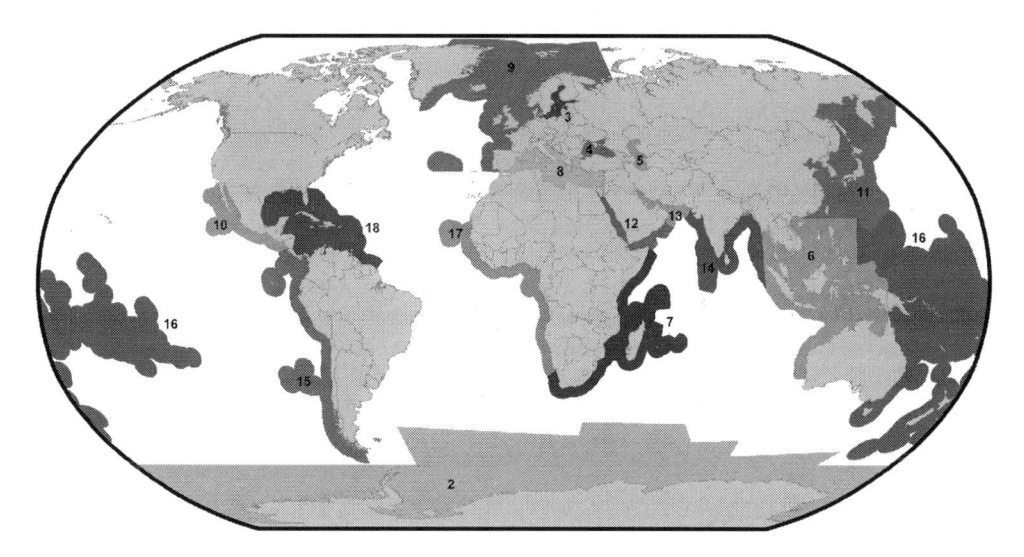

Figure 6.1 United Nations Environment Programme (UNEP) Regional Seas Programmes, including affiliated partners, are as follows:

1 Arctic	7 Eastern Africa	13 ROPME Sea Area
2 Antarctic	8 Mediterranean	14 South Asian Seas
3 Baltic	9 Northeast Atlantic	15 Southeast Pacific
4 Black Sea	10 Northeast Pacific	16 Pacific
5 Caspian	11 Northwest Pacific	17 Western Africa
6 South Asian Seas	12 Red Sea and Gulf of Aden	18 Wider Caribbean

The Convention for the Protection of The Mediterranean Sea Against Pollution (the 'Barcelona Convention') was adopted 16 February 1976 and entered into force on 12 February 1978. Fourteen Mediterranean states and the European Union signed the Convention. The Convention was revised on 9–10 June 1995, as the Convention for the Protection of the Marine Environment and the Coastal Region of the Mediterranean but has not yet entered into force. In addition, the MAP was revised and termed MAP Phase II with 21 participating states focussing on coastal management and biodiversity protection (Vallega, 2002; UNEP, 2005).

The Convention's primary intent is to protect the Mediterranean from pollution through a system of protocols (Article 4). While all types and sources of pollution are covered by the Convention, specific Articles address pollution occasioned by: dumping from ships and aircraft (Article 5), discharges from ships (Article 6), exploration and exploitation of the continental shelf and the seabed and its subsoil (Article 7) and land-based sources (Article 8). Cooperation on pollution emergencies (Article 9), monitoring (Article 10) and scientific cooperation (Article 11) are also treated in distinct articles.

The Convention is amended by the following seven Protocols:

1   Dumping (adopted 1976, entered into force 1978 revised 1995).
2   Prevention and emergencies (adopted 1976, entered into force 1978, revised version adopted 2002, revised version entered into force 2004).

3   Land-based pollution (adopted 1980, entered into force 1983, revised version adopted 1996, revised version entered into force 2008).
4   Specially protected areas and biodiversity (adopted 1982, entered into force 1986, revised version adopted 1995, revised version entered into force 1999).
5   Seabed development and offshore exploration (adopted 1994, entered into force 2011).
6   Hazardous wastes (adopted 1996, entered into force 2008).
7   Integrated coastal zone management (adopted 2008, entered into force 2011).

By uniting states to confront mutual problems, the Convention has achieved success, and there are now 22 contracting parties. Key achievements include the creation of the Mediterranean Strategy for Sustainable Development, which integrates the environment and development (UNEP, 2005), and the Strategic Action Plan for the Conservation of Marine and Coastal Biodiversity in the Mediterranean (SAP BIO) (sapbio.rac-spa.org/).

UNEP Regional Seas Programme: Kuwait/ROPME

The Kuwait Regional Convention for Co-operation on the Protection of the Marine Environment from Pollution ('Kuwait Convention') was adopted on 23 April 1978 by a regional conference convened in Kuwait and attended by Bahrain, Iran, Iraq, Kuwait, Oman, Qatar, Saudi Arabia and the United Arab Emirates. The Convention covers the Persian Gulf and Gulf of Oman. The 8 nations bordering these seas produce 25 per cent of the world's oil and have experienced 3 major wars in the past 25 years and the world's largest oil spill until the DeepWater Horizon accident (de Mora et al., 2010). Furthermore, the Persian Gulf is unique among regional seas in that it is very shallow, averaging 36 metres depth, and in that much of its marine biota lives near the upper range of tolerances to salinity and temperature. The region also has significant exposure to human-induced threats – including externalities of oil and gas production and transportation – alien species and overharvesting of marine resources. Another unique consideration of the Persian Gulf that increases attention on marine water quality is that a significant portion of its coastal residents obtain their water through desalinization (Hamza and Munawar, 2009).

Adopted with the Convention is the Action Plan for the Protection of the Marine Environment and the Coastal Areas of Bahrain, Iran, Iraq, Kuwait, Oman, Qatar, Saudi Arabia and the United Arab Emirates. Further bolstering the protection of the area is the Protocol concerning Regional Co-Operation in Combating Pollution by Oil and other Harmful Substances in Cases of Emergency. Other protocols deal with seabed development and offshore activity (adopted 1989, entered into force 1990), land-based pollution (adopted 1990, entered into force 1993) and hazardous wastes (adopted 1998, entered into force 2003).

Article XVI of the Convention established the Regional Organization for the Protection of the Marine Environment (ROPME) in 1979 to implement the Action Plan and, in 1982, the establishment of the ROPME Secretariat in Kuwait. ROPME, while part of the Regional Seas Programme, operates independently of UNEP. Broadly, the Convention's aims are to prevent damage to the marine environment and protect human health within a context of sustainability.

UNEP Regional Seas Programme: West and Central Africa

The Abidjan Convention for Co-Operation in the Protection and Development of the Marine and Coastal Environment of the West and Central African Region ('Abidjan Convention') was adopted in 1981 and entered into force in 1984 (UNEP, 1981a). Sometimes termed 'WACAF', the Convention has 23 contracting parties and 10 additional ratifying parties. In addition, the

Action Plan for the Protection and Development of the Marine Environment and Coastal Areas of the West and Central African Region and Protocol Concerning Co-Operation in Combating Pollution in Cases of Emergency were also adopted and entered into force at the same time as the Convention.

The primary objectives of the Action Plan are to:

- Undertake environmental assessment and scientific research into coastal and marine areas (Article 4.1).
- Promote methods for sustainable development (Article 4.2).
- Foster regional agreements for the protection and development of marine and coastal areas (Article 4.3).
- Cultivate financial and institutional arrangements for implementing the Action Plan (Article 4.4).

UNEP Regional Seas Programme: Southeast Pacific

The Convention for the Protection of the Marine Environment and Coastal Area of the Southeast Pacific ('Lima Convention') and Action Plan for the Protection of the Marine Environment and Coastal Areas of the Southeast Pacific were adopted 12 November 1981 and entered force in 1986 (UNEP, 1981b). Subsequent protocols and agreements include:

- Agreement on Regional Cooperation in Combating Pollution in the Southeast Pacific by Hydrocarbons and other Harmful Substances in cases of Emergency (signed 1981).
- Complementary Protocol on the Agreement for Regional Cooperation in Combating Pollution in the Southeast Pacific by Hydrocarbons and other Harmful Substances in Cases of Emergency (adopted 1983, entered into force 1987).
- Protocol for the Protection of the Southeast Pacific Against Pollution from Land-Based Sources (adopted 1983, entered into force 1986).
- Protocol for the Conservation and Management of Protected Marine and Coastal Areas of the Southeast Pacific (adopted 1989, entered into force 1995).
- Protocol for the Protection of the Southeast Pacific from Radioactive Pollution (adopted 1989, entered into force 1995).
- Protocol on the Regional Program for the Study of the El Niño phenomenon in the Southeast Pacific (ERFEN) (adopted 1992).
- Proposed Protocol on the Assessment of the Environmental Impact on the Marine Environment and Coastal Areas the Southeast Pacific (signed).
- Proposed Protocol to Prohibit Transboundary Movements of Hazardous Wastes and their Disposal to the Southeast Pacific (signed).

The Convention encompasses the entire Pacific coast of South America and includes Panama, Columbia, Ecuador, Peru and Chile. Its goals are numerous: prevent pollution from all sources (Article 4); prevent, reduce and control erosion from coastal areas (Article 5); cooperate on responding to marine pollution emergencies (Article 6); monitor pollution (Article 7); undertake environmental impact assessment (Article 8); exchange information (Article 9); and cooperate on scientific investigations (Article 10). A Permanent Commission on the South Pacific (CPPS) is responsible for implementing the Convention and Action Plan, and the Secretariat is headquartered in Quito, Ecuador.

UNEP Regional Seas Programme: Red Sea and Gulf of Aden

The Programme for the Environment of the Red Sea and Gulf of Aden (PERSGA) was initiated in 1974 to address coastal development, overfishing, marine transportation and marine emergencies. The Programme Secretariat is headquartered in Jeddah, Saudi Arabia.

The Regional Convention for the Conservation of the Red Sea and Gulf of Aden Environment (Jeddah Convention) was adopted 15 February 1982 and entered into force 20 August 1985 (UNEP, 1982). Parties to the Convention include Djibouti, Egypt, Jordan, Saudi Arabia, Somalia, Sudan and Yemen. In addition to the Jeddah Convention, the regional conference that produced the Convention adopted the Action Plan for the Conservation of the Marine Environment and Coastal Areas in the Red Sea and Gulf of Aden and Protocol Concerning Regional Cooperation in Combating Pollution by Oil and Other Harmful Substances in Cases of Emergency. In September 1985, PERSGA approved the following additional protocols:

- Protocol Concerning the Conservation of Biological Diversity and the Establishment of a Network of Protected Areas in the Red Sea and Gulf of Aden.
- Protocol Concerning the Protection of the Marine Environment from Land-Based Activities in the Red Sea and Gulf of Aden.
- Protocol Concerning the Exchange of Personnel and Equipment in Case of Marine Emergency.

The three most recent protocols are as follows:

1 Protection of Marine Environment from Land-Based Activities (2005).
2 Biodiversity and Establishment of Marine Protected Areas (2005).
3 Technical Cooperation to Borrow and Transfer Experts, Technicians, Equipment and Materials in Cases of Emergency (2009).

Source: www.persga.org

UNEP Regional Seas Programme: Wider Caribbean

The Convention for the Protection and Development of the Marine Environment of the Wider Caribbean Region ('Cartagena Convention') was adopted 24 March 1983 and entered into force 11 October 1986 (UNEP, 1983). Thirty-three island and continental states are party to the Convention, which builds on UNEP's Caribbean Environment Program (CEP) and Caribbean Action Plan adopted in 1981. Twenty-five member states have ratified. The Convention included the Protocol concerning Co-Operation in Combating Oil Spills in the Wider Caribbean Region (entered into force 1986), and is administered by the CEP in Kingston, Jamaica. It is supplemented by the Specially Protected Areas and Wildlife Protocol (entered into force 2000) and the Land-Based Sources of Marine Pollution Protocol (entered into force 2010).

UNEP Regional Seas Programme: Eastern Africa

The Convention for the Protection, Management and Development of the Marine and Coastal Environment of the Eastern African Region ('Nairobi Convention') was adopted in Nairobi on 21 June 1985 and entered into force on 30 May 1996 (UNEP, 1985). It was amended in 2010. Other plans and protocols adopted concurrently and entering into force at the same time were the Action Plan for the Protection, Management and Development of the Marine and Coastal Environment of the Eastern African Region ('EAP'), Protocol Concerning Protected Areas and Wild Fauna and Flora in the Eastern African Region, and the Protocol Concerning

Co-Operation in Combating Marine Pollution in Cases of Emergency in the Eastern African Region. Six states were required to ratify or accede to the Convention before it came into force, and there are currently 10 participating states. UNEP led the initial development of the program.

The Convention is a comprehensive compact covering the protection, management and development of the marine and coastal environment. Nine items comprised a priority list; pollution-related commitments included controlling pollution from ships (Article 5); dumping (Article 6); land-based sources (Article 7); seabed activities (Article 8); airborne sources (Article 9); and in cases of emergencies (Article 11). Additional commitments are related to specially protected areas (Article 10), environmental damage from engineering activities (Article 12), environmental assessment (Article 13) and scientific cooperation (Article 14).

Despite its sensible intentions, the Convention had been criticized for focussing attention on pollution prevention and single-species management at the expense of integrated coastal zone management. More recently, however, marine spatial planning has become a central theme.

UNEP Regional Seas Programme: South Pacific

The Convention for the Protection of Natural Resources and Environment of the South Pacific Region ('Noumea Convention') was adopted in 1986 and entered into force in 1990 (SPREP, 1986). The Convention encompasses the territorial waters and EEZs of 24 state parties, which includes 7,500 islands and approximately one third of the world's languages (Table 6.1). The Convention also contains protocols for dumping and pollution emergencies. It is administered by the Secretariat of the Pacific Community (SPC), headquartered in Anse Vata, Noumea. The South Pacific Regional Environment Programme (SPREP) was established in the late 1970s as a component of UNEP's Regional Seas Programme. The Programme was renamed the Pacific Regional Environment Programme in 2004; however, it is still referred to as 'SPREP' when abbreviated. SPREP left SPC in in 1992 and relocated to Samoa, becoming a fully independent inter-governmental organization in 1993. SPREP has four thematic areas: biodiversity and ecosystems management; climate change; environmental monitoring and governance; and waste management and pollution control.

SPC and SPREP belong to the Council of Regional Organisations (CROP) of the Pacific, chaired by the Pacific Islands Forum Secretariat. Other Pacific CROP agencies involved in environmental matters include the Forum Fisheries Agency (FFA) and the Pacific Islands Development Program (PIDP). Additionally, the Pacific Islands Development Forum, outside of the CROP, focusses on sustainable development.

Table 6.1 State parties to the Convention for the Protection of Natural Resources and Environment of the South Pacific Region (Noumea Convention), including overseas parties

American Samoa	Australia	Cook Islands
Federated States of Micronesia	French Polynesia	Fiji
Guam	Kiribati	Marshall Islands
Nauru	New Caledonia	New Zealand
Niue	Northern Mariana Islands	Palau
Papua New Guinea	Pitcairn Islands	Samoa
Solomon Islands	Tokelau	Tonga
Tuvalu	Vanuatu	Wallis and Futuna
Overseas parties		
France	United States	

UNEP Regional Seas Programme: Black Sea

The 20 states located in the Black Sea watershed input a combined 350 cubic km of river water annually into the Black Sea. The Sea is in turn connected to the world ocean through the shallow, 35km wide Istanbul Strait. As a result, the Black Sea is highly entrained and therefore sensitive to a number of human impacts. Eutrophication/nutrient enrichment, chemical and oil pollution, overfishing, habitat loss, changes in biodiversity and introduced species are all significant threats to the area. The unique nature of the Black Sea called for a convention focussing on its issues; yet those states – Bulgaria, Georgia, Romania, Russian Federation, Turkey and Ukraine – surrounding its waters needed to remain abiding to the 1976 Barcelona Convention.

The Convention on the Protection of the Black Sea Against Pollution ('Bucharest Convention') was adopted 21 April 1992 and entered into force 15 January 1994 (BSC, 1992). Parties to the Convention include Bulgaria, Georgia, Romania, Russian Federation, Turkey and Ukraine and the Convention is supported by a permanent secretariat based in Istanbul, Turkey. It contains three Protocols on the control of land-based pollution, dumping and pollution emergencies. The primary purpose of the Convention is to reduce pollution in the Black Sea, and the Programme developed the 1996 Strategic Action Plan for the Environmental Protection and Rehabilitation of the Black Sea that was updated in 2009 (BSC, 2009).

Discussion questions

- How might the 18 Regional Seas Programmes interact with the 40+ regional fisheries bodies and regional fisheries management organizations?
- Why might a state elect not to join a Regional Seas Programme?
- Some countries, like the United States, are part of Regional Seas Programmes but have not ratified the LOSC. Might this pose certain challenges to those programmes?

Further reading

Akiwumi, P. and Melvasalo, T. (1998) 'UNEP's regional seas programme: Approach, experience and future plans', *Marine Policy*, vol 22, no 3, pp. 229–234.
Black Sea Commission (1992) *Convention on the Protection of the Black Sea Against Pollution*, Istanbul, Turkey, www.blacksea-commission.org/_convention-fulltext.asp, accessed 04 January 2019.
Black Sea Commission (2009) *Strategic Action Plan for the Environmental Protection and Rehabilitation of the Black Sea*, Istanbul, Turkey, www.blacksea-commission.org/_bssap2009.asp, accessed 04 January 2019.
Branch, A. E. and Roberts, M. (2014) *Branch's Elements of Shipping*, Routledge, London.
CBD (1995) *Conservation and Sustainable Use of Marine and Coastal Biological, Diversity*, Montreal, Canada, www.cbd.int/decision/cop/?id=7083, accessed 04 January 2019.
CBD (1998) *COP 4 Decision IV/5: Conservation and Sustainable Use of Marine and Coastal Biological Diversity, Including a Programme of Work*, www.cbd.int/decision/cop/?id=7128, accessed 04 January 2019.
CBD (2000a) *The Jakarta Mandate: From Global Consensus to Global Work*, Montreal, Canada, www.cbd.int/doc/publications/jm-brochure-en.pdf, accessed 04 January 2019.
CBD (2000b) *Cartagena Protocol on Biosafety to the Convention on Biological Diversity*, Montreal, Canada, www.cbd.int/doc/legal/cartagena-protocol-en.pdf, accessed 04 January 2019.
CBD (2006) *COP 8 Decision VIII/22: Marine and Coastal Biological Diversity: Enhancing the Implementation of Integrated Marine and Coastal Area Management*, Montreal, Canada, www.cbd.int/decision/cop/?id=11036, accessed 02 September 2018.
CBD (2008) *COP 9 Decision IX/20: Marine and Coastal Biodiversity*, Montreal, Canada, www.cbd.int/decision/cop/?id=11663, accessed 02 September 2018.

CBD (2010) *Nagoya Protocol on access and Benefit-Sharing*, www.cbd.int/abs/, accessed 02 September 2018.

Davidson, I. C. and Simkanin, C. (2012) 'The biology of ballast water invasions 25 years later', *Biological Invasions*, vol 14, pp. 9–13.

de Mora, S., Tolosa, I., Fowler, S. W., Villeneuve, J.-P., Cassi, R. and Cattini, C. (2010) 'Distribution of petroleum hydrocarbons and organochlorinated contaminants in marine biota and coastal sediments from the ROPME Sea Area during 2005, *Marine Pollution Bulletin*, vol 60, no 12, pp. 2323–2349.

Hamza, W. and Munawar, M. (2009) 'Protecting and managing the Arabian Gulf: Past, present and future', *Aquatic Ecosystem Health & Management*, vol 12, no 4, pp. 429–439.

IMO (1972) *Convention on the Prevention of Marine Pollution by Dumping of Wastes and Other Matter*, IMO, London, treaties.un.org/doc/Publication/UNTS/Volume%201046/volume-1046-I-15749-English.pdf, accessed 02 September 2018.

IMO (2004) *International Convention for the Control and Management of Ships' Ballast Water and Sediments*, IMO, London, www.imo.org/en/About/Conventions/ListOfConventions/Pages/International-Convention-for-the-Control-and-Management-of-Ships%27-Ballast-Water-and-Sediments-(BWM).aspx, accessed 02 September 2018.

Jaffe, D. (1995) 'The international effort to control the transboundary movement of hazardous waste: The Basel and Bamako Conventions', *ILSA Journal of International & Comparative Law*, vol 123.

Osborn, D. and Datta, A. (2006) 'Institutional and policy cocktails for protecting coastal and marine environments from land-based sources of pollution', *Ocean & Coastal Management*, vol 49, nos 9–10, pp. 576–596.

Secretariat of the Pacific Regional Environment Programme (SPREP) (1986) *Convention for the Protection of the Natural Resources and Environment of the South Pacific Region*, Apia, Samoa, www.sprep.org/attachments/NoumeConventintextATS.pdf, accessed 02 September 2018.

UN (1992) *Convention on Biological Diversity*, treaties.un.org/doc/Treaties/1992/06/19920605%2008-44%20PM/Ch_XXVII_08p.pdf, accessed 02 September 2018.

UNEP (1981a) *Convention for Co-Operation in the Protection and Development of the Marine and Coastal Environment of the West and Central African Region and protocol*, Nairobi, Kenya, abidjanconvention.org/media/documents/publications/Abidjan%20Convention%20English.pdf, accessed 02 September 2018.

UNEP (1981b) *Convention for the Protection of the Marine Environment and Coastal Area of the South-East Pacific*, Nairobi, Kenya, sedac.ciesin.org/entri/texts/marine.environment.coastal.south.east.pacific.1981.html, accessed 02 September 2018.

UNEP (1982) *Regional Convention for the Conservation of the Red Sea and Gulf of Aden*, Nairobi, Kenya, www.persga.org/Documents/Doc_62_20090211112825.pdf, accessed 02 September 2018.

UNEP (1983) *Convention for the Protection and Development of the Marine Environment in the Wider Caribbean Region*, Nairobi, Kenya, www.cep.unep.org/cartagena-convention/text-of-the-cartagena-convention, accessed 02 September 2018.

UNEP (1985) *Convention for the Protection, Management and Development of the Marine and Coastal Environment of the Eastern African Region*, Nairobi, Kenya, wedocs.unep.org/bitstream/handle/20.500.11822/21167/UNEP-DEPI-EAF.CP.7.Inf4a%20-en%20Amended%20Nairobi%20Convention.pdf?sequence=1&isAllowed=y, accessed 02 September 2018.

UNEP (1989) *Basel Convention on the Control of Transboundary Movements of Hazardous Wastes and their Disposal*, Nairobi, Kenya, treaties.un.org/doc/Treaties/1992/05/19920505%2012-51%20PM/Ch_XXVII_03p.pdf, accessed 02 September 2018.

UNEP (1995a) *Amendment to the Basel Convention on the Control of Transboundary Movements of Hazardous Wastes and their Disposal*, Nairobi, Kenya, treaties.un.org/doc/Treaties/1995/09/19950922%2012-54%20PM/Ch_XXVII_03_ap.pdf, accessed 02 September 2018.

UNEP (1995b) *Global Programme of Action for the Protection of the Marine Environment from Land-Based Activities*, Nairobi, Kenya, www.gpa.unep.org, accessed 02 September 2018.

UNEP (2005) *Mediterranean Strategy for Sustainable Development: A Framework for Environmental Sustainability and Shared Prosperity*, Nairobi, Kenya, mio-ecsde.org/epeaek09/basic_docs/unep_mssd_eng.pdf, accessed 02 September 2018.

UNEP (2012) *State of the Mediterranean Marine and Coastal Environment 2012 Highlights for Policy Makers*, Nairobi, Kenya, wedocs.unep.org/bitstream/handle/20.500.11822/375/unepmap_soehighlights_2012_eng.pdf, accessed 02 September 2018.

UNEP (2018) *Proceedings of the United Nations Environment Assembly at its third session*, Nairobi, Kenya, papersmart.unon.org/resolution/uploads/k1709367.english.pdf

UNESCO (1972) *World Heritage Convention*, whc.unesco.org/archive/convention-en.pdf

UNESCO (2011) *Navigating the Future of Marine World Heritage*, no 28, whc.unesco.org/en/series/28/

Vallega, A. (2002) 'The regional approach to the ocean, the ocean regions, and ocean regionalisation? A post-modern dilemma', *Ocean & Coastal Management*, vol 45, nos 11–12, p. 721.

Verlaan, P. A. and Khan, A. S. (1996) 'Paying to protect the commons: Lessons from the Regional Seas Programme', *Ocean and Coastal Management*, vol 31, nos 2–3, pp. 83–104.

World Commission on Environment and Development (1987) *Our Common Future*, Oxford University Press, Oxford, UK.

Chapter 7

Addressing climate change and its impacts on the world ocean

International efforts to mitigate and adapt to a changing planet

Key topics

While the effects of climate change on marine environments are not fully understood, climate change has significant potential implications to coastal and marine environments, including:

- Rising sea levels (0.09–0.88m by 2100) and their impact on coastal areas and island nations.
- Melting of Polar sea ice resulting in changes to marine trophic structure. In addition, the loss of sea ice reduces the albedo effect (where incoming solar radiation is reflected back into space).
- Changes to global ocean circulation patterns and the effects on terrestrial climates, marine trophic structure and commercial fisheries.
- Increased ocean acidification, resulting in diminished shell formation for calcareous organisms (organisms with shells).
- Increased severity of extreme weather events due to ocean warming.
- Increased frequency of severe weather events in places historically unaccustomed to them.
- Migration of biological communities towards Polar areas as a result of increasing ocean temperatures.
- Release of seabed methane deposits as a result of increasing ocean temperatures.

The domestic and international responses to climate change are to limit the amount of greenhouse gasses (GHGs) input into the atmosphere (mitigation); build or maintain resilience in ecosystems (protection); adapt human societies to a changing planet (adaptation); and repair damage that has already occurred (repair/rehabilitation/restoration).

The primary international agreements governing mitigation and adaptation are the United Nations Framework Convention on Climate Change (UNFCCC), Kyoto Protocol, and Paris Agreement.

Introduction

Since oceans cover 73 per cent of the earth's surface and comprise a volume of 1.3 billion square km, it is no surprise that the oceans are important in structuring and regulating the global climate. As well as storing and exchanging gasses such as CO_2, oceans store and release heat, transfer heat horizontally and vertically and provide water for the atmosphere.

There is no doubt that the earth's climate changes over time and that cyclical changes transpired long before humans dominated the planet. Global climate changes have been responsible for mass extinctions in the past and the earth's climate will continue to vary, likely resulting in future mass extinctions. During the Quaternary period, sea levels deviated as much as 85m, which inhibited the evolution of established marine communities in coastal and shelf environments. Unquestionably, humans have impacted global temperatures; since the 1980s, there has been considerable debate on differentiating the natural and anthropogenic contributions to climate change. Recent evidence, however, concludes that markedly hotter summers since the 1950s is human-induced, suggesting a significant role for humankind in climate change more generally (Hansen et al., 2012).

Human activities that influence change include the release of CO_2 through the burning of fossil fuels and large-scale deforestation – which decreases the removal of CO_2 from the atmosphere. The global atmospheric CO_2 concentration in 1750, prior to the start of the industrial revolution, was estimated at 227 parts per million (ppm). Concentrations have since increased to 340 ppm in 1980 to 404 ppm in 2017. At present trends and without significant changes to government regulation and individual human behaviours, cumulative emissions of CO_2 will reach 565 ppm by the mid-21st Century (Le Quéré et al., 2016).

This chapter reviews the present state of, and impacts to, the global marine environment from the anthropogenic addition of greenhouse gasses to the atmosphere. Following this is a discussion on the primary international mechanisms that govern the mitigation and adaptation of climate change effects on the marine environment.

Impacts of climate change on the global ocean

While the specific impacts of climate change on marine environments are discussed below, an overall scientific consensus has emerged, which concludes that a) the global ocean drives global climate and provides valuable services to humanity with respect to climate regulation, b) oceans are currently affected by climate change and many ecological and human communities are at high risk of impacts before 2100, c) significant CO_2 reductions are immediately required to prevent further damage to ocean ecosystems and human settlements, and d) the longer the delay in reducing CO_2 emissions the opportunities for protection, adaptation and remediation diminish (Gattuso et al., 2015).

Physical impacts and corresponding impacts on marine biota

The primary physical impacts on ocean systems as a result of climate change is increasing temperatures that result in rising sea levels, loss of sea ice, changes in ocean stratification and circulation (mixing), reduced oxygen, and increased storm severity (and in some places, frequency). Many of these impacts are additive and interact to negatively impact marine and terrestrial biological and human communities.

The IPCC concluded that the overall average temperature of the surface layers (first 75m) of the global ocean increased by 0.11°C per decade since 1971 and that it is virtually certain that the first 700m of the ocean surface has warmed during this period. Furthermore, the Panel concluded that areas deeper than 700m have warmed over the past 150 years (IPCC, 2014).

Temperatures increase differently across the planet with corresponding differential impacts to marine areas. For example, the Arctic has warmed nearly twice as fast as the rest of the earth over the past couple of decades with significant implications for sea ice cover. The more rapid increase in Arctic warming is a result of reduced sea ice reducing the albedo (reflectivity) of the Arctic's surface resulting in additional near-surface warming, establishing a positive feedback loop of less ice and greater temperatures. The current sea ice contraction has not been exceeded in at least 1.5 millennia; over the past 100 years, over two million square km of sea ice has been lost.

Temperature increases have also affected global circulation patterns. Perhaps most notably, climate change has been demonstrated to affect the El Niño Southern Oscillation (ENSO) cycle. As discussed in Chapter 2, the ENSO is a periodic variation in which the waters of the central and eastern Pacific are either warmer (El Niño) or cooler (La Niña) than normal. Both El Niño and La Niña cause changes to precipitation patterns resulting in floods, droughts, severe storms (including cyclones and hurricanes). For example, the 1997–98 El Niño caused $US 35–45 billion in damage and claimed an estimated 23,000 lives worldwide. The La Niña of the following year claimed 25,000–50,000 lives due to landslides in Venezuela and displaced over 200 million people in Asia. The surface water warming caused by El Niño directly stresses the health of coral reefs throughout the Pacific, thus impacting reef fisheries and all the other ecological services that coastal peoples receive from healthy coral reefs. Climate change models have demonstrated that increased ocean warming will increase the frequency and severity of ENSO events.

The implications on marine flora and fauna from increasing ocean temperatures are broadly understood to be changes in species ranges and range shifts to find cooler waters, damage to fixed species (such as corals), increases in disease and increased storm severity leading to damage to coastal ecosystems. While certain species will benefit in the short term from small increases in temperature, the rate of change will likely exceed the ability for the broader marine community to adapt cohesively, with consequences for ocean-dependent human societies.

In response to warming temperatures, mobile species can either move poleward or deeper to avoid temperature stress. Ocean communities (both plankton and nekton) may move up to 400km per decade in response to ocean warming. However, sessile (immobile) communities such as coral reefs are unable to move in response to temperature changes and are thus at the mercy of ocean temperatures. Most tropical reef forming corals form between 20°C and 30°C and are found between 30°N and 30°S degrees latitude in all oceans. Excessive temperatures can cause corals found in photic (lighted) zones to expel their symbiotic photosynthetic zooxanthellae resulting in greatly reduced fitness or death.

Susceptibility of marine organisms to disease and reduced fitness has been linked to increased ocean temperatures for many different types of species. Many types of disease and pathogens affecting corals, for example, have been shown to increase with warming. Polar bears and other marine species that utilize sea ice are negatively affected by sea ice loss due to Arctic warming. Warming can also create opportunities for non-native or invasive species to colonize areas previously inhospitable to the invaders. Temperatures affect the sex ratios of offspring for certain organisms, including some species of sea turtles, fish and copepods (Chapter 2). Consequently, increased temperatures reduce fitness by unbalancing the proportion of males and females in a population.

Changes in temperature can also affect the timing of spring blooms creating a 'mis-match' between pulses of primary productivity and the food webs that have evolved to consume these pulses. It has been shown in the Arctic that spring phytoplankton blooms are occurring up to 20 days earlier than historical normal, which can cause havoc on marine food webs, especially when consumers have evolved with migratory patterns to take advantage of these blooms but now arrive too late.

Other physical impacts of climate change on the ocean include changes to ocean stratification and circulation (collectively captured as 'mixing'), which can disrupt or alter nutrient pathways between nutrient-rich, oxygen-poor deep oceans and nutrient-poor, oxygen-rich shallow seas where photosynthesis occurs. With less mixing in the open ocean, nutrients in shallower depths can become depleted, reducing overall production. A final physical manifestation of increasing ocean temperatures is the projected increase in the severity of storms (which is approximately an exponential function of temperature) and their impacts on coastal ecosystems.

Chemical impacts and corresponding impacts on marine biota

The primary **chemical impacts** of climate change on the composition of seawater are ocean acidification and hypoxia. The global ocean is a net sink for atmospheric CO_2, and both organic and inorganic carbon are captured in ocean sediments. Atmospheric CO_2 diffuses into the ocean through direct contact, resulting in increases in forms of dissolved organic carbon. Since the industrial revolution in the 1800s, oceans have absorbed an estimated 30–40 per cent of the increase in CO_2 produced by human activity, much of which has been sequestered in the coastal ocean and continental shelf sediments (another quarter used by plants and the atmosphere accepts the remainder). The cost of this service to humanity, however, is a decrease in the pH (acidification) of the global ocean. The ocean, which is historically slightly alkaline, has become 35 per cent more acidic since 1750 and has been estimated to have sequestered approximately 525 billion tonnes of CO_2 during this time. At the current trajectory without significant reductions in CO_2 increases, the ocean is expected to become approximately 125 per cent more acidic by 2100. While the rate of CO_2 sequestering by the global ocean slows as pH drops, the ocean has a massive capacity for additional CO_2 to be captured.

The primary impact of ocean acidification is on those organisms that produce calcium carbonate ($Ca\ CO_2$) shells and skeletons. While different organisms that rely on calcification experience different responses to decreasing pH under varying ocean conditions, overall those organisms relying on calcification must spend more energy on shell development, which means less energy devoted to reproduction, defence etc. Studies in ocean areas throughout the world have shown decreases in shell thickness, particularly over the past several decades. Fewer such organisms could mean that the global 'carbon pump' is weakened, with the ocean sequestering less carbon in its sediments. The effects of increasing pH on marine ecosystems are still largely unknown, but already some unexpected results have emerged, such as some fish becoming less fearful of (or even attracted to) predators and the slightly increased transmission of sound in an already noisy ocean.

Further contributing to ocean acidification is the loss of marine vegetated habitats (e.g. macroalgae, mangroves, salt marshes and seagrasses) as a result of human activities and coastal developments. While marine vegetated habitats occupy only 0.2 per cent of the ocean surface, they are responsible for sequestering half of the CO_2 entering the ocean. This benefit to humans, however, is compromised by the loss of 25–50 per cent of these communities worldwide.

Hypoxic areas, also termed oxygen minimum zones (Chapter 2), are oxygen-deficient regions of the ocean caused when oxygen is consumed (by bacterial decay) faster than it can be replaced through ocean circulation. Situated on continental shelves often near larger rivers or human population centres, over 400 of these 'dead zones' are known and, until recently, were understood to be caused solely by eutrophication, or excess nutrients (e.g. fertilizer) flowing from terrestrial to marine environments. Climate change appears to be exacerbating hypoxic areas. A warming, and therefore more energetic, ocean results in more rapid ocean circulation resulting in additional nutrients being upwelled from the deep ocean to shallower water where they trigger increases in primary production. Not all biological material is able to be consumed in the surface waters, and therefore this matter sinks into deeper oceans where bacteria use the remaining available oxygen for decomposition. Newly identified hypoxic areas have been identified in upwelling systems off the west coast of North America and Africa.

Impacts of climate change on marine biological communities and human societies

Under our current rate of emissions, most marine organisms evaluated will have a very high risk of impacts by 2020 and many more by 2050. As discussed above, impacts of a warming, more acidic ocean may include: increasing hypoxic areas, coral bleaching, species range shifts, changes to phenology (life cycles and life histories), changes in behaviours affecting survival, reduced organism body size, changes to community composition and food webs, increased noise, storm frequency and severity affecting coastal vegetation and loss of habitats (e.g. sea ice, coastal vegetation).

Socio-economic impacts of climate change

While the public discussion and international climate negotiations generally focus on the socio-economic impacts of terrestrial climate, the impacts of humanity of a warming, more acidic ocean are measurable and forecasted to impact food security, employment, coastal settlements and infrastructure. For example, 'worst case' climate models conclude that up to 4.6 per cent of the world's population will experience annual inundation due to sea level rise at a cost of up to 9.3 per cent of GDP. Low-lying islands in some countries (e.g. Kiribati and Tuvalu) may no longer be habitable (due to saltwater intrusion into their drinking water and/or storm surges), rendering both the possibility of climate change refugees and the loss of national territorial seas and exclusive economic zones (which according to LOSC rely upon land baselines).

International law related to climate change

The global response to both climate change mitigation (reducing the releases of anthropogenic greenhouse gasses [GHGs]) and adaptation (preparing for the impacts of climate change) has significant marine policy implications for coastal, fisheries, public safety and biodiversity management. Accordingly, an appreciation of the international approaches to mitigating and adapting to climate change and how these approaches affect the marine environment is essential.

United Nations Framework Convention on climate change

While climate change was recognized by the UN General Assembly as a potential threat to the global environment and human population as early as 1989, the first stand-alone international agreement on climate change was the United Nations Framework Convention on Climate

Change (UNFCCC). It opened for signature 9 May 1992 at the UNCED in Rio de Janeiro and entered into force on 21 March 1994 (UNFCC, 1992). Presently, the Convention has 195 parties.

The Convention's objective (Article 2) is 'stabilization of greenhouse gas concentrations in the atmosphere at a level that would prevent dangerous anthropogenic interference with the climate system'. Furthermore, Article 2 states that 'such a level should be achieved within a time-frame sufficient to allow ecosystems to adapt naturally to climate change, to ensure that food production is not threatened, and to enable economic development to proceed in a sustainable manner'.

Article 3 of the Convention outlines the principles intended to guide parties, which include:

- More developed states will lead in the mitigation of and adaptation to climate change.

    (3.1)

- Less developed states, and especially those vulnerable to climate change, will be given special consideration.

    (3.2)

- Scientific uncertainty should not be used to postpone decisions.

    (3.3)

- Parties have a right to sustainable development and should use economic cooperation and trade to enable sustainable development and to address climate change.

    (3.4 and 3.5)

Article 4.1 of the Convention commits all parties to the following key actions:

- Establish inventories of greenhouse gas emissions by source.

    (4.1a)

- Address and report on anthropogenic emissions.

    (4.1b)

- Develop and share mitigation technologies and practices.

    (4.1c)

- Promote the conservation and enhancement of sinks and reservoirs.

    (4.1d

- Prepare climate change adaptation plans.

    (4.1d)

- Incorporate climate change considerations into national policies.

    (4.1e)

- Promote and share knowledge on climate science.

    (4.1f, 4.1g)

- Promote, and cooperate on, education and awareness related to climate change.

    (4.1i)

Article 4.2 of the Convention commits developed state parties listed in Annex I of the Convention to the specific strategic actions:

- Develop and implement national policies on limiting greenhouse gas emissions and protecting and enhancing greenhouse gas sinks and reservoirs.

(4.2a)

- Reduce the emission of greenhouse gasses to 1990 levels.

(4.2b)

- Use best available scientific information.

(4.2c)

The directives in Articles 4.3–4.10 outline the specific responsibilities of developed parties in assisting less developed parties in their efforts to mitigate and adapt to climate change.

The remainder of the Convention is summarized below:

- Article 5: Commits parties to undertake research and monitoring and to assist developing states to this end.
- Article 6: Commits parties to education, training and public awareness.
- Article 7: Establishes a Conference of the Parties and outlines the duties of the COP.
- Article 8: Establishes a Secretariat and its functions.
- Article 9: Establishes a subsidiary body to provide scientific and technological advice.
- Article 10: Establishes a subsidiary body to implement the Convention.
- Article 11: Establishes a financial mechanism to fund grants or the transfer of technology.
- Article 12: Identifies reporting requirements for parties to the Convention.
- Articles 13–16: Provides housekeeping details common to most international agreements.

Kyoto Protocol to the United Nations Framework Convention on Climate Change

The Kyoto Protocol was adopted on 11 December 1997 and entered into force on 16 February 2005. While the UNFCCC only encourages parties to stabilize and reduce greenhouse gas emissions, the Kyoto Protocol committed 37 developed parties to 5 per cent reductions to their 1990 levels of greenhouse gas emissions over the 2008–2012 period (UNFCC, 1997). Certain states have greater than 5 per cent GHG reduction targets including Iceland (10 per cent) and Australia (8 per cent).

LDSs were not bound by the Protocol because of their disproportionately small contributions of greenhouse gasses relative to MDSs. This principle is stated in the Protocol as 'common but differentiated responsibilities'. The United States did not ratify the Protocol, and Canada withdrew from the Protocol in 2011.

The Protocol functioned in two ways: First, by setting limits on the overall amounts of greenhouse gasses that a state may emit, it monetized emissions (creates an economic value). Emissions that were previously unregulated now have economic value. This created the opportunity to buy and sell emissions domestically for the industrialized state parties to the Convention. Second, the Protocol created market-based mechanisms where states may meet their emissions targets by reducing emissions in other states. The intent of these mechanisms was to reduce GHG emissions in LDSs where the costs of reduction are likely less than developed nations and the benefits to LDSs greater.

The Protocol contained three market-based mechanisms:

1. Emissions trading (Article 17): Allowed states that have exceeded their GHG reduction targets to sell the difference between their targets and actual emissions to other states. Known as the 'carbon market', emissions trading also allowed states to consider changes to land use/land cover or changes in practices that reduce CO_2 as equivalent to direct reductions on emissions.
2. Clean development mechanism (CDM, Article 12): Allowed states to undertake emissions reduction projects in other states in order to assist with meeting their own GHG targets (e.g. funding a switch from fossil fuels to renewable energy for electricity generation).
3. Joint implementation (Article 6): Allowed states with emission reductions targets to trade emissions between each other.

Greenhouse gasses covered in the Protocol include:

- Carbon dioxide (CO_2).
- Methane (CH_4).
- Nitrous oxide (N_2O).
- Hydrofluorocarbons (HFCs).
- Perfluorocarbons (PFCs).
- Sulphur hexafluoride (SF_6).

Some states have explored ocean fertilization and direct injection of CO_2 into the deep oceans as potential methods to sequester CO_2 and meet their GHG reduction targets. Ocean fertilization generally proposes the addition of biologically available iron or urea (nitrogen) to increase primary production, while deep injection proposes to inject liquefied CO_2 into the deep oceans or in geological structures (Freestone and Rayfuse, 2008). To some researchers/lawmakers, the UNFCCC's direction to consider stabilization of climate through the use of sinks suggests that ocean fertilization may be an avenue for states to meet their reduction targets. Ocean fertilization, however, may be considered a 'pollutant' under the London Dumping Convention and potentially may contravene the LOSC if fertilization potentially harms marine living resources.

Subsequent to some high-profile ocean fertilization experiments in the high seas, under the International Maritime Organization, parties to the London Protocol adopted a framework by which to assess such activities. In 2013, an amendment to the London Protocol was adopted to regulate placement of matter for ocean fertilization and other marine geo-engineering activities (LP.4(8)). Perhaps due to the negative or inconclusive results of the earlier experiments, however, no new ones have come forward, thus the assessment framework and amendment have not yet been put to the test.

Paris Agreement

The Paris Agreement was adopted on 12 December 2015. All 197 UNFCCC members have either signed or acceded to the Paris Agreement. As of February 2019, 194 states plus the European Union have signed the Agreement (= 195 signatories); 184 states plus the EU (= 185 parties), representing more than 88 per cent of global greenhouse gas emissions, have ratified or acceded to the Agreement, including China, the United States (which has stated it will be leaving) and India (UNFCC, 2019). The Agreement nests under the UNFCCC and aims to strengthen the global response to the threat of climate change through:

- Holding the increase in the global average temperature to below 2°C but preferably to limit the increase to 1.5°C over pre-industrial levels.
- Improving climate adaptation and fostering low carbon economic growth.
- Aligning financing to support low carbon economic growth and climate-resilient development.

Unlike the Kyoto Protocol, which sets legally binding emissions targets for states party to the Protocol, the Paris Agreement allows states to voluntarily establish their own 'nationally determined contributions' (NDCs, Article 4) that they are encouraged to revise every five years. States may adjust their NDCs anytime and report their NDCs and progress towards their targets to UNFCCC Secretariat. The NDC Partnership Office was established to assist states with meeting their NDCs. The rationale for switching from enforceable to voluntary reduction targets was to encourage states to be aspirational in setting emissions reduction (shifting from a 'stick' to a 'carrot'), allow greater flexibility in approach and to broaden the emission reduction focus from industry, where it had been concentrated under Kyoto, to the entire economy. It is still too early to judge the success of this new approach. However, according to a 2018 special report by the IPCC, the 2°C target of the Agreement will likely translate into considerably greater ecological and economic disruption than the 1.5°C aspirational value (IPCC, 2018); therefore, to avoid considerable costs and harm to humans and the environment, NDCs will have to be strengthened.

Early climate change agreement meetings (UNFCCC conferences of the parties [COPs]) barely mentioned the ocean, focussing on land and the atmosphere. However, from 2017 to 2018, Fiji as President of COP23 did much to raise awareness of the interconnectedness of all the realms on the earth, launching the Ocean Pathway Partnership. As a result, ocean issues have gained exposure at the UNFCCC meetings, though to date mainly though side events. It is hoped by oceanic countries that a future COP will formally recognize the role of the ocean, thus freeing up considerable financing to address marine climate change issues.

On the question of whether sea level rise could affect the sovereignty of states over their maritime jurisdictions, in 2012, the International Law Association created a Committee to examine the issue, which continues to meet. In 2018, it released its first full report, which included a resolution that, *inter alia*, endorsed,

> on the grounds of legal certainty and stability, provided that the baselines and the outer limits of maritime zones of a coastal or an archipelagic State have been properly determined in accordance with the 1982 Law of the Sea Convention, these baselines and limits should not be required to be recalculated should sea level change affect the geographical reality of the coastline.
>
> (ILA, resolution 5/2018)

As that the ILA is not an international treaty organization, such resolutions are advisory at best and may not affect the behaviours of states interested in usurping marine natural resources possibly lost to a state through climate change. Pacific Island states are especially concerned and have been organising a regional response through the Pacific Islands Forum, including an interstate agreement to respect each other's maritime boundaries, which it is hoped will carry some legal weight in future international dispute resolution, as may arise.

Discussion questions

- In what ways does the world ocean regulate climate?
- What might be the most effective ways to make international shipping – as a major marine contributor of GHGs – more climate friendly?
- What are 'blue carbon' initiatives, and how might they assist with mitigating and adapting to climate change?
- How might sea level rise affect low-lying states' rights to marine and seabed resources? Outline the arguments that could be made for and against their continued access.

Further reading

Freestone, D. and Rayfuse, R. (2008) 'Ocean iron fertilization and international law', *Marine Ecology Progress Series*, vol 364, nos 213–218, pp. 227–233.

Gattuso, J.-P., Magnan, A., Billé, R., Cheung, W. W. L., Howes, E. L., Joos, F., Allemand, D., Bopp, L., Cooley, S. R., Eakin, C. M., Hoegh-Guldberg, O., Kelly, R. P., Pörtner, H.-O., Rogers, A. D., Baxter, J. M., Laffoley, D., Osborn, D., Rankovic, A., Rochette, J., Sumaila, U. R., Treyer, S. and Turley, C. (2015) 'Contrasting futures for ocean and society from different anthropogenic CO2 emissions scenarios', *Science*, vol 349, no 6243, pp. 45.

Hansen, J., Sato, M. and Ruedy, R. (2012) 'Perception of climate change', *Proceedings of the National Academy of Sciences*, vol 109, no 37, pp. 2415–2423, doi:10.1073/pnas.1205276109

IPCC (2014) *Climate Change 2014: Synthesis Report*, Contribution of Working Groups I, II and III to the Fifth Assessment Report of the Intergovernmental Panel on Climate Change, IPCC, Geneva, Switzerland, 151 pp.

IPCC (2018) *Summary for Policymakers: In: Global Warming of 1.5°C: An IPCC Special Report on the Impacts of Global Warming of 1.5°C above Pre-Industrial Levels and Related Global Greenhouse Gas Emission Pathways, in the Context of Strengthening the Global Response to the Threat of Climate Change, Sustainable Development, and Efforts to Eradicate Poverty*, Meteorological Organization, Geneva, Switzerland, 32 pp.

Le Quéré, C., Andrew, R. M., Canadell, J. G., Sitch, S., Korsbakken, J. I., Peters, G. P., Manning, A. C., Boden, T. A., Tans, P. P., Houghton, R. A., Keeling, R. F., Alin, S., Andrews, O. D., Anthoni, P., Andrews, O. D., Barbero, L., Bopp, L., Chevallier, F., Chini, L. P., Ciais, P., Currie, K., Delire, D., Doney, S. C., Friedlingstein, P., Gkritzalis, T., Harris, I., Hauch, J., Haverd, V., Hoppema, M., Goldewijk, K. K., Jain, A. K., Kato, E., Körtzinger, A., Landschützer, P., Lefèvre, N., Lenton, A., Lienert, S., Lombardozzi, D., Melton, J. R., Metzl, N., Millero, F., Monteiro, P. M. S., Munro, D. R., Nabel, J. E. M. S., Nakaoka, S., O'Brien, K., Olsen, A., Omar, A. M., Ono, T., Pierrot, D., Poulter, B., Rödenbeck, C., Salisbuty, J., Schuster, U., Schwinger, J., Séférian, R., Skjelvan, I., Stocker, B. D., Sutton, A. J., Takahashi, T., Tian, H., Tilbtook, B., van der Laan-Luijkx, I. T., van der Werf, G. R., Viovy, N., Walker, A. P., Wiltshire, A. J. and Zaehle, S. (2016) 'Global carbon budget 2016', *Earth System Science Data*, vol 8, pp. 605–649.

UNFCCC (1992) *United Nations Framework Convention on Climate Change*, Bonn, Germany, treaties.un.org/doc/Treaties/1994/03/19940321%2004–56%20AM/Ch_XXVII_07p.pdf, accessed 04 January 2019.

UNFCCC (1997) *Kyoto Protocol to the UN Framework Convention on Climate Change*, Bonn, Germany, unfccc.int/resource/docs/convkp/kpeng.pdf, accessed 04 January 2019.

UNFCCC (2019) *Paris Agreement*, Bonn, Germany, treaties.un.org/pages/ViewDetails.aspx?src=TREATY&mtdsg_no=XXVII-7-d&chapter=27&clang=_en, accessed 26 February 2019.

Chapter 8

International fisheries policy
Sustaining global fisheries for the long term

Key topics

- Over 1 billion people depend on seafood as their primary source of protein, and fishing is estimated to employ 34 million people in full- or part-time jobs, producing over 81 million tonnes of seafood annually. The first-sale value of the world's fisheries is $US 100 billion.
- Marine fisheries – and fisheries in areas beyond national jurisdiction in particular – have historically been poorly managed or not managed at all. Overall, 31 per cent of the world's fisheries are overexploited, 20 per cent of the global fisheries catch consists of unwanted by-catch; it is estimated that illegal, unreported and unregulated fishing may comprise up to 30 per cent of the global fisheries catch.
- Overfishing has largely been eliminated in the United States, with good progress being made in Canada and the European Union.
- Fisheries may be managed under single-species, multi-species or ecosystem-based regimes. The ecosystem approach to fisheries management is currently being advocated by many states and the FAO as an approach that attempts to balance ecological and human well-being objectives through good governance.
- The FAO is responsible for reporting on the status of the world's fisheries, providing information to decision-makers and supporting regional fisheries management organizations.
- A number of binding and non-binding agreements have been developed to govern fishing and whaling.
- Rights-based approaches confer a private right to a public good. Marine property rights may be conceived of as tangible objects (e.g. territory, fish, oil and gas reserve) or intangible benefits or income streams (e.g. right to fish) recognized by some level of government. These approaches are used primarily in fisheries management, and currently 15 per cent of the world's catch is managed under this scheme.

Introduction

From a western perspective, the importance of marine fisheries to global food security is often overlooked. In North America and many parts of Europe, seafood is simply another source of protein for consumers to choose. Elsewhere, over 1 billion people depend on seafood as their primary source of protein, and 3.1 billion people rely on seafood for at least 17 per cent of their protein. Fishing has recently been estimated to employ 34 million people in full- or part-time jobs, producing 81.4 million tonnes of seafood in 2014. The first-sale value of the world's fisheries is estimated at $US 100 billion.

Concern over the sustainability of marine capture fisheries has increased over the past two decades. The Food and Agriculture Organization of the United Nations (FAO) currently reports that 31.4 per cent of the world's fisheries are now overexploited, depleted or recovering. Indeed, 57.4 per cent are fully exploited and only 20 per cent under exploited or moderately exploited (FAO, 2018). Global catches peaked in 1996 and, for the past 15 years, have been relatively stable at a level approximately 10 per cent less than 1996 (FAO, 2018). For context, in 1974, the corresponding percentages were 10, 50 and 40 per cent, respectively. However, the situation in some parts of the world is improving. In the United States, the proportion of fish stocks assessed as overfished is now 16 per cent, or half of what it was in 1997. In Argentina, Chile and Peru, the proportion of overfished stocks has declined from 75 per cent in 2000 to 45 per cent in 2011 (Hilborn, 2016).

In national jurisdictions, many stocks are still much less abundant than historically, and full recovery is likely many years away. North Sea cod stocks of Atlantic cod, for example, estimated at 6 per cent of historical levels, are depressed to such a degree that recently fishers have been unable to harvest enough to meet their allowable catches, and each year further reductions are recommended.

The status of many high seas stocks is less well known. However, the North Atlantic has a longer management history than most other regions. According to the Northwest Atlantic Fisheries Organization (NAFO) stock advice for 2018, of the seven stocks listed, only one was considered to be healthy (fished at maximum sustainable yield). However, of the other six, all but two were seen to be no longer overfished, and most stocks are stable or recovering.

Through an examination of regionally based on-average catches from 2005–2009, the following regions contribute the largest proportions of the global catch:

- Northwest Pacific (25 percent).
- Southeast Pacific (15 percent).
- Western Central Pacific (14 percent).
- Northeast Atlantic (11 percent).
- Eastern Indian Ocean (8 percent).

Source: FAO (2018)

A closer look at certain aspects of marine capture fisheries shows that 50–70 per cent of pelagic predators (i.e. tunas, swordfish) have been removed by fishing. Fishing pressure continues to shift towards lower trophic levels as apex predators decline; this has been termed 'fishing down marine food webs' (Pauly et al., 1998). Furthermore, the global fishing fleet is far larger than what necessity dictates, especially since technological innovations increase the ability to catch fish. This surplus fishing capacity is underwritten by $US 30–40 billion in annual subsidies by most fishing nations, thus providing no incentive to reduce fishing efforts (Sumaila et al., 2016). In addition, evidence suggests that fishing efforts from port-based fisheries have increased over the past three decades at upwards of 3 per cent per decade. As such, more fishers are chasing fewer fish; if one were to imagine the global fishing fleet as a country, it would be the 18th largest oil-consuming nation on earth, equivalent to the Netherlands (Tyedmers et al., 2006).

Another serious problem is the unintended harvest of species. An estimated 20 per cent of the global fisheries catch constitutes unwanted by-catch that is discarded. Shrimp fisheries produce the largest by-catch and small pelagic fisheries the least. By-catch also consists of the incidental take of endangered marine mammals, turtles and seabirds, although recent advances in gear technology are reducing these impacts.

This chapter first provides a brief overview of how marine fisheries are managed. This is followed by a discussion on the laws and policies of international fisheries and the organizations that oversee marine fisheries. Lastly, rights-based approaches to ocean management are presented.

An overview of fisheries management

There was little perceived need for fisheries management prior to the 20th Century. Marine fisheries (including marine mammals) were thought to be inexhaustible, so there was little impetus to allocate resources towards their management. Most fisheries (including nearshore ones) were 'commons' resources, which were open to anyone with the equipment to exploit them. With the advent of the steam engine, followed by the internal combustion engine, fisheries easily accessible from land were quickly depleted, and states realized the need for management. A suite of national and international legislation and conventions was developed to enforce fisheries regulations. These included the International Convention for the Regulation of Whaling (1946) and the first agreements under Law of the Sea negotiations (1958) (Chapter 4). Two of the earlier regional regulatory instruments were the 1949 International Commission for the Northwest Atlantic Fisheries and the 1959 North East Atlantic Fisheries Convention, which were replaced by NAFO in 1978 and the North-East Fisheries Commission (NEAFC) in 1982. A number of other conventions (discussed in Chapter 6), including MARPOL (1973/78), the London Dumping Convention (1972) and Protocol (1996), and the Convention for the Prevention of Marine Pollution by Dumping of Wastes and Other Matter (1993), were designed to protect marine environments – and therefore fisheries – from land- and marine-based pollutants.

The purpose of this section, however, is not to chronicle the plight of specific fisheries or dwell on the failure (or success) of fisheries management. Rather, this section presents some of the principles of fisheries management and how these have evolved since the early 20th Century. While humans have utilized almost all marine taxa for food, fuel and medicine, fishing efforts have traditionally targeted finfish (e.g. class/order Agnatha, Chondrichthyes, Osteichthyes), crustaceans (e.g. class/order Amphipoda, Decapoda, Euphausiacea, Mysidacea), molluscs (e.g. class/order Bivalvia, Cephalopoda, Gastropoda) and marine mammals (e.g. class/order Odontoceti, Mysticeti, Pinnipedia, Sirenia). Fisheries for marine algae are important in many regions, and many species from other marine taxa (e.g. Echinodermata) are also intensely harvested (see Chapter 2).

Fisheries employ a variety of techniques and can be categorized in a number of ways (Box 8.1). However, four types of fisheries are generally recognized: 'food', 'industrial', 'shellfish' and 'recreational'. Food fisheries are generally operated by and for local peoples, and – while declining over the past 30 years – they continue to be vital to both developed and developing nations, particularly small island states. Food fisheries generally harvest nearshore finfish and shellfish, with some remaining subsistence whaling by Indigenous peoples in Arctic areas. Industrial fisheries have been the fastest growing in recent decades, due to improvements in vessels, fishing gear and technology (e.g. satellite images, global positioning systems). Industrial fisheries are the sole harvesters on the high seas and are increasingly harvesting pelagic invertebrates (e.g. squid, krill). Shellfish fisheries, on the other hand, target invertebrates that are primarily on or within the benthos. Finally, recreational fisheries operate throughout the world, at a variety of scales ranging from personal single-boat operations to commercially organized fleets with 'mother ships' and processing facilities. There are limited statistics on recreational fisheries and their impacts; however, one study in the United States suggested that recreational landings for species of concern accounted for about one quarter of the total (Coleman et al., 2004).

Box 8.1 Types of fisheries based on gear types

- Bottom trawl. A fishing net is dragged across the seafloor (benthic trawling) or immediately above the seafloor (demersal trawling). Bottom trawling targets shrimp and various types of groundfish. It has been widley criticized as environmentally damaging.
- Cable longline. This gear consists of heavy wire for the mainline instead of monofilament. It has been used to target mako and blue sharks.
- Drift gillnets. A gillnet is a panel of netting suspended vertically in the water by floats, with weights along the bottom. Fish are entangled in the net. Drift gillnet gear is anchored to a vessel and drifts along with the current. It is usually used to target swordfish and some shark species.
- Dynamite. Explosives are used to concuss vertebrate fish to facilitate harvesting. Most frequently applied in tropical coral reef environments. This destructive method has been banned in most places.
- Harpoon. A larger spear, which may contain explosives, that is thrown or launched at individual animals. The harpoon fishery mainly targets swordfish and marine mammals.
- Midwater trawl. A fishing net is towed, suspended in the pelagic zone, targeting schooling fish. This gear is associated with the largest fisheries by catch volume.
- Pelagic longline. Pelagic longline gear consists of a main horizontal line that has shorter lines with baited hooks attached to it. The gear is used at various depths and may be anchored to the bottom, depending on the species being targeted.
- Poison. Poison, most often sodium cyanide or bleach, are used to kill or stun fish. Most frequently applied in tropical coral reef environments. This method is being phased out in most places.
- Purse seine. A purse seine is an encircling net that is closed by means of a purse line threaded through rings on the bottom of the net. This gear is effective in catching schooled tunas. 'Coastal' purse seiners are smaller vessels that fish close to the shore. They mainly harvest coastal pelagic species (sardines, anchovies, mackerel). 'Large' purse seine gear is used in major high seas fisheries, such as tuna.
- Recreational fisheries. The recreational fisheries for resident and migratory species consist of private vessels and charter vessels usually using hook-and-line gear.
- Trolling. Use of one or more baited fishing lines towed through the water column behind a boat. Trolling targets pelagic species such as albacore tuna.

To date, fisheries management has essentially been limited to assessment of individual species, or a sub-group of them, known as a 'stock'. The basic goal of fishery management is to estimate the amount of fish that can be harvested (total allowable catch – TAC) so as to impose catch limits (Box 8.2) while maintaining viable fish populations. These biological estimates may be modified by political, economic and social considerations. As a result, overly conservative management resulting in under-utilized fisheries production (under-harvesting) seldom occurs. Overly aggressive management, on the other hand, is likely to result in overharvesting and, in extreme cases, severely reduced or commercially extirpated populations.

Fisheries management is generally composed of four components: the fish that are harvested; the species that are not harvested; the environment or habitat where the fish live; and the human use and interactions with the fish and their habitat. Each of these aspects of fisheries

> **Box 8.2 Catch limits**
>
> **Total allowable catch** (TAC) is a management measure that limits the total output from a fishery by setting the maximum weight or number of fish that can be harvested over a period of time. TAC-based management requires that landings be monitored and that fishing operations stop when the TAC for the fishery is met. A TAC is based on stock assessments (discussed below) and other indicators of biological productivity, usually derived from both fishery-dependent (catch) and fishery-independent (biological survey) data. Data collected from fishermen, processors or dockside sampling can be combined with at-sea observations and independent fishery survey cruises to provide information about the total biomass, age distribution and number of fish harvested. Typically, the TAC is determined on an annual basis but can be partitioned across areas and seasons. To the extent that a TAC is well estimated and enforced, it can sustainably control total fishing mortality on a stock (e.g. Pacific halibut off the west coast US and Canada).
>
> **Trip limits and bag limits** are measures that limit the amount of harvest of a species in any given trip. Trip limits are commonly applied in recreational fisheries and sometimes in commercial fisheries to space out the landings over time and avoid the accidental overharvesting when the fish happen to be easier to catch (e.g. spawning or transiting closer to shore). They may be accompanied by a limit on the frequency of landings. Trip/bag limits have the advantage of being relatively simple to enforce at port side.
>
> **Individual fishing quotas (IFQs)** are a fishery management tool that allocates a certain portion of the TAC to individual vessels, fishermen or other eligible recipients, based on initial qualifying criteria (see Chapter 12). Used throughout the world, IFQs are used in the United States to manage Alaska halibut and sablefish, wreckfish and surf clam/ocean quahog fisheries.
>
> **Individual vessel quotas (IVQs)** are used in a number of fisheries worldwide, including some Canadian and Norwegian fisheries. IVQs are similar to IFQs, except that they divide the TAC among vessels registered in a fishery, rather than among individuals (see below).

management has its own set of theories, concepts and methods to address these issues. While the ideal ecological unit of fisheries management is a biological population, for practical reasons fisheries are more often managed on the basis of 'stocks', with administratively defined spatial boundaries. Sometimes more than one species or population is included in a stock because they are harvested together as though they were one species. In other cases, a stock may include a sub-population, and/or different species, for convenience.

Fisheries managers have a number of tools and techniques to manage a stock. These techniques can be broadly separated into input controls, an indirect form of control since they do not limit the catch, and output controls, which directly limit the catch.

Input controls are designed to limit the fishing effort and/or the efficiency of fishing. Input controls, often adopted when a fishery is first managed, include restrictions on gear types, number and size of vessels, the area fished, the time fished, the time of year or the numbers of fishers. They have been applied to both commercial and recreational fisheries. Two popular input controls are licenses and license endorsements. They may be used to certify fishermen or vessels or used as a management measure to limit the number and types of fishers or vessels participating in the fishery. The licensing system is designed to limit fishing capacity and effort, but their effect on either is indirect. For instance, if licenses do not stipulate a maximum number of days at sea, or a maximum storage capacity, the effort and capacity of the fleet can drift upwards. This problem arises when the factors regulated are not a full reflection of fishing power. Poorly considered restrictions (e.g. vessel length, or limited seasons) can lead to outcomes that are economically inefficient (e.g. too many vessels fishing too few fish), fuel inefficient and unseaworthy (e.g. too-wide vessels) and dangerous to fishermen (e.g. fishing during storms). Input controls have the advantage of being generally inexpensive to enforce.

Output controls directly limit catch and therefore a significant component of fishing-induced mortality (which also includes mortality from by-catch, ghost fishing and habitat degradation due to fishing). Output controls can be used to set catch limits for an entire fleet or fishery, such as a TAC (see Box 8.2 for this and other types of limits). They can also be used to set catch limits for specific vessels (trip limits, IVQs), owners or operators (IFQs). In this case, the sum of the catch limits for individuals or vessels equals the TAC for the entire fishery. Output controls rely on the ability to monitor total catch. This can be achieved by either measuring total landed catch with reliable landings records, port-sampling data and some estimates of discarded or unreported catch, and/or measuring the catch with (full or partial) at-sea observer coverage and verifiable logbook data.

Stock assessment

One of the most important aspects of fisheries management is the ability to describe the condition or status of a stock, also known as stock assessment. Not only do stock assessments provide information on the size, age structure and health of a stock, but they inform recommendations on the management of the stock. There are two components of stock assessment. The first is to study the biology and life history of the species, and the second is to understand the effects and impacts of fishing on the stock. For many stocks throughout the globe, there is little information available on the life history of the species, so fisheries management decisions are often made based on information obtained from the landings (catch) of fish from the stock. This is a classic dilemma in fisheries management: Since fisheries are often established without baseline information on the natural state and variability of the stock prior to fishing, management decisions must either be based on patterns in the landings of fish or must attempt to build a 'picture' of the stock prior to exploitation from which to base management decisions.

In a perfect world with perfect information, however, full and accurate stock assessments would be made using the following information:

- The number of fishers and types of gear used in the fishery (e.g. longline, trawl, seine).
- Annual catches by gear type.
- Annual effort expended by gear type.
- Geographic location of the effort of individual vessels.
- Age structure of the fish caught by gear type.
- Ratio of males to females in the catch.
- How the fish are marketed (preferred size, etc.).
- Value of fish to the different groups of fishermen.
- Timing and location of best catches.

The biological information would include:

- The age structure of the stock.
- The age at first spawning (maturity).

- Fecundity (average number of eggs each age fish can produce).
- Ratio of males to females in the stock.
- Natural mortality (the rate at which fish die from natural causes).
- Fishing mortality (the rate at which fish die from being harvested).
- Growth rate of the fish.
- Spawning behaviour (frequency, time and place).
- Habitats of recently hatched fish (larvae), of juveniles and of adults.
- Migratory habits.
- Food habits for all ages of fish in the stock.
- Habitat variables that are changeable over time (e.g. shifts in water temperature due to climate change, availability of oxygen in in river deltas due to eutrophication).

When the above information is collected by examining the landings of fishermen, it is called **fishery-dependent data**. When the information is collected by biologists through their own sampling program, it is called **fishery-independent data**. Both methods contribute valuable information to the stock assessment, but each have limitations when considered in isolation. Scientific methods of sampling (catching) fish may not reflect how fish are actually caught in the fishery and, for reasons of consistency over time, may not be adapted. Commercial data, on the other hand, often lack the thoroughness of scientific data and are restricted only to those areas that happen to be available to fishing through regulated openings, longstanding fishing habits, proximity to port and amenities.

In practice, very little of the above information is available, particularly for newer fisheries that target poorly understood species (often high seas or deep-sea species). Depending on how much information is available on the stock, there are a number of ways to manage a fishery, which include estimating populations of stocks based on the ease of catching fish, protecting fish until they can spawn, ensuring a proper age structure in the stock and developing models of population viability. The following paragraphs briefly discuss some of these methods.

The simplest stock assessments are made by calculating **catch per unit effort** (CPUE) using a combination of the history of landings (catches) for the stock and the level of effort expended in harvesting the stock. CPUE, therefore, is essentially an indicator of stock abundance. When new fisheries are established, CPUE tends to be high as catches are high while the effort required to catch the fish is low. As more fishers participate in the fishery, or technological improvements result in the ability to harvest more fish, CPUE will at first continue to increase, instilling a sense of optimism in fishermen and managers alike. It then will either stabilize in a well-managed fishery or begin to decline in a poorly managed fishery. At first glance, CPUE is relatively simple and intuitive but in practice has some serious limitations and therefore is no longer solely used as a management tool. The primary shortcoming is that the fish stock is often biologically overfished by the time that reliable estimates of CUPE trends have been generated that account for the ever-improving experience, techniques and technology that have grown up alongside the fishery. Moreover, other problems hamper the efficiency of CUPE: insufficient information on landings, inadequate information on fishing effort and insufficient understanding of the effects of technological advances that confound comparisons with past effort data.

A potentially more reliable way to assess a stock is to determine at what age(s) fish spawn and then to structure the fishery in such a way that fish are not harvested before they can recruit (reproduce). The objective is to protect fish until they are old enough to spawn. The harvest of fish before they can recruit and replace themselves is termed **recruitment overfishing** and has had serious consequences on many fisheries. Once spawning age(s) is determined, the fishery can be managed through specifications on gear type (e.g. mesh size in nets) or size limits. Nevertheless, protection against recruitment fishing does not protect a stock from being overharvested for more fish can still be removed than can recruit. Additionally, most larger, older fish are disproportionately more fecund than smaller younger fish in relation to body sizes, thus the commercially valuable big fish are also biologically valuable to the health of the stock. Fishing only large fish can also have the undesirable effort of genetically selecting for smaller, earlier spawners.

Another method to estimate stocks is to develop a mortality and **spawning potential ratio** (SPR). An SPR is the number of eggs that could be produced by an average recruit over its lifetime when the stock is fished divided by the number of eggs that could be produced by an average recruit over its lifetime when the stock is unfished. In other words, SPR compares the spawning ability of a stock in the fished condition to the stock's spawning ability in the unfished condition. SPR can also be calculated using either the biomass (weight) of the entire adult stock, the biomass of mature females in the stock, or the biomass of the eggs they produce. These measures are called **spawning stock biomass** (SSB); when they are put on a per-recruit basis, they are called **spawning stock biomass per recruit** (SSBR). SPRs are based on knowledge of the age structure of a stock collected from either estimating ages from the lengths and weights of a fish or from the examination of the ear bones ('otoliths') from fish, termed otoliths which – like trees – develop (usually, but not always, annual) growth rings that can be counted.

Current approaches and difficulties with fisheries management

Many definitions of 'fishery management' exist, but the FAO defines fishery management as:

> The integrated process of information gathering, analysis, planning, consultation, decision-making, allocation of resources and formulation and implementation, with enforcement as necessary, of regulations or rules which govern fisheries activities in order to ensure the continued productivity of the resources and the accomplishment of other fisheries objectives.
> (FAO, 1997)

The human history of fishing is relatively straightforward and tends to repeat itself. As humans begin to over-exploit a species, they may (a) exploit the same species somewhere else, (b) exploit less preferred species locally, and (c) increase local resource production through interventions (e.g. hatcheries) and aquaculture (Lotze, 2004). Thus, because fish stocks are limited, if effort is left uncontrolled, mortality will increase until the fishery becomes economically non-viable and/or the population collapses.

The previous section provided an overview of how fisheries function and the methods used to set catch limits. This section will explore approaches to how fisheries are currently managed and the consequences and limitations of these approaches.

Single-species approach to fisheries

When managed, most fisheries have historically operated under the principle of single-species management, where each target stock or population is managed independently of other targeted stocks, which may overlap in space and time (Larkin, 1977). Thus, fishery management advice under this model is provided on a stock-by-stock basis – a model under which most species continue to be managed today. Under moderate fishing pressure, the single-species approach is an acceptable management strategy; it has been demonstrated to work well for terrestrial wildlife management and, if properly applied, for the management of certain marine species (e.g. whales – see below). However, as discussed in the previous section, global fishing pressures are currently at the highest they have ever been, and narrowly focussed species-by-species management has the potential to seriously undermine the structure and function of entire marine ecosystems.

The purpose of this section is to review the single-species approach and its strengths and weaknesses and show how this approach has been improved over the past two decades to encompass communities and ecosystems.

Single-species fishery management seeks to control fishing catch in order to avoid over-exploitation of a stock and to ensure that fishers are provided with a suitable return on their investment. Without good single-species management from the onset, situations can arise where fishers progressively make less and less money, limiting managers' abilities to restrict their activities further. Meanwhile the fishery becomes biologically over-exploited through overfishing, resulting in fish being caught at a size before they have realized their full growth and spawning potential, potentially being driven to local extinction ('extirpation'). Therefore, the goal of single-species management is to invoke technical conservation measures early to prevent over-exploitation and limited management options down the road. This can be actualized by protecting a proportion of the fish (ideally across all age classes) from the outset or making the fishery sufficiently 'inefficient' that the zero profit level is reached before the stock is over-exploited (Cochrane, 2002).

The fundamental objective of traditional single-species management is to manage a fishery to **maximum sustainable yield** (MSY), which assumes that every fish stock generates 'surplus production' and that this production can be safely reduced (by fishing) to a biomass necessary in order to maintain the stock at some sustainable level. Since the early 20th Century, proponents of MSY have argued that harvesting less than MSY was an inefficient use of fish and harvesting more than MSY was an inefficient use of effort. For these reasons, MSY was adopted as the primary fisheries' management tool from the 1940s until the 1970s. In addition to MSY, many fisheries managers believed the stocks could not become extinct as the cost of harvesting the few remaining fish would be such that fisheries would shift to other, more readily available species. Economic theory, however, did not reflect reality as fishers often did continue to pursue weakened stocks to ecological and economic extinction. Furthermore, overly simplistic fisheries models, limited by computing power and data, often did not account for minimum density of biomass required for successful spawning ('allee effect'), nor for the geographic disparities hidden within statistics summarized across fishing areas. Although overall numbers may have appeared to support MSY, local overfishing could still cause localized collapses, triggering a domino effect of displaced fishing effort and further localized declines.

Traditional applications of MSY usually result in biomass reductions of the target stock of 30–50 per cent of unfished levels (Mace, 2001, 2004). Given that managing multiple stocks to MSY in the same geographic (food web) area will inevitably affect food web structure, fisheries managers have begun to realize that MSY is an upper limit that should not be exceeded, rather than a target to be aimed for. Hence, they have begun stipulating a fishing mortality less than MSY, often re-expressed as 'F_{MSY}'.

Regardless of the improvements to the MSY concept since its introduction in the 1930s, the following assumptions underpin single-species management (summarized by Babcock and Pikitch, 2004):

- The objective of management is to maximize the long-term average yields of the fishery.
- A population biomass level exists that will maximize the long-term average yield.
- Fish growth, natural mortality and fecundity are constant and do not change over time, irrespective of the abundance of other species, environmental changes or the effects of fishing.
- The total fishing mortality can be controlled by regulation of the fishery.

With the failure of conventional MSY calculations as a fisheries management tool, the concept of optimum sustained yield (OSY) was developed. OSY is essentially MSY modified to reflect relevant economic, ecological or social considerations and is more of a management goal than an empirically derived number. Although OSY offers flexibility in meeting multiple management considerations, the vagueness of how to calculate OSY has led to difficulties and debates.

A further improvement on the single-species approach is **multispecies fishery management**, where decisions are made in consideration of other target species under harvest in the same area and may include food web ('trophic') considerations in management decisions.

Current fishery management practices in most countries and international agencies are based on the single-species approach and generally follow the model below (adapted from Hilborn, 2004):

- Single-species stock assessments are undertaken for each stock or stock complex to set maximum sustained yield (MSY) or some other maximum threshold such as F_{MSY}.
- Regulations that determine allowable time, area, gear and catch limits are recommended by fishery managers. These are ultimately established through a political process.
- A centralized management agency responsible for science, decision-making and enforcement is responsible for operating the fishery.
- Stakeholders are involved in decision-making, either through policy, legal or political means.

In reality, the traditional assumptions that underpin the single-species management model have resulted in the following characteristics (adapted from Pavlikakis and Tsihrintzis, 2000; Marasco et al., 2007):

- A disproportionate focus on short-term economic objectives and the maintenance of a single species.
- Removal of humans (and other species) as ecosystem components from fishery models and decision-making.
- Political, economic and social values being either discounted or ignored.
- Needs of commercial fishing stakeholders taking precedence over the broader public interest and other sectors.
- Science and management occurring at local scales, where scaling up to regional, national or international interests is difficult or unwarranted.
- A focus on single-species population ecology rather than community ecology in conjunction with oceanography, fisheries economics, broader ecology and other species' biology.
- Modern tools, such as geographic information systems (allowing for spatial analyses), oceanographic models (e.g. inter-annual variability) and economic techniques (e.g. behavioural models and forecasting), not being applied.

Single-species fishery management is primarily based on developing stock assessment models to establish reference points that define overfishing limits based on (spawning) stock biomass and fishing mortality. When these limits are exceeded, protocols are triggered to reduce fishing, either through input controls (e.g. number and size of vessels, time allotted to fish, gear restrictions) or output controls (e.g. amount of fish that can be caught).

Properly applied, the single-species approach is a viable model for the management of certain fisheries and can provide significant economic and ecological value to the sustainable management of a fishery. Single-species approaches can work where co-existing fisheries have minimal economic and ecological impact on one another, stock status can be determined, life histories are well known, the effects of environment changes on stocks are understood and compliance with regulations can be enforced. Notable examples of species that can be managed under this approach include certain marine mammals, such as those covered by the International Whaling Commission's Revised Management Procedure (see below), and nearshore shellfish stocks where stock status and environmental conditions can be easily determined. In such successful situations, flexibility is still required to quickly adapt to unanticipated variables such as temperature shifts, disease outbreaks, harmful algal blooms and natural disasters.

Most fisheries, however, do not meet the conditions for being managed under the single-species approach. In certain circumstances, scientifically well-conducted single-species management approaches can still lead to over-exploitation of a stock or economic failure of the fishery because of external interactions. That said, failures of the approach have generally not been the fault of the science and management but rather due to data limitations, poor senior decision-making under economic and political pressures and failures of political will. For example, a review of the scientific advice provided to European decision-makers on 18 fish stocks in 2002 found that scientists provided the correct advice 53 per cent of the time, provided the wrong advice resulting in a detriment to the fishery 23 per cent of the time and 24 per cent of the advice consisted of 'false alarms' where catch reductions were recommended but later found to be unnecessary (Frid et al., 2005). Thus, scientific advice on single-stocks was either correct or neutral to the health of the stock 77 per cent of the time. More recently, a review of the advice provided by the International Commission for the Conservation of Atlantic Tunas (ICCAT) found that, since 1970, decision-makers chose to take no action on advice 40.4% of the time, ignored the advice 20.6% of the time and followed advice just 39.0% of the time (Galland et al., 2018).

Perhaps the most contentious aspect of the single-species approach is the reliance on MSY or MSY variants in order to set fishery harvests. Many jurisdictions continue to use stock assessment models to attempt to determine MSY or the rate of fishing morality F_{MSY} that a stock can sustain. As discussed earlier in this chapter, over 30 per cent of the world's fisheries are overfished, suggesting that the current approach is failing. This failure is a result of either how MSY is calculated, uncertainty in the knowledge of how marine ecosystems function or political interference that sets allocations above MSY. Many stock assessment models used to calculate MSY result in typical biomass reductions of 50–70 per cent below unfished levels, which has been demonstrated to be biologically unfeasible for slow growing, late maturing species such as sharks, rays and deep-sea species (Hirshfield, 2005; Norse et al., 2012). As such, setting catches below MSY results in larger stock sizes that should, in turn, cause fewer stock collapses and more favourable outcomes (Hilborn, 2004). In addition, traditional application of the MSY concept has led to presumptions that all mature female fish are of equal importance. In actuality, older females have a disproportionate spawning potential, including healthier eggs, and are critical to maintaining population viability. The MSY concept, moreover, ignores ecosystem effects of high harvest rates on fast growing, high biomass species, where populations appear to be resilient to high levels of fishing but are important sources of food for other species as well (Hirshfield, 2005).

Much of the difficulty in managing to MSY is related to uncertainty in single-species stock assessment models, which can be sufficient reason for decision-makers to ignore these models. In some cases, incorrect assessments leading to rapid fishery declines has jolted confidence in the models. Another difficulty with the models is that they have been weak in providing assistance to governments attempting to design regulatory systems to achieve fishery targets (Pauly et al., 2002). Uncertainty has been further exacerbated by funding limitations: Models have lacked input data, scientific analyses and applications across spatial and temporal scales.

There are a number of recent and proposed improvements to the single-species models, some of which are now enshrined in national and international law. Most suggested improvements fall under the guise of establishing marine protected areas (MPAs) or applying ecosystem approaches (see Chapter 12) and improving fishery management regulations. While useful, they are generally used in specific circumstances and have limited scope to address the broader challenges of fisheries management. A broader consideration of fundamental improvements to the approach fall under the following types (adapted from Hilborn, 2004; Symes, 2007):

- Eliminating the approximately $US 30 billion in annual subsidies for fishing fleets.
- Reducing target fishing moralities through adopting a more precautionary approach to setting harvest levels.
- Establishing no fishing areas over a significant (e.g. 30 per cent) area of the world ocean either through MPAs or other area-based management tools.
- Eliminating destructive fishing practices, including prohibiting bottom trawling of previously untrawled or lightly trawled areas and reducing by-catch.
- Evolving the current command and control fisheries management model to a system of co-management where responsibilities are shared between stakeholders.
- Establishing new forms of marine tenure where individuals or groups of fishers are guaranteed a specific share of a future catch, thus removing incentives to over-capitalize, overfish in the short term and better facilitating alignment of economic interests with long-term conservation goals.
- Establishing more incentive-based (versus regulatory-based) management approaches.

Additional improvements to the single-species approach may include consideration of the species of interest within the broader ecosystem context. Examples of recent efforts towards these ends include: modifying stock assessment models to account for density dependence for the target species that arises from predation from another species (community effects); considering time-dependent disease and predation effects when modelling natural mortality; incorporating perceived regime shifts in defining biological reference points; and incorporating considerations for low productivity or endangered species that may be affected by harvest of the target species in stock models (Marasco et al., 2007).

In conclusion, improvements on the single-species approach – including more precautionary re-expressions of MSY that also account for uncertainty (e.g. recruitment failure) in the dynamics of stocks and populations – is a step in the right direction, but this has not yet in practice been proven to result in more sustainable fisheries management outcomes. This is most likely due to political interference and illegal fishing. Similarly, the multispecies approach has also shown little promise that it is a sustainable tool for fisheries management. Since 1986, the United States Northeast multi-species groundfish management plan has consisted of 24 target species managed collectively. A 20-year review of the plan concluded that indicators of population and community health (e.g. spawning stock biomass) were not improving as projected and that illegal harvest was responsible for 12–24 per cent of total harvest (King and Sutinen, 2010). There is no shortage of national and international fishery conservation laws, conventions, agreements and policies that should be sufficient to control overfishing and ensure that community and ecosystem considerations are incorporated into decision-making (see below). Furthermore, without accepted rules about how the (imperfect) models are interpreted and applied, they will remain subjected to the political whims of governments, which are rarely willing to fundamentally reform fisheries policies (Symes, 2007).

Guerry (2005) eloquently summarized the reasons for the failure of traditional species-based approaches as follows: fragmented ocean governance where the fishery as a commons resource has led to jurisdictions competing for the resources; inability to maintain ecosystem elements, such as water quality or spawning habitat necessary to sustain successful fisheries; inability to manage diverse, non-fisheries related impacts, including pollution, habitat loss, overharvesting, climate change and introduced species; and lack of recognition of connections between ecosystem structure, function and services.

The ecosystem approach to fisheries management

The ecosystem approach to fisheries management (EAFM) stems from the realization that single- (and multi-) species approaches to fisheries management have, regardless of the reasons, resulted in a situation where most of the world's fish stocks are fully exploited or overfished (FAO, 2018). The term 'EAFM' is used here as the successor to earlier terms, including the ecosystem approach to fisheries (EAF) and ecosystem-based fisheries management (EBFM). This section will first explore the EAF concept as it emerged out of the challenges with multispecies fisheries management followed by the EBFM, which invokes human well-being and good governance on equal footing as ecological well-being.

The evolution of fisheries management from single species to EBFM has taken over 50 years (Figure 8.1) and is the result of many international agreements discussed later in this chapter.

The concept of EAF is fundamentally a re-expression of the principles of sustainable development, which have been enshrined in international conventions since the 1972 UNCHE (Chapter 6). General concepts and definitions of the EAF converge around the need either to alter existing practices or to develop a new fishery management paradigm that explicitly recognizes the interrelationships and dependences within food webs and human activities, as well as the need for effective relationships between science and policy making. The EAF is predicated on the assumption that an improved understanding and management of stock interactions, stock – prey relationships, and stock-habitat requirements will result in more accurate fishery assessment models (Christie et al., 2007). Activities and concepts that roll up into the EAF include: establishing a sustainable fishery resource for future generations; inclusion of humans in ecosystems; emphasizing ecosystem sustainability over ecosystem products; understanding the dynamic nature of marine systems; establishing clear management goals and objectives; precautionary; and adaptive management.

The overall goal of the EAF is to achieve ecologically sound resource conservation that is responsive to the reality of ecosystem processes. The EAF objectives typically include (modified from Pikitch et al., 2004):

- Avoiding degradation of, and potentially restoring, ecosystems as measured by indicators of environmental quality.
- Maintaining ecosystem structure, process and function at the community and ecosystem level.
- Obtaining and maintaining long-term socio-economic benefits without compromising the long-term sustainability of the ecosystem.
- Generating knowledge of ecosystem processes sufficient to understand the likely consequences of human actions and changing conditions.

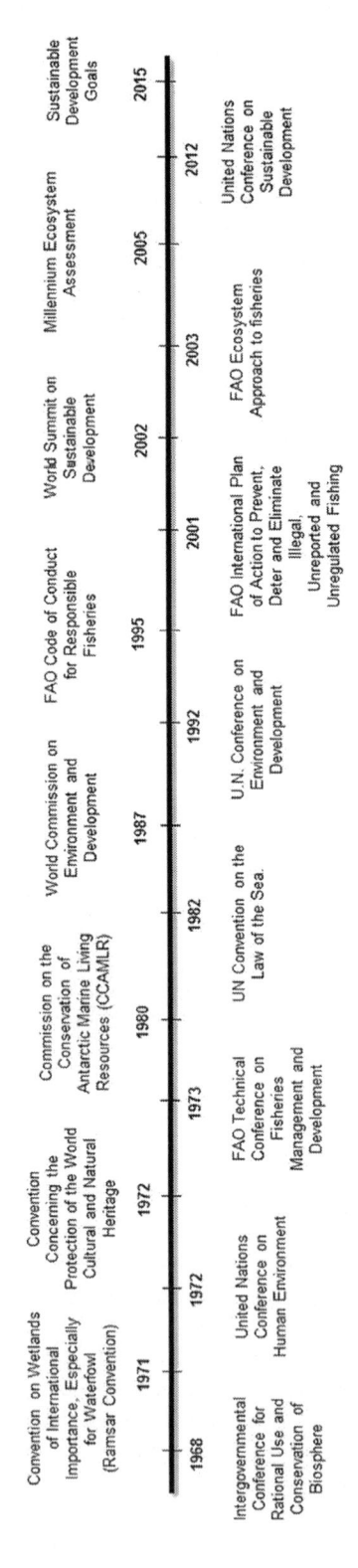

Figure 8.1 International initiatives that contributed to the ecosystem approach to fisheries management

Source: Reproduced with permission from P. Ramirez-Monsalve et al. (2016)

Elements of the EAF are summarized as follows (adapted from Marasco et al., 2007; Sissenwine and Murawski, 2004):

- Ensuring that broader societal goals are taken into account.
- Employing geographic (spatial) representation.
- Recognizing the importance of climatic-oceanic conditions.
- Emphasizing food web interactions and pursuing ecosystem modelling and research.
- Incorporating improved habitat information (for target and non-target species).
- Expanding monitoring and ecosystem assessments.
- Acknowledging and responding to higher levels of uncertainty.
- Employing adaptive management.
- Accounting for ecosystem knowledge and uncertainties.
- Considering multiple external influences.
- Avoiding decisions that can cause irreversible (long-term) harmful effects on ecosystems and human communities.

Perspectives on the EAF range from the EAF being simply an incremental extension of current fisheries management approaches to a complete redesign of marine management (Pitcher et al., 2009). Regardless of which perspective is applied, the EAF necessitates that short-term socio-economic benefits will be reduced due to decreased harvests. This is due to the application of the precautionary principle and consideration of the long-term ecosystem effects of fishing. Furthermore, uncertainty cannot be used as an excuse to maintain the status quo; rather, it should be recognized as a reason to be more precautionary.

There have been many criticisms of the EAF. Perhaps the most significant issue with the EAF is that, given humanity's dismal record with single-species management, the likelihood of successfully implementing the EAF is low, especially given that many of the same fishery scientists, politicians and stakeholders responsible for single-species declines are now in charge of implementing the EAF. Compounding this problem is the lack of a general theory of the functioning of marine ecosystems, which limits the ability to explain and predict even simplified single-species impacts on marine systems. As such, the expectation that the EAF will be able to cope with the increased uncertainty due to the additional need to model community and ecosystem considerations will severely limit the usefulness of the tool (Cury et al., 2005; Valdermarsen and Suuronen, 2003). Others argue that improvements in single-species approaches, such as the application of F_{MSY}, negate the need to implement EAF. While these criticisms are valid, many of the primary issues to be addressed under EAF, including by-catch, indirect effects of harvesting and interactions between biological and physical components of ecosystems, are tractable and in some cases are already being addressed by conventional management (Sissenwine and Murawski, 2004).

With a few exceptions that will be discussed below, the EAF to date has consisted of further application of marine protected areas, limits on destructive fishing practices and efforts to reduce by-catch (Shelton, 2009). The EAF, being relatively new, has not yet realized its full potential primarily because the sustainability objectives have not yet been clearly articulated for most fisheries in a way that can be solved through models or equations. The MSY (or F_{MSY}) concept has not been quantitatively translated into the context of ecosystem objectives, and therefore setting EAF-informed MSYs has not been undertaken for most fisheries. In addition, the realization that the EAF, if applied properly and with a precautionary focus, will invariably lead to reduced fishery targets has led to bureaucratic inertia by governments and stakeholder groups unwilling to accept more precautionary catch allocations.

The most comprehensive application of the EAF involves the integration of the target stock with community dynamics and environmental variables before determining catch limits and management measures. While this approach requires a level of inventory and scientific understanding that may not currently exist, it does attempt to identify the surplus production required to satisfy ecosystem needs (Goodman et al., 2002). Additional challenges of this application of the EAF are addressing the high levels of uncertainty as a result of integrating biological and oceanographic processes into stock assessment models. These are likely to have limited predictive power, given most currently available data and understanding of marine ecosystem function (Jennings and Revill, 2007). For example, Georges Bank, off the east coast of North America, is one of the world's most studied ecosystems, yet the proper application of the EAF has been hampered by a lack of sufficient data (Froese et al., 2008).

Regardless of what EAF entails for a particular application, it faces a number of implementation challenges. Among the things that the EAF will need to develop are long-term ecosystem-related objectives, meaningful indicators that clearly demonstrate when a threshold has been triggered or a limit exceeded, and a stronger scientific foundation to inform EAF decisions (Cury et al., 2005).

The science gaps to fully operationalize the EAF can be summarized as follows (modified from Frid et al., 2006):

- Understanding the relationships between hydrographic regimes and fish stock dynamics.
- Understanding and inventorying habitat distributions.
- Establishing the desired fisheries role for marine protected areas and incorporating that into their design.
- Understanding ecological dependencies within marine communities.
- Developing predictive capabilities in complex systems (choosing correct scales).
- Incorporating uncertainty into management advice and direction.
- Understanding the genetics of target and non-target stocks, populations and species.
- Understanding the response of fishers to management measures.

Regardless of how the EAF is used, a number of management options have been identified, many of which are currently in practice and discussed elsewhere in this book (modified from Pikitch et al., 2004):

- Marine habitat delineation and protection at multiple spatial and temporal scales.
- Identification and protection of essential fish habitat.
- Marine protected areas.
- Marine spatial planning and zoning.
- Reducing impacts of fisheries on endangered species.
- The reduction of by-catch.
- Managing several target species.
- Managing forage species critical for fisheries and other animals.
- Precautionary approach.
- Adaptive management.
- New analytical models and management tools.
- Integrated management plans with other fisheries and other sectors.

The concept of EAFM builds on prior efforts to implement the EAF with an increased focus on the governance and oversight aspects of fisheries management. Informed by earlier attempts to implement EAF and in light of the experiences with EAF implementation, EAFM attempts to manage fisheries by finding a balance between ecological and human well-being through good governance. As such, EAFM is foremost a policy-driven process that can only successfully operate if informed by science and stakeholders both. Furthermore, EAFM approaches build management strategies for entire ecological and social systems rather than for individual components.

The seven principles of EAFM, as set out in the FAO Code of Conduct for Responsible Fisheries, include:

1. Good governance.
2. Appropriate scale.
3. Increased participation.
4. Multiple objectives.
5. Cooperation and coordination.
6. Adaptive management.
7. Precautionary approach.

The management benefits of an EAFM include (modified from FAO (2014)):

- Consideration of links between ecosystems and fisheries.
- More effective resource use planning.
- Identifying and negotiating trade-offs between different stakeholder priorities, balancing human and ecological needs.
- Increased stakeholder participation.
- Balancing fish production with conservation of biodiversity and habitat protection.
- Resolving or reducing conflicts between stakeholders.
- Recognition of cultural and traditional values in decision-making.
- Enabling of larger-scale, longer-term issues to be recognized and incorporated into fisheries and coastal resource management.

New Zealand's approach to fisheries management closely aligns to the principles of the EAFM. The state's journey started with the adoption of an ITQ system in 1986 followed by the *Fisheries Act 1996* and substantial incremental improvements since this time. While New Zealand has not formally adopted the EAFM, 73 per cent of New Zealand fisheries were above their management targets, defined as a biomass level or fishing mortality rate (Cryer et al., 2016).

While EAFM approaches have not been applied long enough to sufficiently assess their performance, a number of studies review the performance of EAFM given the longer application of this approach to real-world management decisions. Pitcher et al. (2009) evaluated the performance of the EAFM using a number of criteria. Only two states were rated as 'good' (Norway, USA) while four countries (Iceland, South Africa, Canada, Australia) were rated as 'acceptable', leaving over half of the countries with failing grades.

However, there are a number of EAFM successes, including Alaskan groundfish management and the Convention for the Conservation of Antarctic Marine Living Resources (CCAMLR – Chapter 9). CCAMLR has demonstrated that the EAFM and precautionary approach can be applied in high seas environments. Under the treaty, the scientific committee has developed a number of innovative methods to manage prey species in order to protect dependant predators, limit by-catch and develop precautionary-based protocols prior to the exploitation of new fisheries. Advice from the CCAMLR scientific committee is almost always followed, and the quality of science is high (Constable, 2004). Continuing challenges to implement the EAFM

Table 8.1 Traditional fisheries management compared with EAFM

	Existing approaches	*EAFM*
Species considered	Mainly target species.	Key species in the ecosystem, particularly those impacted by fishing.
Management objectives	Relate mainly to target species and conventionally focussed on biological objectives for maximizing sustainable yield.	Multiple objectives covering the fisheries, ecosystem goods and services and socio-economic considerations.
Scale	Addresses fisheries management issues at the stock/fishery scale.	Addresses the key issues at the appropriate spatial and temporal scales. These are often nested (local, national, sub-regional, regional, global)
Data and information used	Mainly scientific data focussing on target species.	Broader knowledge base (both scientific and traditional) that emphasizes learning by doing (adaptive management).
Assessment methods	Largely stock assessment for key target species.	Multispecies and ecosystem assessments through indicators.
Management intervention	Mainly control of fishing.	Broad-based incentives (including ecosystem tools such as Marine Protected Areas (MPAs)). Links with Integrated Coastal Zone Management (ICM) and sustainable economic development more broadly.
Planning	Usually in the form of a Fisheries Management Plan that considers target species.	The EAFM plan that considers the fishery, ecosystem and human systems and governance.
Stakeholders	Fishers, fishing industry/communities.	Broader stakeholders: people affected by or who affect EAF management
Sectors	Sectoral, i.e. focuses mainly on fisheries sector issues.	Deals more explicitly with the interactions of the fishery sector with other sectors, e.g. coastal development, tourism, aquaculture, navigation, offshore petroleum, renewable energy industries.
Policy and decision-making	Largely at the government level. Addresses mainly corporate (fisheries sector) interests.	Participatory with major stakeholders. Addresses the interests and aspirations of a broader stakeholder community.
Participation	Top-down (command and control) approaches typify conventional fisheries management.	Participatory approaches, e.g. various forms of co-management are a key feature of EAFM.
Compliance and enforcement	Operates through regulations and penalties for non-compliance.	Encourages compliance with regulations through incentives (as well as penalties).

Source: Modified from www.fao.org/3/a-i3778e.pdf

under the treaty exist, ranging from gaining consensus decisions across a diversity of states, data limitations, controlling fishing effort and incorporating climate change into ecosystem and stock assessments.

In conclusion, the EAF and the EAFM concepts have been increasingly part of the suite of fishery management tools for nearly two decades. In particular, they have been successfully applied to ecosystems and human communities with certain characteristics. Those that are generally bottom-up structured and those where their EEZs fall within developed nations (CCAMLR excepted) have fared well. Success has also come where there is genuine stakeholder and political will to move away from traditional, single-species management and towards a more holistic approach to ocean conservation. The degree and speed of uptake of the EAF/EAFM will depend on the willingness of fishing industries and communities to bear the (sometimes high) short-term costs associated with reducing fishing effort and the willingness of governments to buffer and manage the transition moving towards sustainability.

History of international fisheries law and policy

International efforts to regulate fishing are not new: France and Britain's first collective attempt to manage fisheries in the North Sea was in 1839. This was followed by the multi-lateral 1881 North Sea Fisheries Conference focussed on the orderly exploitation of cod, haddock and plaice stocks. While overfishing was understood to be occurring at the time of the Conference, it was not formally part of the Conference. Instead, the Conference addressed dispute resolution and avoiding gear conflicts.

The International Council for the Exploration of the Sea (ICES) was established in 1902; by 1934, size limits, closed areas to assist fisheries and mesh sizes had been recommended.

The first conservation-based international agreement was the North Pacific Fur Seal Convention (signed 7 July 1911 and entered into force on 14 December 1911). In this Convention, Great Britain (for Canada), Japan, Russia and the United States agreed to halt the pelagic (sea-based) harvesting of Northern fur seals and sea otters (Table 8.2).

In 1937, in recognition of the rapid depletion of European inshore fisheries since World War I, the British government convened the first of three conferences on overfishing. The conference to implement the 1934 ICES recommendations was held, but due to World War II, the convention was never brought into force. The second convention in 1943 also failed to reach consensus on any issues. However, in 1946, the London International Overfishing Conference was convened and attended by 12 European nations to address overfishing in the North Sea and waters adjacent to the British Isles. The resulting Convention for the Regulation of Meshes of Fishing Nets and the Size Limits of Fish (the 'Overfishing Convention') applied only to the Northeast Atlantic and did not enter into force until 1953. By this time, however, the Overfishing Convention was inadequate to address emerging fisheries issues (e.g. herring), and the parties (now numbering 14) to the Convention signed the Northeast Atlantic Fisheries Convention in 1959 which, in turn, established the North East Atlantic Fisheries Commission (NEAFC), a body extant today (see below).

A companion convention to the Northwest Atlantic, the International Convention for the Northwest Atlantic Fisheries was enacted in 1950. This established the International Commission for the Northwest Atlantic Fisheries (ICNAF) with the authority under Article VIII to prescribe 'an overall catch limit for any species of fish'.

The widespread introduction in the mid-seventies of exclusive economic zones (EEZs) and the adoption of the LOSC in 1982 provided a new framework for the better management of marine resources. More recently, advances in fishing technology (e.g. cold storage, GPS, sonar) have opened up the previously inaccessible 60 per cent of the oceans that are outside national EEZs. The high seas' contribution to overall marine catches increased from 9 per cent in the

1950s to 15 per cent today. This increase necessitated the establishment of regional fisheries management organizations that are discussed below (Cullis-Suzuki and Pauly, 2010). Furthermore, the ability to access the high seas has resulted in the growth of **illegal, unreported and unregulated** (IUU) fishing (Sumaila et al., 2006). It is estimated that IUU fishing may comprise up to 30 per cent of the global fisheries catch (Borit and Olsen, 2012).

Organizations that oversee international fisheries management

The Food and Agriculture Organization of the United Nations

In 1945, the United Nations established the FAO as a specialized agency to improve agricultural productivity. Better productivity would ensure the world's population would have access to sufficient and nutritious food, improve rural quality of life and contribute to the world's economy. Currently, the FAO is headquartered in Rome and has 191 member states, two associate members and one member organization (the European Union). The FAO has staff located in over 130 states and is governed by the FAO Council, which oversees the activities of the FAO and sets programs and budgets.

The FAO's Fisheries and Aquaculture department is one of seven FAO departments. It was established to improve the conservation and utilization of the world's fisheries in inland and marine waters. The department is broadly responsible for the following activities:

- Reporting on the status of the world's wild fisheries and aquaculture operations.
- Collecting, analysing and summarizing environmental and socio-economic information on the world's wild fisheries and aquaculture operations.
- Supporting national and regional (multi-national) fisheries management organizations through the provision of scientific, policy, technical and legal advice.

The Fisheries and Aquaculture department is overseen by the Committee on Fisheries (COFI), which is a subsidiary body of the FAO Council. The COFI reviews and approves the fisheries and aquaculture-related work of the FAO and makes recommendations on pertinent fisheries and aquaculture issues. The COFI also provides a venue for the negotiation of international agreements.

Key long-term fisheries and aquaculture related FAO activities include:

- Implementing the Code of Conduct for Responsible Fisheries (see below).
- Implementing the various International Plans of Action on the management of sharks, incidental seabird catches, fishing overcapacity and IUU fishing (see below).
- Administering the Global Partnerships for Responsible Fisheries (FishCode) to improve the socio-economic and nutritional status of developing nation populations.
- Administering the Species Identification and Data Programme (FAO FishFinder) to improve fish identification.
- Administering GLOBEFISH, which provides analysis and information on the international fish trade and promoted market access, including managing the FAO Fishery Country Profiles and FAO Fishing Areas (Figures 8.2 and 8.3).
- Developing the Global Record of Fishing Vessels, Refrigerated Transport Vessels and Supply Vessels (Global Record) in order to provide a single database of global fishing vessels to deter and eliminate IUU fishing.

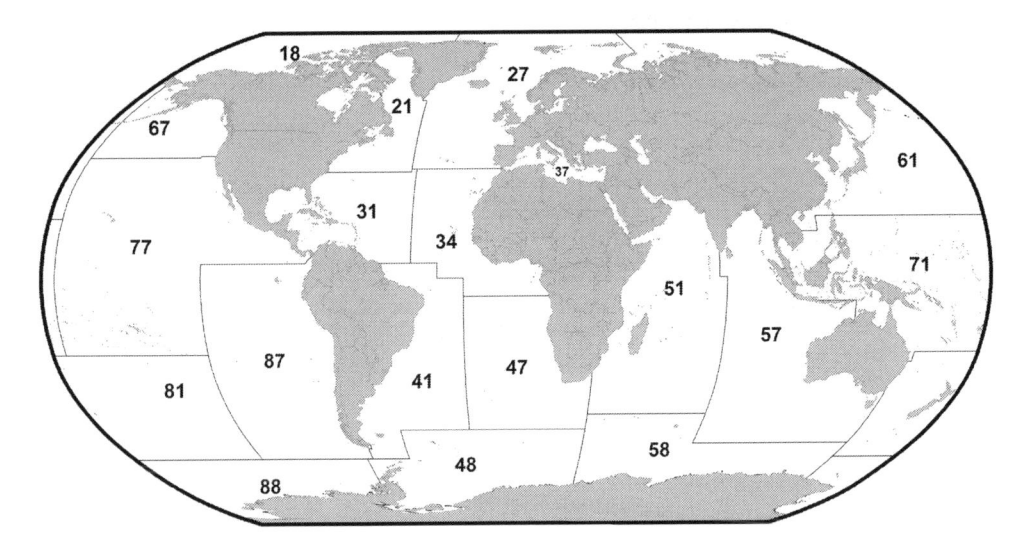

Figure 8.2 The Food and Agriculture Organization of the United Nations Major Fishing Areas. Fishing Area names are as follows:

Atlantic Ocean and adjacent seas
18 Arctic Sea
21 Atlantic, Northwest
27 Atlantic, Northeast
31 Atlantic, Western Central
24 Atlantic, Eastern Central
37 Mediterranean and Black Sea
41 Atlantic, Southwest
47 Atlantic, Southeast
Indian Ocean
51 Indian Ocean
57 Indian Ocean

Pacific Ocean
61 Pacific, Northwest
67 Pacific, Northeast
71 Pacific, Western Central
77 Pacific, Eastern Central
81 Pacific, Southwest
87 Pacific, Southeast
Southern Ocean
48 Atlantic, Antarctic
58 Indian Ocean, Antarctic
88 Pacific, Antarctic

Source: www.fao.org/fishery/area/search/en

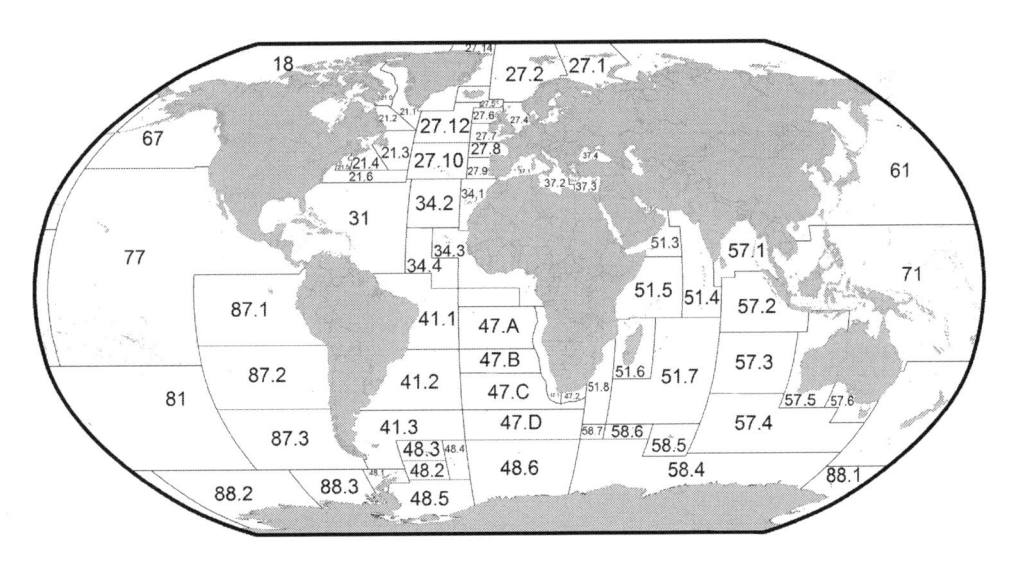

Figure 8.3 The Food and Agriculture Organization of the United Nations Major Fishing Subareas

Source: www.fao.org/fishery/area/search/en

Regional fisheries bodies and regional fisheries management organizations

Globally applicable agreements are important for setting the overall direction for fisheries policy with respect to access, conservation, equity and the beneficial use of fisheries. However, given the diversity of marine ecosystems and the states that depend on them, global agreements have limited use in establishing management regimes for individual regions.

This limitation was recognized during the LOSC negotiations and, while the freedom of fishing is guaranteed to all nations (including landlocked as per Articles 87 and 116), these rights are constrained by both the need to conserve resources on the high seas (Articles 117–120) and the requirement to establish regional or sub-regional organizations to manage stocks found in the EEZs of two or more coastal states (Articles 63–67). Furthermore, the IJC concluded in the 1974 *Fisheries Jurisdiction Case* that the freedom to fish doctrine was 'replaced by a recognition of a duty to have due regard of other states and the needs of conservation for the benefit of all'. Subsequent agreements, most notably Articles 8–13 of the 1995 UN Fish Stocks Agreement (discussed below), direct signatories to establish **regional fisheries management organizations** (RFMOs) as the means to achieve fisheries objectives outlined in the LOSC, UNCED and UNCBD (Chapter 4).

While treaties on the management of 'fisheries' (including marine mammals) date back to 1911, the International Pacific Halibut Commission (IPHC – see Table 3.2), established in 1923, was the first true regional fisheries body. This was followed in the 1950s with the establishment of a number of tuna-based regional fisheries bodies (Table 8.2) to provide scientific advice and management. While regional fisheries bodies established before the 1995 UN Fish Stocks Agreement may have had both an advisory and management mandate, more recently 'regional fisheries bodies' are assumed to have an advisory mandate. RFMOs, meanwhile, set fisheries quotas binding state parties to the agreements (Drankier, 2012).

RFMOs are legally mandated fisheries management bodies operating in areas beyond national jurisdiction (ABNJ) as well as sometimes within national jurisdictions depending on the nature of the agreement (Figure 8.4). RFMOs may include 'regional economic integration organizations'

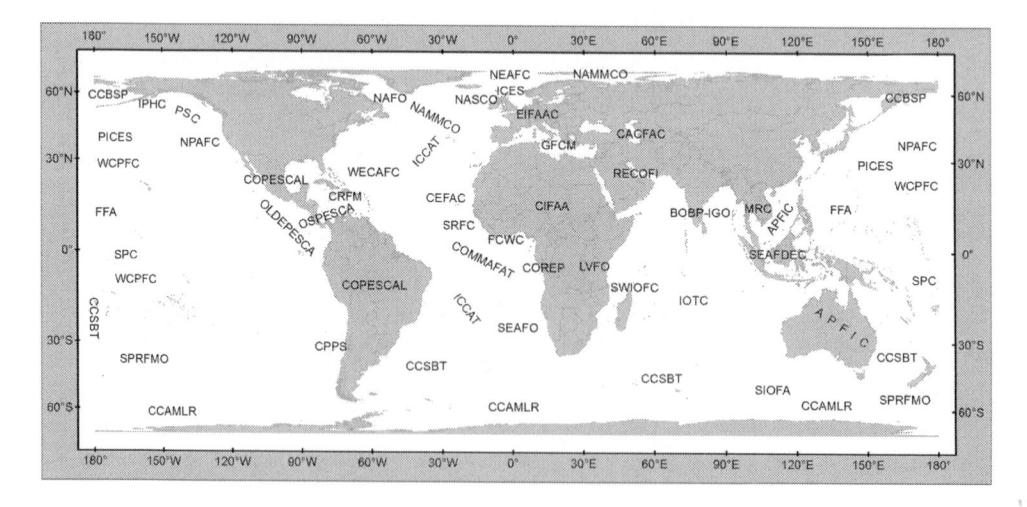

Figure 8.4 Regional Fisheries Management Organizations. Acronyms are listed in Table 8.1

Source: www.fao.org/fishery/topic/14908/en

such as the European Union and semi-sovereign states such as Taiwan. Besides fisheries management, RFMOs gather and exchange information, address broader resource management issues of significant import and nurture scientific cooperation. RFMOs also provide a forum for international cooperation on the management of straddling fish stocks (see Box 8.3).

Most RFMOs follow the current organizational model of international agreements and have a decision-making body and secretariat. The decision-making body may be a Commission, Committee, Council, Meeting of Ministers or any other forum selected by signatories. RFMOs lack supranational powers, meaning that states have to agree to be bound by RFMO decisions. This is one reason why some question whether RFMOs are meeting their conservation mandates (Cullis-Suzuki and Pauly, 2010). Certain RFMOs, particularly those established to manage tuna fisheries, have been criticized for ignoring scientific advice and setting unsustainable harvest levels (McCarthy, 2013).

Box 8.3 The case of deep-sea fisheries

While there is no agreed-upon definition of 'deep sea' fisheries, various RMFOs define them as starting at 400–500 metres depth beyond marginal seas and continental shelves and comprising various habitat types (e.g. seamounts, fjords, slopes) that support these fisheries. Most deep-sea fisheries occur in depths of less than 1,500 metres, although a few are exploited to depths of 2,000 metres or more. Commercially viable deep-sea species include sharks, crabs and finfish. Deep-sea fisheries occur both within and beyond national jurisdictions where they may not be as well researched or managed. Furthermore, deep-sea species are generally slow growing, long-lived and reproduce intermittently, making them vulnerable to large-scale fishing practices (Norse et al., 2012).

The management of ABNJ deep-sea fisheries falls mainly to RFMOs with overarching policies and guidance provided by the UN General Assembly and FAO. The LOSC (see Chapter 4) articles 87.2, 117, 118, 119 and 194.5 provide some guidance to deep-sea fisheries. In addition, a number of UN and FAO agreements (discussed below) address fishing in ABNJ, including the 2008 FAO International Guidelines for the Management of Deep-sea Fisheries in the High Seas that arose out of the 2006 UN General Assembly resolution 61/105 (later augmented by the 2009 UNGA resolution 64/72). The FAO Guidelines are voluntary and, while a binding treaty on deep-sea fisheries has been proposed, these discussions are in their infancy. It should be noted that on 15 June 2015, the UN General Assembly resolved that a binding legal instrument be developed to protect marine biodiversity in ABNJ; however, the impact, if any, of this new instrument currently under negotiation on deep-sea fisheries remains unclear. There have been a series of recurring UN General Assembly resolutions to implement the FAO guidelines on vulnerable marine ecosystems, complete impact assessments and to utilize best available science in making conservation and management decisions.

Of the 40 plus RFMOs, 8 have competencies to manage deep-sea fisheries and whose decisions are binding. The future of deep-sea fishery management likely rests on the continued work of RFMOs, further supported by the new multilateral treaty to protect biodiversity in ABNJ.

Source: Adapted from Oanta (2018)

Table 8.2 List of current marine regional fisheries management organizations (RFMOs) grouped by purpose and listed by year of establishment

Name by type	Year of establishment	Coverage of fish stocks	Membership	Decisions binding on parties
Regional management				
International Pacific Halibut Commission (IPHC)	1923	Halibut	Closed	Yes
Asia-Pacific Fisheries Commission (APFIC)	1948	All	Open	No
General Fisheries Council for the Mediterranean (GFCM)	1949	All	Open	Yes
Inter-American Tropical Tuna Commission (I-ATTC)	1949	Tuna	Closed	No
Permanent Commission for the South Pacific (CPPS)	1952	All	Closed	Yes
International Commission for the Conservation of Atlantic Tunas (ICCAT)	1966	Tuna	Open	Yes
International Baltic Sea Fisheries Commission (IBSFC)	1973	All	Closed	Yes
Joint Technical Commission of the Maritime Front (CTMFM)	1973	All	Closed	Yes
Joint Norwegian-Russian Fisheries Commission (JointFish)	1974	Cod, haddock, capelin, king crab	Closed	Yes
Northwest Atlantic Fisheries Organization (NAFO)	1978	All	Closed	Yes
North east Atlantic Fisheries Commission (NEAFC)	1980	All	Closed	Yes
Commission for the Conservation of Atlantic Marine Living Resources (CCAMLR)	1980	All	Open	Yes
North Atlantic Salmon Conservation Organization (NASCO)	1983	Anadromous	States of origin	No
Regional Commission of Fisheries of Gulf of Guinea (COREP)	1984	All	Open	Yes
Pacific Salmon Commission (PSC)	1985	Anadromous	States of origin	Yes
Ministerial Conference on Fisheries Cooperation among African States Bordering the Atlantic (COMHAFAT-ATLAFCO)	1991	All	Open	Yes
North Atlantic Marine Mammal Commission (NAMMCO)	1992	Marine mammals	Closed	No
North Pacific Anadromous Fish Commission (NPAFC)	1993	Anadromous	States of origin	No
Commission for the Conservation of Southern Bluefin Tuna (CCSBT)	1994	Tuna	Closed	Yes
Convention on the Conservation and Management of the Pollock Resources in the Central Bering Sea (CCBSP)	1995	Pollock	Open	Yes
Indian Ocean Tuna Commission (IOTC)	1996	Tuna	Open	Yes

Name by type	Year of establishment	Coverage of fish stocks	Membership	Decisions binding on parties
Regional Commission for Fisheries (RECOFI)	2001	All	Open	Yes
South East Atlantic Fisheries Organization (SEAFO)	2003	All (except tuna)	Open	Yes
Caribbean Regional Fisheries Mechanism (CRFM)	2003	All	Open	Yes
Western and Central Pacific Fisheries Commission (WCPFC)	2004	Tuna and other highly migratory	Open	Yes
South Indian Ocean Fisheries Agreement (SIOFA)	2006	All except tuna		Yes
Fishery Committee of the West Central Gulf of Guinea (FCWC)	2006	All	Open	Yes
South Pacific Regional Fisheries Management Organization (SPRFMO)	2012	All	Open	Yes
Central Asian and Caucasus Regional Fisheries and Aquaculture Commission (CACFish)	2010	All	Open	Yes
Southwest Indian Ocean Fisheries Commission (SWIOFC)	2004	All	Open	Yes
North Pacific Fisheries Commission (NPFC)	2015	All with exceptions	Open	Yes
Coordination and development				
Committee for the Eastern Central Atlantic Fisheries (CECAF)	1967	All	Closed	No
Western Central Atlantic Fisheries Commission (WECAFC)	1973	All	Open	No
Southeast Asian Fisheries Development Center (SEAFDEC)	1967	All	Open	No
South Pacific Forum Fisheries Agency (FFA)	1979	Tuna	Closed	No
Latin American Organization for the Development of Fisheries (OLDEPESCA)	1984	All	Closed	No
Sub-Regional Fisheries Commission (SRCF)	1985	All	Closed	No
Regional Convention on Fisheries Cooperation among African States Bordering the Atlantic Ocean (ASBAO)	1995	All	Closed	No
Bay of Bengal Programme Inter-Governmental Organisation (BOBP-IGO)	2003	Small scale fisheries	Open	No
Scientific research				
International Council for the Exploration of the Sea (ICES)	1902	All	Closed	No
North Pacific Marine Science Organization (PICES)	1992	All	Closed	No

Regional Fishery Management Organizations are defined as organizations established through international treaties with a primary focus on fisheries issues. The International Whaling Commission is sometimes considered a RFMO and is discussed below.

Adapted from Sydnes (2001) and www.fao.org/fishery/rfb/search/en

Recent international agreements affecting fisheries

International fisheries agreements negotiated since the LOSC entered into force in 1982 include the following 'soft law' agreements:

- 1995 FAO Code of Conduct for Responsible Fisheries and its associated FAO international plans of action (IPOAs).
- Various resolutions arising from the 1998 Jakarta Mandate under the Convention on Biological Diversity. Notably, this includes the decisions to protect ecologically or biologically significant areas (EBSAs), beginning with Decision IX/20 and its three annexes, adopted in 2008.

'Hard law' agreements include:

- 1993 (in force 2003) FAO Agreement to Promote Compliance with International Conservation and Management Measures by Fishing Vessels on the High Seas.
- 1995 (in force 2001) Agreement for the Implementation of the Provisions of the United Nations Convention on the Law of the Sea of 10 December 1982 Relating to the Conservation and Management of Straddling Fish Stocks and Highly Migratory Fish Stocks. (Fish Stocks Agreement.)
- 2009 Port State Measures Agreement (in force 2016).

The following sections discuss the key global fisheries agreements negotiated since the World War II. There is a particular focus on the management of fisheries under the LOSC. Though many of the following agreements include discussions on aquaculture, since these activities largely occur within states' internal waters they will not be discussed here.

International Convention for the Regulation of Whaling

Mechanization and technical improvements to the world's whaling fleets throughout the 19th and 20th Centuries resulted in the near extinction of many cetacean species. For example, the blue whale (*Balaenoptera musculus*) was reduced to less than 1 per cent of its historical population levels and perhaps as few as 5,000 individuals existed prior to the 1986 moratorium on their harvest. Similarly, the western Pacific grey whale (*Eschrichtius robustus*) population continues to be approximately 130 animals and remains listed as 'critically endangered' by the IUCN (www.iucn.org/wgwap/).

Recognizing that many whale (cetacean) populations had become precariously low as a result of commercial whaling, the International Convention for the Regulation of Whaling (ICRW) was convened in Washington, D.C in 1946 to conserve whale stocks to ensure the development of an orderly whaling industry (Box 8.4).

The Convention was formally amended only once, in 1956, through a Protocol that designated helicopters or aircraft as 'whale catchers'. The annual catch limits by species and geographic area are published annually in the 'Schedule' to the Convention. The Schedule also prescribes capture methods, the locations of capture and rules for IWC Sanctuaries and reporting requirements for whaling states. The Schedule is reviewed and approved annually at the meetings of the Commission.

There are currently 89 member states party to the Convention as well as a number of non-governmental organizations who may attend IWC meetings. The Convention established the International Whaling Commission, whose secretariat is headquartered in Cambridge, England.

Box 8.4 Whales species covered by the International Convention for the Regulation of Whaling

Of the 120+ species of cetaceans, currently only 21 species of 'great whales' are covered by the ICRW. Cetaceans covered by the convention either reside in ABNJ or migrate between states and, as such, are 'shared between nations'.

Baleen whales (order Mysticeti – use baleen plates to filter food):

Blue whale (*Balaenoptera musculus*)
Bowhead whale (*Balaena mysticetus*)
Bryde's whale (*Balaenoptera edeni, B. brydei*)
Fin whale (*Balaenoptera physalus*)
Grey whale (*Eschrichtius robustus*)
Humpback whale (*Megaptera novaeangliae*)
Minke whale (*Balaenoptera acutorostrata, B. bonaerensis*)
Pygmy right whale (*Caperea marginata*)
Right whale (*Eubalaena glacialis, E. australis*)
Sei whale (*Balaenoptera borealis*)

Toothed whales (order Odontoceti)

Cuvier's beaked whale (*Ziphius cavirostris*)
Shepherd's beaked whale (*Tasmacetus shepherdi*).
Baird's beaked whale (*Berardius bairdii*)
Arnoux's whale (*Berardius arnuxii*)
Southern bottlenose whale (*Hyperoodon planifrons*)
Northern bottlenose whale (*Hyperoodon ampullatus*).
Killer whale (*Orcinus orca*)
Pilot whale (*Globicephala melaena*) (*G. macrorhynchus*).
Sperm whale (*Physeter macrocephalus*)

The Commission is comprised of the Scientific, Technical, Finance and Administration and Conservation Committees. The Scientific Committee recommends research necessary for the regulation of whaling, reports on the status and trends of whale stocks and recommends actions to increase whale populations. The work of the Scientific Committee is used by the Commission to develop regulations for the member states.

Changes to the Schedule require the consent of three quarters of the member states. These modifications become effective 90 days after they are approved at the annual meeting of the commission. A change to the Schedule is non-binding on signatory states if formally objected to within 90 days. If one state objects within the 90-day period, additional time is provided for other states to object.

The IWC's original mandate to regulate whaling in order to foster the development of the whaling industry has changed significantly over the past decades. The IWC, while still retaining its original mandate, has evolved into a conservation organization, much to the displeasure of whaling states. This shift began in 1972 where the UNCHE called on the IWC to establish a 10-year moratorium on commercial whaling along with a recommendation to undertake additional research efforts. In the early 1970s, this call coincided with the establishment of global anti-whaling campaigns by Greenpeace and other environmental organizations.

In 1982, the IWC, at its annual meeting, agreed to a 'Comprehensive Assessment' or a pause in commercial whaling. The goal of the Assessment was to undertake population assessments and other research such that improved catch limits could be set to avoid the further decline of harvested populations. The Comprehensive Assessment, better known as the 'moratorium', was later defined by the IWC Scientific Committee as 'an in-depth evaluation of the status of all whale stocks in the light of management objectives and procedures . . . that . . . would include the examination of current stock size, recent population trends, carrying capacity and productivity' (www.iwcoffice.org/conservation/estimate.htm). Stated differently, the moratorium (Paragraph 10e of the Schedule) sets commercial whaling catch limits to zero for all species in all areas, irrespective of their conservation status while a comprehensive assessment of stock status is undertaken. Thus, the moratorium, adopted in 1982 and implemented in 1985, is a de facto global whale sanctuary. The moratorium applies only to commercial harvesting activities, and a number of nations, including Japan, Norway and Iceland continued whaling under scientific research permits where the products from this 'research whaling' were sold commercially. Norway resumed commercial whaling in 1994 and Iceland in 2006. In 2018, Japan announced its withdrawal from the IWC and will cease 'scientific' whaling in Antarctic waters but plans commercial whaling in its own waters in 2019.

Currently, the IWC is nearly evenly divided between pro- and anti-whaling states. This, of course, makes progress on any IWC business nearly impossible. Owing to the number of states previously encouraged by Japanese foreign aid to support whaling motions, its withdrawal could tip the balance in favour of conservation.

Convention on Fishing and Conservation of the Living Resources of the High Seas (1958)

The primary purpose of the four 1958 Geneva Conventions, culminating in the 1982 LOSC, was to codify the geographical limits of, and obligations in, the territorial seas. Yet, the Conventions also laid the groundwork for future fisheries agreements. In particular, the 1958 Conventions bestowed full control of fisheries within a state's territorial seas and continental shelf with no duties of conservation. Beyond territorial waters, the 'freedom of the high seas' doctrine also applied to fishing, and the only stipulation (albeit large) was that nations must consider the interests of other nations when fishing in the high seas.

The 1958 Convention on Fishing and Conservation of the Living Resources of the High Seas (the 'Convention on Fishing'), was signed on 19 April 1958 and entered into force on 20 March 1966 (UN, 1958). The Convention, for the first time in international law, recognized that fisheries could and were being over-exploited. The Convention also recognized that international cooperation was the appropriate vehicle for the sustainable management of high seas fish stocks.

While the Convention was the first attempt at managing fish stocks in the high seas, it was widely perceived to have a number of shortcomings. It seemed to favour the rights of fishing nations over coastal states, and it was unable to accommodate the growing number of nations claiming sovereignty out to 200nmi. Also, in lieu of the ability of coastal states to enforce compliance, the dispute resolution mechanisms appeared hopelessly convoluted. As a result, only 38 states have ratified the Convention. In addition, the concept of managing to MSY has failed, resulting in continued over-exploitation of fisheries.

In 1974, the International Court of Justice ruled that Iceland could claim a 12nmi zone seaward of the coastline where the state fully controlled fishing rights. The court also recognized that Iceland had 'preferential fishing rights' beyond the 12nmi seaward boundary to an unspecified distance so that its fisheries could be preserved (ICJ, 1975). This decision necessitated the establishment of marine jurisdictional arrangements and was pivotal in the development of the zoning contained in LOSC.

Fisheries management and the Convention on international trade in endangered species of wild Fauna and Flora (1973)

The Convention on International Trade in Endangered Species of Wild Fauna and Flora (CITES) was signed on 3 March 1973 and entered into force on 1 July 1975. It was amended on 22 June 1979 (the 'Bonn Amendment'). The Convention currently has 183 parties (CITES, 1973).

The purpose of CITES is to regulate the trade in wild animals and plants to prevent their over-exploitation. The Convention functions by assigning species of concern to one of the following three lists:

1. Appendix I: All species threatened with extinction that are, or may be, affected by trade.
2. Appendix II: All species that may become threatened with extinction that are, or may be, affected by trade.
3. Appendix III: All species which any state identifies as being subject to regulation within its jurisdiction for the purpose of preventing or restricting exploitation and needing the cooperation of other Parties in the control of trade.

Approximately 5,800 plants and more than 30,000 animals are currently listed in the CITES Appendices. While CITES has no authority over the operation of marine fisheries, the Convention has the ability to contribute to fisheries management objectives by listing species of commercial importance. Listed marine species on Appendix I of current or past commercial importance include all marine turtles, all beaked whales, all dugongs, all coelacanths and a number of other seals, dolphins, porpoises, sturgeon and fish. Appendix II lists thousands of marine species, including the Caribbean queen conch (*Strombus gigas*), whale shark (*Rhincodon typus*), great white shark (*Carcharodon carcharias*) and basking shark (*Cetorhinus maximus*). Proposals for CITES to add additional commercially important species, including threatened or endangered shark and tuna species, has seen partial success, with several sharks listed under Appendix II in 2014, followed by rays and more sharks in 2017. Pressure for CITES to list additional commercial species, such as Bluefin tuna, continues to increase.

Fisheries management and the United Nations Convention on the Law of the Sea

It is important to note that while the LOSC is not a prescriptive convention on fisheries management it has fundamentally changed the way global fisheries are managed. For example, the establishment of the 200nmi EEZs resulted in approximately 95 per cent of the world's fisheries harvest to fall within national jurisdiction, either within internal waters, the territorial sea or archipelagic waters. The LOSC sets the legal framework for the conservation, management and research with respect to marine living resources and is the foundation for subsequent international fisheries management agreements. Broadly, the LOSC attempts to balance several considerations, including: coastal states' sovereign rights to manage their fisheries against providing opportunities for other states to fish in their territorial seas and EEZs; optimum utilization against long-term sustainability; and freedom to fish the high seas against the requirement to cooperate with other nations to avoid over-exploitation.

The specific LOSC Articles relevant to fisheries management are summarized below.

Part II (Territorial Sea and Contiguous Zone)

- Article 19: Fishing activities are not considered innocent passage and may only be conducted with the agreement of the coastal state.

Part IX (Enclosed or semi-enclosed seas)

- Article 123: Coordinate the management, conservation, exploration and exploitation of the living resources of the sea.

Part V (EEZs), Articles 61–68

- Article 61: The coastal state determines the allowable catch of the living resources in its EEZ and is required to ensure that these resources are not over-exploited.
- Article 62: Coastal states will promote the objective of optimum utilization of living resources within their EEZs and provide other states the opportunity to harvest surplus catch. Nationals of other states fishing within another state's EEZ must abide by the laws and regulations of the coastal state.
- Article 63: States will cooperate on the conservation and development of fish stocks that overlap two or more states' EEZs.
- Article 64: States that fish for highly migratory species (listed in Annex I), whether in or out of a state's EEZ, must cooperate through an international organization to ensure the conservation and optimum utilization of the stock.
- Article 65: States will cooperate to conserve marine mammals.
- Article 66: States in whose rivers anadromous stocks originate have the primary interest in, and responsibility for, such stocks.
- Article 67: States in whose waters catadromous species spend the greater part of their life cycle shall have responsibility for the management of these species and shall ensure the ingress and egress of migrating fish.

Part VII (high seas), Section 2, Articles 116–120

- Article 116: All states have the right for their nationals to engage in fishing on the high seas.
- Article 117: All states have a duty to the conservation of the living resources of the high seas.
- Article 118: States shall cooperate with each other in the conservation and management of living resources in the areas of the high seas including the establishment of sub-regional or regional fisheries organizations.
- Article 119: States will manage for MSY on the high seas based on best scientific evidence. States will share information and not discriminate again fishers from any state.
- Article 120: States will cooperate to conserve marine mammals on the high seas.

The LOSC has been complemented by the Agreement to Promote Compliance with International Conservation and Management Measure by Fishing Vessels on the High Seas (Compliance Agreement – discussed below) intended to improve the regulation of fishing vessels on the high seas by strengthening 'flag-state responsibility'. Parties to the agreement must ensure that they maintain an authorization and recording system for high seas fishing vessels and that these vessels do not undermine international conservation and management measures. The agreement aims to deter the practice of 're-flagging' vessels with the flags of states that are unable or unwilling to enforce such measures.

The 2009 Port State Measures Agreement (in force 2016) puts in place a series of port-side inspection and reporting requirements to combat illegal and unreported fishing.

Fisheries management and the UN conference on environment and development

As discussed in Chapter 6, the purpose of the 1992 UNCED was to reconcile environment and development concerns in the context of disparities between nations' abilities to implement sustainable development. The conference launched Agenda 21 aimed at preparing the world for the challenges of the following century and contained a program of action for sustainable development to be undertaken by the world's nations. Agenda 21 contained Chapter 17 titled 'Protection of the Oceans', which addresses: integrated management and sustainable development of coastal areas; marine environmental protection; sustainable use and conservation of marine living resources of the high seas; sustainable use and conservation of marine living resources under national jurisdiction; addressing critical uncertainties for the management of the marine environment and climate change; strengthening international, including regional, co-operation and co-ordination; and sustainable development of small islands (UNEP, 1992).

In addition, UNCED resulted in the UNFCCC and CBD, which are discussed below.

Fisheries management and the United Nations Convention on Biological Diversity

As discussed in Chapter 6, the 1992 CBD came into force in 1993 and is the global umbrella convention for the protection of terrestrial and marine living resources (UN, 1992). The Convention addresses biodiversity at the genetic, species, and ecosystem or seascape levels and directs signatories to establish systems of protected areas and to protect endangered species. A 1995 meeting in Jakarta, Indonesia, led to the Jakarta Ministerial Statement (the 'Jakarta Mandate') on the implementation of the CBD, which includes protection of marine biodiversity within sustainable fisheries practices (Sinclair et al., 2002). A program of action for the Jakarta Mandate was developed in 1998 with a focus on integrated coastal and marine management, sustainable use of marine resources, establishing marine reserves (protected areas) and addressing alien species (Cochrane and Doulman, 2005).

In 2008, as part of establishing protected areas, the CBD agreed to identify marine 'biologically or ecologically significant areas' (EBSAs). To date, over 300 have been identified through a series of regional workshops, found to meet one or more of the following seven scientific criteria:

1 Uniqueness or rarity.
2 Special importance for life history stages of species.
3 Importance for threatened, endangered or declining species and/or habitats.
4 Vulnerability, fragility, sensitivity, or slow recovery.
5 Biological productivity.
6 Biological diversity.
7 Naturalness.

FAO agreement to promote compliance with international conservation and management measures by fishing vessels on the high seas

The Agreement to Promote Compliance with International Conservation and Management Measures by Fishing Vessels on the High Seas (1993 FAO Compliance Agreement) was approved on 24 November 1993 by Resolution 15/93 of the Twenty-Seventh Session of the FAO Conference. The Agreement entered into force on 24 April 2003. The Agreement originated from discussions at the 1992 UNCED for the need to effectively implement the LOSC. The Compliance Agreement was drafted to advance the objectives for marine living resources in the LOSC and articulate clear responsibilities, particularly for flag states, on ensuring that vessels flying their flag are in compliance with regional and sub-regional high seas fisheries agreements and that information on high seas fishing vessels is provided to the FAO (FAO, 1995a). The Agreement closes a legal loophole where vessels could change their state of registration ('flag') to avoid being bound by international agreements that had been signed by their previous flag state. The intent of the Agreement is that state parties will transpose the Agreement into national legislation (Freestone et al., 2001).

1995 United Nations agreement for the implementation of the provisions of the United Nations Convention on the Law of the Sea of 10 December 1982 relating to the conservation of straddling Fish Stocks and highly migratory Fish Stocks (UN Fish Stocks Agreement)

The 1995 UN Fish Stocks Agreement (UNFSA) entered into force on 11 December 2001. Besides ensuring the long-term conservation and sustainable use of straddling fish stocks and highly migratory fish stocks in ABNJ (Article 2) the Agreement provides specific language related to the protection of the marine environment (Articles 5 and 6). As with the FAO Code of Conduct, the intent of the Agreement is that state parties will transpose its objectives into national legislation (Freestone et al., 2001). The Agreement contains some similar general measures as the FAO Code of Conduct; however, it goes further by also mandating the creation of RFMOs, discussed above.

Key Articles of the agreement include:

- Article 3: Agreement applies to straddling fish stocks and highly migratory stocks outside of national jurisdiction (Article 3.1), requires states to apply conservation principles outlined in Article 5 (Article 3.2) and commits signatories to assist developing states in implementing the Agreement (Article 3.3).
- Article 5: Requires signatories to: sustainably manage straddling and highly migratory fish stocks and promote optimum utilization (Article 5.a); base management decisions on best available science and pursue maximum sustained yield (Article 5.b); apply the precautionary approach (Article 5.c); consider the ecosystem effects of fishing and manage other fish stocks in an appropriate manner (Article 5.d, e); minimize pollution, waste, discards, incidental catch by lost or abandoned gear, and catch of non-target species (Article 5.f); protect marine biodiversity (Article 5.g); prevent overfishing and excess fishing capacity (Article 5.h); consider artisanal and subsistence fishers (Article 5.i); collect and share data (Article 5.j); conduct scientific research (Article 5.k); and implement monitoring and compliance (Article 5.l).
- Article 6: Requires signatories to adopt the precautionary approach for straddling fish stocks and highly migratory fish stocks in order to protect the living marine resources and preserve the marine environment.

- Article 7: Commits signatories to cooperate on management measures for straddling fish stocks and highly migratory fish stocks. It also provides specific direction on how states should cooperate with each other.
- Article 8: Commits signatories to cooperate and resolve disputes through sub-regional or regional fisheries management organizations or other arrangements.
- Articles 9–13: Provides further specifics on the operation of sub-regional or regional fisheries management organizations or arrangements.
- Article 14: Commits states to collect and provide information and cooperation in scientific research.

FAO Code of Conduct for Responsible Fisheries

The FAO Code of Conduct for Responsible Fisheries was adopted on 31 October 1995 and originated from a recommendation from the 1992 International Conference on Responsible Fishing, held in Cancun (Mexico). Unlike other fisheries agreements discussed in this Chapter, the Code is voluntary and provides the principles and standards applicable to the conservation, management and development of all fisheries. The Code also addresses the capture, processing and trade of fish and fishery products. Fishing operations, aquaculture, fisheries research and the integration of fisheries into coastal area management fall under the Code's umbrella as well (FAO, 1995b).

While voluntary, the Code is an ambitious effort to integrate a number of binding principles from the LOSC, the FAO Compliance Agreement and the UN Fish Stocks Agreement. To address all aspects of fishing, a much broader set of operating principles and best practices to address all aspects of fishing is exploited. The Code applies to inland fisheries and also addresses aquaculture development (Hoscha et al., 2011).

Article 7 (Fisheries Management) is the most pertinent Article to fisheries management and commits signatories to adopt measures for the long-term conservation and sustainable use of fisheries resources. The management scheme calls for effective compliance and enforcement, transparent science and decision-making, reductions in excess fishing capacity, protection of aquatic habitats and endangered species, minimizing by-catch and implementation of the precautionary approach. Specifically, the agreement states that there is a, 'conscious need to avoid adverse impacts on the marine environment, preserve biodiversity, and maintain the integrity of marine ecosystems'.

Article 8 (Fishing Operations) describes the duties of flag and port states with respect to: monitoring; compliance and enforcement; health and safety standards; search and rescue; education and training; recording keeping and reporting; vessel and gear markings; insurance; and accident reporting. This Article also directs states to prohibit the use of dynamite, poison and other destructive fishing practices; minimize by-catch and discards; utilize methods to limit the loss of fishing gear; develop and utilize selective fishing gear; reduce energy consumption and minimize pollution; provide port-based services to minimize pollution and introduction of alien species; and consider the use of artificial reefs and aggregation devices to increase stock populations.

Additional Articles address the following subject areas:

- Article 9: Aquaculture development.
- Article 10: Integration of fisheries into coastal area management.
- Article 11: Post harvest practices and trade.
- Article 12: Fisheries research.

Importantly, the Code of Conduct introduced and reinforced key principles of sustainable fisheries management. Among these principles was use of traditional ecological knowledge (6.4), the precautionary approach (6.5 and 7.5), an ecosystem-based approach to fisheries management (6.2, 6.4, 6.6, 6.8, 7.2.2, and 7.3.3.), environmentally responsible gear and practices (6.6 and 8.5) and international cooperation (7.1.4 and 7.1.5.). Also, the Code identifies the FAO Secretariat to monitor the application of the Code and to report on developments to the Committee of Fisheries (4.2) (Juda, 2002).

The Code engendered a global discussion on fisheries management and informed the development of 'hard law' agreements including the FAO Compliance Agreement, the UN Fish Stocks Agreement and subsequent amendments to the Code of Conduct (discussed below).

The Code was amended by the Rome Declaration on the Implementation of the Code of Conduct for Responsible Fisheries, agreed to on 11 March 1999 at a FAO Ministerial Meeting on Fisheries (FAO, 1999). The Declaration specifically recognized the following:

- Article 2: The growing amount of IUU fishing activities being carried out, including fishing vessels flying 'flags of convenience'.
- Article 3: The importance of ensuring the contribution of sustainable aquaculture to food security, income and rural development.
- Article 5: The need for further technical and financial support for certain states to apply the Code and Plans of Action.
- Article 6: The need to give greater consideration to the development of more appropriate ecosystem approaches to fisheries development and management.

To date, aspects of the Code have proved particularly challenging, particularly with respect to the amount and quality of information necessary to implement the Code. Hoscha et al. (2011) found, in a review of the implementation of the Code in nine states, that its goals have been hampered by socio-economic considerations, administrative inertia, faltering political will and short-sighted leadership. Furthermore, they found that little process had been made on combating IUU fishing and addressing fishing overcapacity.

The Code has also provided the basis for the development of four international plans of action (IPOA) for the following, which are administered by the FAO (the two primary plans are discussed below):

- IPOA for the conservation and management of sharks (1999).
- IPOA for reducing the incidental catches of seabirds (1999).
- IPOA for the management of fishing capacity (1999).
- IPOA to prevent, deter and eliminate IUU fishing (2001).

The Agreement on Port State Measures to Prevent, Deter and Eliminate Illegal, Unreported and Unregulated Fishing

Illegal, unreported and unregulated (IUU) fishing may comprise up to 30 per cent of the global fisheries catch, with a landed value of between $US 10–23 billion (Agnew et al., 2009; Österblom and Bodin, 2012). IUU fishing may defeat efforts to sustainably manage regional fisheries, manage seafood safety through traceability (chain of custody) and can place fishers in dangerous situations or without adequate compensation (Sumaila et al., 2006). With an annual landed value greater than $US 5 million, the Patagonian toothfish is perhaps the most heavily exploited IUU species. Moreover, as a result of the use of longline fishing techniques for toothfish, bycatch impacts on albatross (*Diomedidae spp.*) and white-chinned petrels (*Procellaria aequinoctialis*) are substantial (Sumaila et al., 2006; Österblom and Bodin, 2012).

To address IUU, the European Union has adopted the EU IUU Regulation (1005/2008) and EU Control Regulation (1224/2009) to describe and mandate the development of a new seafood traceability program. It encompasses all components of the seafood production cycle (Borit and Olsen, 2012). In addition, the CCAMLR has demonstrated to be an effective regional body for addressing IUU fishing of the Patagonian toothfish (Österblom and Bodin, 2012).

However, it was not until the introduction of the FAO's 2009 Port State Measures Agreement (in force 2016) that a significant step forward was taken in battling IUU globally. The Agreement on Port State Measures to Prevent, Deter and Eliminate Illegal, Unreported and Unregulated Fishing (its full name) specifies Integration and coordination at the national level (Article 5), international cooperation and exchange of information among parties (Article 6), designated ports where inspections will be carried out (Article 7) and importantly, the ability to deny a vessel entry to a port and its provisioning services if the vessel does not have authorization to fish, or there is evidence that it has been involved in IUU activities (Article 11). Articles 12 to 15 concern port-side inspections, and Article 18 the possible denial of services if inspections reveal evidence of IUU. The Agreement also contains articles for training of inspectors, the requirements of developing states and the resolution of disputes.

Reykjavik Declaration on Responsible Fisheries in the Marine Ecosystem

Following the development of the FAO Code of Conduct, the FAO prepared a background paper in 2001 titled 'Towards Ecosystem-Based Fisheries Management' for the Reykjavik Conference on Responsible Fisheries in the Marine Ecosystem (FAO, 2003). The key messages of this document are captured as the 2001 Reykjavik Declaration on Responsible Fisheries and its subsequent Operational Guidelines on the ecosystem approach to fisheries (FAO, 2003) that defines the EAF as:

> An ecosystem approach to fisheries strives to balance diverse societal objectives, by taking into account the knowledge and uncertainties about biotic, abiotic and human components of ecosystems and their interactions and applying an integrated approach to fisheries within ecologically meaningful boundaries.
>
> (FAO, 2003)

Private ocean rights

A frequent theme throughout this text is the risks of open access (see Chapter 3) where individuals, corporations and states have few incentives to produce, regulate, manage or monitor the production of marine-derived goods and services efficiently in ocean areas which are shared with others. Open marine access leads to multifarious, undesirable effects: rapid depletion of resources due to 'gold rush' behaviour, low profits due to overcapitalization and excessive costs, dangerous working conditions due to low profit margins and competition and poor product quality (Deacon, 2009). Most discussion related to the perils of open access centres on the rapid depletion of fisheries due to circumstances in which fishers have few economic incentives to consider the long-term health or productivity of fish stocks. Yet, other sectors – transportation, energy development, mining and others – often lack incentives to operate in a manner that minimizes environmental harm and maximizes long-term opportunities.

The proliferation of declarations of EEZs over the past several decades has resulted in 95 per cent of the world's fisheries and nearly all of the world's known offshore hydrocarbon resources falling within national jurisdictions. The establishment of EEZs has been a significant step towards economic rationality and enclosing the commons. However, many of the most important global fisheries, including tuna, occur primarily outside of EEZs.

Introduction to marine property rights

Marine property is frequently conceived of as a tangible object (e.g. territory, fish, oil and gas reserve), but it can also be an intangible benefit or income stream (e.g. right to fish) recognized by some level of government (Edwards, 1994). Marine property rights span a continuum, ranging from weak usage rights – such as those granted to companies to explore for oil and gas deposits – to the full ownership of ocean resources found in some fisheries management systems (Box 8.5). Property rights have different meanings depending on whether the resource is found within a state's (or multiple state's) EEZs or ABNJ. With the exception of some privately held estuarine and intertidal areas in certain states, marine environments are held in trust by states on behalf of their citizens within territorial waters and EEZs. Outside of these areas, ABNJ are held in trust for the benefit of all states and managed cooperatively by the international organizations discussed in previous chapters.

Rights-based fishery management systems

As discussed in Chapter 6, throughout human history humans have managed to overfish nearly any area or stock that was accessible with available technology. The fundamental reasons for the repeated and unchecked exploitation of fish stocks are open access (the rule of capture), weak

Box 8.5 Types of marine property rights

Types of rights	Characteristics	Examples
Proprietary (possessory)	Physical control of the property to use it (or not use it), to change its form and substance and to transfer all rights through the sale or some rights through the rental. Responsible for damages to other's interests and may claim damages if others harm or damage the property.	Collects rents, royalties from leases. States have proprietary rights held in trust for their citizens.
Use	Rights to use for specific purposes at specific times and places.	Fishing rights, oil and gas exploration and development rights, offshore energy installations, aquaculture.
Exclusionary	Right to exclude others from the property.	Exclusive rights to explore and exploit (e.g. energy and minerals). May exclude others for safety concerns.
Disposition	Right to transfer, sell, exchange.	Individual transferrable fishing rights. Rights to resell use or exclusionary rights.

Adapted from Young (2007)

governance (decision-making), and insufficient enforcement (Hannesson, 2011). These reasons are especially relevant to high seas fisheries where fish stocks are truly a global 'commons'. The problem is not limited to the high seas. This predicament also exists within many state's EEZs if the state cannot control or enforce access.

As early as 1954, it was understood that economic inefficiencies and over-exploitation are inevitable consequences of fish as a 'commons' resource. Gordon (1954) defined the commons as 'free goods for the individual and scarce goods for society'. Under open access, fisheries would experience 'market failure', meaning fishing becomes unprofitable as fishing effort increases and fish stocks decrease (Gordon, 1954). Gordon was an early proponent of treating fish as a private resource; but it was not until 1979, when Iceland began limiting entry to its herring fishery using license limitations, that modern rights-based fishing emerged (Chu, 2009).

Approaches to rights-based fishing can broadly be separated into approaches that:

- Limit entry to a fishery by determining who is allowed to fish and where they can fish (license limitation, see Box 8.5).
- Limit who is allowed to utilize processing facilities (landing limitation).
- Limit who is allowed access to goods and services (e.g. fuel, supplies) that support fishing.
- Conferring ownership of a portion or all of a particular fish stock upon an individual, company or community (privatization, see Box 8.5).

Interestingly, rights-based fishing systems are not always established by governments. North American lobster fisheries informally control the entry to a fishery in Maine through 'harbour gangs' or a 'berth' system in Nova Scotia where rights to a specific fishing area are held informally by local fishers. These systems evolved from the Scottish 'kindness' system where access to nearshore ocean areas was granted by private landowners (Acheson, 2015).

The most prominent and frequently used rights-based approaches are those that establish a total allowable catch (TAC) for a fish stock and then allocate a portion of the catch (known as 'catch shares') to individuals, vessels, corporations or communities. Types of catch share approaches include:

- Individual transferrable quotas (ITQs).
- Individual vessel quotas (IVQs).
- Community development quotas (CDQs).
- Enterprise Allocations (EAs – allocated to fishing corporations).
- Territorial user rights in fisheries (TURFs – allocated to residents of a geographical area).

ITQs are the most frequently used catch share approaches, and the term 'ITQ' has become synonymous with all types of catch shares. Currently, ITQs have been introduced in over 121 fisheries in 18 states with some evidence that they can be successful. Hilborn et al. (2003) found that by-catch declined an average of 40 per cent in a study of 10 fisheries in the United States and Canada after they moved to management under catch shares. Chu (2009) found improvements in 12 of 20 stocks after ITQs were introduced. Despite these promising statistics, catch share programs continue to be the exception in fisheries management. A paltry 1 per cent of the world's commercial fisheries and 15 per cent of the world's catch are managed under the scheme (Deacon, 2009). One reason for this is that in order to stop cheating, ITQ systems require robust monitoring and enforcement, which is more expensive than enforcing a seasonal opening. Also, the enforcement infrastructure (observes and cameras) have been seldom practical in the developing world.

Box 8.6 Successful examples of catch share programs

The Peruvian anchovy fishery

The Peruvian anchovy (*Engraulis ringens*) fishery comprises nearly 10 per cent of the global fisheries landings, supports approximately 1,200 fishing vessels and 140 factories, employs over 18,000 people and represents a minimum of 7 per cent ($US 2 billion) of Peru's foreign exchange earnings. The fishery was also recently ranked as the world's most sustainable fishery in a study developed by the University of British Columbia Fisheries Centre.

The fishery operates under a TAC set annually by government with scientific advice from the Peruvian Marine Research Institute. Since 1992, the government has addressed overfishing by limiting the issuance of new licenses. In 2008, it adopted a system of Individual Vessel Quotas (IVQ) to divide the TAC among vessels registered in the fishery. Each day during the fishing season, the government publishes a list of vessels permitted to fish. In addition, fishing plants are prohibited from receiving fish from vessels not authorized to fish.

The Peruvian government has developed a legislated management scheme to control domestic and foreign access by designating geographical boundaries, limiting the number of boats and their fishing capacity and maintaining biological productivity. These statutes are modified as necessary by ministerial resolutions relating to closed areas, periods or quotas.

The northeast Pacific halibut fishery

The Canadian and United States populations of pacific halibut (*Hippoglossus stenolepis*) have been managed by the International Pacific Halibut Commission (IPHC) since 1923. The Canadian fishery moved to an IVQ system in 1991 and the United States to an individual fishing quota (IFQ – same as an ITQ) system in 1995.

The Canadian halibut fishing season was six days long in 1990. Today the season is 10 months in duration, which has resulted in 95 per cent of the catch sold fresh (not frozen), yielding higher market prices. The longer season also produced stability for fish processors, greater crew safety (through shorter shifts and not needing to fish during stormy weather), a smaller fleet fishing more weeks per vessel, and a product price increase of $US 3.30 per kg at the wholesale level (DFO, 2008).

The United States halibut fishing season was reduced from 150 days in 1970 to 16 days in 1979. Overcapitalization in the fishery – characterized by an increase in the size and number of fishing vessels – occasioned this drastic measure. After the establishment of IFQs, the vessel fleet was reduced by 35 per cent through consolidation and the fishing season was increased to nine months. Additionally, the halibut price paid to fishers increased 10.5 per cent per year for the first eight years of the ITQ program. The program added further efficacious measures: it instituted vessel size quotas so that larger operators would be prohibited from dominating the fishery, it limited transferability to fishers with experience in commercial fishing and it introduced use caps on individual owners and vessels discourage consolidation.

Source: DFO (2008), Schreiber (2012), and Carothers (2013)

It is important to note that catch share programs do not bestow full property rights to fishers. Full property rights would require that the holder of a catch share have an exclusive, inalienable right in perpetuity that could be sold, leased or rented. While some catch share programs, such as those practiced in some New Zealand fisheries, approach private rights, most programs occupy a place between open access and full private rights. So long as the TAC is set correctly, catch shares can promote sustainability by providing fishers with an incentive to conserve the fish resource. The incentive to fish responsibly is based in part on the security, exclusivity and transferability of the fishing right.

There has been considerable debate as to whether catch shares are applicable to high seas fisheries such as tuna, swordfish and Patagonian toothfish. While certain ITQs and other catch share programs have successfully established sustainable fisheries within EEZs or between EEZs or high seas areas (straddling stocks), there has been little progress on launching rights-based systems in high seas areas (Hannesson, 2011). As discussed previously, while nearly all the world's fisheries occur within EEZs, certain fisheries including cod (Grand Banks), turbot (Barents Sea) and Alaska Pollock (North Pacific) occur outside of EEZs. RFMOs are responsible for managing high seas fisheries, yet this remains problematic. The open access characteristic of the high seas makes developing and enforcing regulations difficult, especially when dealing with states that are not inclined to participate in RFMO processes. Two solutions are offered by Hannesson (2011) to establish private rights in high seas fisheries: First, states could extend their national EEZs to capture globally important fisheries that currently reside in the high seas. Second, corporations could be established that would 'own' fish on the high seas where states are shareholders (Carothers, 2013). At this time, however, neither approach has much traction at the international level. In the western central Pacific, where tuna are managed in national waters using a variant of a time quota system ('vessel day scheme'), countries collectively agreed through the relevant RFMO (the WCPFC) in 2008 to close fishing in the 'high seas pockets', thus keeping the revenue stream within national jurisdictions and avoiding a different access regime on the high seas.

Discussion questions

- Why is an ecosystem approach to fisheries management often considered advantageous over traditional single- and multispecies approaches?
- Explain some of the challenges of taking an ecosystem approach to fisheries management.
- How might a changing climate necessitate changes to how marine fisheries are managed?
- What are some of the challenges and barriers to implementing fisheries management in ABNJ?

Further reading

Acheson, J. M. (2015) 'Private land and common oceans analysis of the development of property regimes', *Current Anthropology*, vol 56, no 1, pp. 28–55.

Agnew, D. J., Pearce, J., Pramond, G., Peatman, T., Watson, R., Beddington, J. R. and Pitcher, T. J. (2009) 'Estimating the worldwide extent of illegal fishing', *PLoS One*, vol 4, no 2, p. e4570.

Babcock, E. A. and Pikitch, E. K. (2004) 'Can we reach agreement on a standardized approach to ecosystem-based fishery management?', *Bulletin of Marine Science*, vol 74, no 3, pp. 685–692.

Borit, M. and Olsen, P. (2012) 'Evaluation framework for regulatory requirements related to data recording and traceability designed to prevent illegal, unreported and unregulated fishing', *Marine Policy*, vol 36, pp. 96–102.

Carothers, C. (2013) 'A survey of US halibut IFQ holders: Market participation, attitudes, and impacts', *Marine Policy*, vol 38, pp. 515–522.

Christie, P., Fluharty, D. L., White, A. T., Eisma-Osorio, L. and Jatulan, W. (2007) 'Assessing the feasibility of ecosystem-based fisheries management in tropical contexts', *Marine Policy*, vol 31, no 3, pp. 239–250.

Chu, C. (2009) 'Thirty years later: The global growth of ITQs and their influence on stock status in marine fisheries', *Fish and Fisheries*, vol 10, pp. 217–230.

CITES (1973) *Convention on International Trade in Endangered Species of Wild Fauna and Flora*, www.cites.org/eng/disc/text.php, accessed 04 January 2019.

Cochrane, K. L. (2002) 'A fishery manager's guidebook: Management measures and their application', in *FAO Fisheries Technical Paper*, vol 424, FAO, Rome.

Cochrane, K. L. and Doulman, D. J. (2005) 'The rising tide of fisheries instruments and the struggle to keep afloat', *Philosophical Transactions of the Royal Society of London B*, vol 360, no 1435, pp. 77–94.

Coleman, F. C., Figueira, W. F., Ueland, J. S. and Crowder, L. B. (2004) 'The impact of United States recreational fisheries on marine fish populations', *Science*, vol 305, pp. 1958–1960.

Constable, A. J. (2004) 'Managing fisheries effects on marine food webs in Antarctica: Trade-offs among harvest strategies, monitoring, and assessment in achieving conservation objectives', *Bulletin of Marine Science*, vol 74, no 3, pp. 583–605.

Cryer, M., Mace, P. M. and Sullivan, K. J. (2016) 'New Zealand's ecosystem approach to fisheries management', *Fisheries Oceanography*, vol 25, suppl 1, pp. 57–70.

Cullis-Suzuki, S. and Pauly, D. (2010) 'Failing the high seas: A global evaluation of regional fisheries management organizations', *Marine Policy*, vol 34, pp. 1036–1042.

Cury, P. M., Mullon, C., Garcia, S. M. and Shannon, L. J. (2005) 'Viability theory for an ecosystem approach to fisheries', *ICES Journal of Marine Science*, vol 62, no 3, pp. 577–584.

Deacon, R. T. (2009) *Creating Marine Assets: Property Rights in Ocean Fisheries*, PERC Policy Series, No 43, Bozeman, Montana.

DFO (2008) *Employment Impacts of ITQ Fisheries in Pacific Canada*, Fisheries and Oceans Canada, Ottawa, Canada.

Drankier, P. (2012) 'Marine protected areas in areas beyond national jurisdiction', *The International Journal of Marine and Coastal Law*, vol 27, pp. 291–350.

Edwards, S. F. (1994) 'Ownership of renewable ocean resources', *Marine Resource Economics*, vol 9, pp. 253–273.

FAO (1995a) *Agreement to Promote Compliance with International Conservation and Management Measures by Fishing Vessels on the High Seas*, Rome, www.fao.org/docrep/meeting/003/x3130m/X3130E00.HTM, accessed 04 January 2019.

FAO (1995b) *Code of Conduct for Responsible Fisheries*, Rome, www.fao.org/docrep/005/v9878e/v9878e00.HTM, accessed 04 January 2019.

FAO (1997) 'Fisheries management', in *FAO Technical Guidelines for Responsible Fisheries*, vol 4, FAO, Rome, www.fao.org/3/a-i1146e.pdf, accessed 04 January 2019.

FAO (1999) *International Plan of Action for the Management of Fishing Capacity*, Rome, www.fao.org/docrep/006/x3170e/x3170e04.htm, accessed 04 January 2019.

FAO (2001) *International Plan of Action to Prevent, Deter and Eliminate Illegal, Unreported and Unregulated Fishing*, FAO, Rome, www.fao.org/3/a-y1224e.pdf, accessed 20 April 2019.

FAO (2003) 'The ecosystem approach to fisheries', in *FAO Fisheries Technical Paper*, vol 443, FAO, Rome, www.fao.org/3/a-y4773e.pdf, accessed 04 January 2019.

FAO (2014) *Essential EAFM: Ecosystem Approach to Fisheries Management Training Course*, FAO, Rome, www.fao.org/3/a-i3778e.pdf, accessed 08 January 2019.

FAO (2018) *The State of World Fisheries and Aquaculture 2018*, FAO, Rome, www.fao.org/documents/card/en/c/I9540EN, accessed 24 July 2018.

Freestone, D., Gudmundsdotti, E. and Edeson, W. (2001) *Legislating for Sustainable Fisheries*, World Bank, Washington, DC.

Frid, C. L. J., Paramor, O. A. L. and Scott, C. L. (2005) 'Ecosystem-based fisheries management: Progress in the NE Atlantic', *Marine Policy*, vol 29, no 5, pp. 461–469.

Frid, C. L. J., Paramor, O. A. L. and Scott, C. L. (2006) 'Ecosystem-based management of fisheries: Is science limiting?', *ICES Journal of Marine Science*, vol 63, no 91, pp. 567–1572.

Froese, R., Stern-Pirlot, A., Winker, H. and Gascuel, D. (2008) 'Size matters: How single-species management can contribute to ecosystem-based fisheries management', *Fisheries Research*, vol 92, nos 2–3, pp. 231–241.

Galland, G. R., Nickson, A. E. M., Hopkins, R. and Miller, S. K. (2018) 'On the importance of clarity in scientific advice for fisheries management', *Marine Policy*, vol 87, pp. 250–254.

Goodman, D., Mangel, M., Parkes, G., Quinn, T., Restrepo, V., Smitch, T. and Stokes, K. (2002) *Scientific Review of the Harvest Strategy Currently Used in the BSAI and GIA Groundfish Fishery Management Plans*, North Pacific Fishery Management Council, Anchorage, AL.

Gordon, H. S. (1954) 'The economic theory of a common property resource: The fishery', *Journal of Political Economy*, vol 62, no 2, pp. 124–142.

Guerry, A. D. (2005) 'Icarus and Daedalus: Conceptual and tactical lessons for marine ecosystem-based management', *Frontiers in Ecology and the Environment*, vol 3, pp. 202–211.

Hannesson, R. (2011) 'Rights based fishing on the high seas: Is it possible?', *Marine Policy*, vol 35, pp. 667–674.

Hilborn, R. (2004) 'Ecosystem-based fisheries management: The carrot or the stick?', *Marine Ecology Progress Series*, vol 274, pp. 275–278.

Hilborn, R. (2016) 'Marine biodiversity needs more than protection', *Nature*, vol 535, pp. 14–17.

Hilborn, R., Branch, T. A., Ernst, B., Magnusson, A., Minte-Vera, C. V., Scheuerell, M. D. and Valero, J. L. (2003) 'State of the world's fisheries', *Annual Review of Environment and Resources*, vol 28, pp. 359–399.

Hirshfield, M. F. (2005) 'Implementing the ecosystem approach: Making ecosystems matter', *Marine Ecology Progress Series*, vol 300, pp. 253–257.

Hoscha, G., Ferrarob, G. and Faillerc, P. (2011) '1995 FAO code of conduct for responsible fisheries: Adopting, implementing or scoring results?', *Marine Policy*, vol 35, no 2, pp. 189–200.

International Court of Justice (1975) *Fisheries Jurisdiction (United Kingdom of Great Britain and Northern Ireland v. Iceland)*, www.icj-cij.org/en/case/55/judgments, accessed 04 January 2019.

Jennings, S. and Revill, A. S. (2007) 'The role of gear technologists in supporting an ecosystem approach to fisheries', *ICES Journal of Marine Science*, vol 64, pp. 1525–1534.

Juda, L. (2002) 'Rio plus ten: The evolution of international marine fisheries governance', *Ocean Development and International Law*, vol 33, pp. 109–144.

King, D. M. and Sutinen, J. G. (2010) 'Rational noncompliance and the liquidation of Northeast groundfish resources', *Marine Policy*, vol 34, no 1, pp. 7–21.

Larkin, P. A. (1977) 'An epitaph for the concept of maximum sustainable yield', *Transactions of the American Fisheries Society*, vol 106, pp. 1–11.

Lotze, H. K. (2004) 'Repetitive history of resource depletion and mismanagement: The need for a shift in perspective', *Marine Ecology Progress Series*, vol 274, pp. 282–285.

Mace, P. M. (2001) 'A new role for MSY in single-species and ecosystem approaches to fisheries stock assessment and management', *Fish and Fisheries*, vol 2, pp. 2–32.

Mace, P. M. (2004) 'In defence of fisheries scientists, single-species models and other scapegoats: Confronting the real problems', *Marine Ecology Progress Series*, vol 274, pp. 285–291.

Marasco, R. J., Goodman, D., Grimes, C. B., Lawson, P. W., Punt, A. E. and Quinn, T. J. (2007) 'Ecosystem-based fisheries management: Some practical suggestions', *Canadian Journal of Fisheries and Aquatic Sciences*, vol 64, no 6, pp. 928–939.

McCarthy, M. (2013) 'Is this the end of the bluefin tuna?', www.independent.co.uk/environment/nature/is-this-the-end-of-the-bluefin-tuna-1040246.html, accessed 04 January 2019.

Norse, E. A., Brooke, S., Cheung, W. W. L., Clark, M. R., Ekeland, I., Froese, R., Gjerde, K. M., Haedrich, R. L., Heppell, S. S., Morato, T., Morgan, L. E., Pauly, D., Sumaila, R. and Watson, R. (2012) 'Sustainability of deep-sea fisheries', *Marine Policy*, vol 36, no 2, pp. 307–320.

Oanta, G. A. (2018) 'International organizations and deep-sea fisheries: Current status and future prospects', *Marine Policy*, vol 87, pp. 51–59.

Österblom, H. and Bodin, Ö. (2012) 'Global cooperation among diverse organizations to reduce illegal fishing in the Southern Ocean', *Conservation Biology*, vol 26, no 4, pp. 638–648.

Pauly, D., Christensen, V., Dalsgaard, J., Froese, R. and Torres, F., Jr. (1998) 'Fishing down marine food webs', *Science*, vol 279, pp. 860–863.

Pauly, D., Christensen, V., Guenette, S., Pitcher, T. J., Sumaila, U. R., Walters, C. J., Watson, R. and Zeller, D. (2002) 'Towards sustainability in world fisheries', *Nature*, vol 418, no 6898, pp. 689–695.

Pavlikakis, G. E. and Tsihrintzis, V. A. (2000) 'Ecosystem management: A review of a new concept and methodology', *Water Resources Management*, vol 14, no 4, pp. 257–283.

Pikitch, E. K., Santora, C., Babcock, E. A., Bakun, A., Bonfil, R., Conover, D. O., Dayton, P., Doukakis, P., Fluharty, D., Heneman, B., Houde, E. D., Link, J., Livingston, P. A., Mangel, M., McAllister, M. K., Pope, J. and Sainsbury, K. J. (2004) 'Ecosystem-based fishery management', *Science*, vol 305, no 5682, pp. 346–347.

Pitcher, T. J., Kalikoski, D., Short, K., Varkey, D. and Pramod, G. (2009) 'An evaluation of progress in implementing ecosystem-based management of fisheries in 33 Countries', *Marine Policy*, vol 33, no 2, pp. 223–232.

Ramìrez-Monsalve, P., Raakjaer, J., Nielsen, K. N., Laksa, U., Danielsen, R., Degnbol, D., Ballesteros, M. and Degnbol, P. (2016) 'Institutional challenges for policy-making and fisheries advice to move to a full EAFM approach within the current governance structures for marine policies', *Marine Policy*, vol 69, pp. 1–12.

Schreiber, M. A. (2012) 'The evolution of legal instruments and the sustainability of the Peruvian anchovy fishery', *Marine Policy*, vol 36, pp. 667–674.

Shelton, P. A. (2009) 'Eco-certification of sustainably managed fisheries: Redundancy or synergy?', *Fisheries Research*, vol 100, no 3, pp. 185–190.

Sinclair, M., Arnason, R., Csirke, J., Karnicki, Z., Sigurjonsson, J., Skjoldal, H. R. and Valdimarsson, G. (2002) 'Responsible fisheries in the marine ecosystem', *Fisheries Research*, vol 58, no 3, pp. 255–265.

Sissenwine, M. P. and Murawski, S. (2004) 'Moving beyond "intelligent tinkering": Advancing an ecosystem approach to fisheries', *Marine Ecology Progress Series*, vol 274, pp. 291–295.

Sumaila, U. R., Alder, J. and Keith, H. (2006) 'Global scope and economics of illegal fishing', *Marine Policy*, vol 30, no 6, pp. 696–703.

Sydnes, A. K. (2001) 'Regional fishery organizations: How and why organizational diversity matters', *Ocean Development & International Law*, vol 32, pp. 349–372.

Symes, D. (2007) 'Fisheries management and institutional reform: A European perspective', *ICES Journal of Marine Science*, vol 64, no 4, pp. 779–785.

Tyedmers, P., Watson, R. and Pauly, D. (2006) 'Fueling global fishing fleets', *Ambio*, vol 34, pp. 635–638.

UN (1958) *Convention on Fishing and Conservation of the Living Resources of the High Seas*, United Nations, Treaty Series, vol 559, p. 285, legal.un.org/ilc/texts/instruments/english/conventions/8_1_1958_fishing.pdf, accessed 04 January 2019.

UN (1992) *Convention on Biological Diversity*, www.cbd.int/doc/legal/cbd-en.pdf, accessed 04 January 2019.

UN (1995) *The United Nations Agreement for the Implementation of the Provisions of the United Nations Convention on the Law of the Sea of 10 December 1982 Relating to the Conservation and Management of Straddling Fish Stocks and Highly Migratory Fish Stocks*, www.un.org/depts/los/convention_agreements/texts/fish_stocks_agreement/CONF164_37.htm, accessed 04 January 2019.

UNEP (1992) *Protection of the Oceans, All Kinds of Seas, Including Enclosed and Semi-Enclosed Seas and Coastal Areas and the Protection Rational Use and Development of Their Living Resources*, UNEP, Nairobi, Kenya, www.un-documents.net/a21-17.htm, accessed 04 January 2019.

Valdermarsen, J. W. and Suuronen, P. (2003) 'Modifying fishing gear to achieve ecosystem objectives', in M. Sinclair and G. Valdimarsson (eds.), *Responsible Fisheries in the Marine Ecosystem*, FAO, Rome.

Young, O. (2007) 'Rights, rules, and common pools: Solving problems arising in human/environment relations', *Natural Resources Journal*, vol 47, no 1, pp. 1–16.

Chapter 9

Marine transportation and safety policy

The operation and regulation of international shipping

Key topics

- More than any other activity, shipping and port development has enabled the global economy.
- Currently, 90 per cent of global trade is carried on 91,000 ships that move 8.4 billion tonnes of goods annually. Over 1.4 million seafarers work on ships registered in over 150 nations, and over 2 billion passengers are moved annually on ships.
- Thirty-five states control 95 per cent of the world's shipping fleet. Of these states, Greece, Japan, Germany, the United States and China control approximately 50 per cent of the world's shipping fleet.
- The LOSC assures the right of innocent passage and transit passage; immunity of warships; freedom of the high seas and use of the high seas for peaceful purposes; rights of all states to participate in shipping; prohibitions on carrying slaves; the necessity of combating piracy; and the right of all states to fish.
- The LOSC also articulates the rights and duties of coastal and flag states as well as the rights of states to determine the conditions for granting of their nationality to ships and standards for manning and training crews.
- Shipping is primarily regulated by the International Maritime Organization (IMO) and the International Labour Organization (ILO).
- The IMO's authority only applies to international shipping. Member states are expected to implement and enforce the 40 IMO conventions, protocols, as well as the 800+ codes and recommendations.
- The Comité Maritime International (CMI), established in 1897, is the oldest international maritime organization, established to oversee the codification of maritime law, commercial practices and the promotion of national associations of maritime law.
- The major marine shipping conventions are the International Convention for the Safety of Life at Sea (SOLAS), International Convention on Standards of Training, Certification and Watchkeeping for Seafarers (STCW), Maritime Labour Convention (MLO), and International Convention for the Prevention of Pollution from Ships (MARPOL 73/78).
- Current global shipping issues include addressing piracy, climate change, seafarer safety and greenhouse gas emissions.

Introduction

Until recently, archaeologists believed that humans (*Homo sapiens*) and their recent ancestors (*Homo erectus*) did not cross deepwater straits until approximately 50,000 years ago. However, recent evidence suggests that *Homo erectus* may have crossed the Wallace Line (an imaginary boundary separating Australian and Asian floras and faunas based on marine geographic separation) up to 800,000 years ago and that Australia was colonized more than 60,000 years ago. Evidence from southeast Asia also indicates that humans were frequently travelling between the Moluccas, Borneo and Bali as long as 33,000 years ago (Bellwood, 1997; Gibbons, 1998).

It is likely that by 4,000 years BCE, extensive trade networks existed along Asia's major waterways and that sea trade linked Asia, India and the Middle East to trade pearls, herbs, spices, pepper, sesame oil and sugar. During this time, Egypt and other Mediterranean states used rowing boats averaging 10 metres in length to trade in copper, papyrus, wool, cedar, linens and dyes.

The first recorded instances of the use of long rowing boats were documented in Egypt in the time of Sesostris III (3,700–4,100 years ago). Supporting these written records was the discovery of six boats, averaging 10m in length, next to the pyramid of Sesostris III. Until the appearance of the modern galley in the 16th Century, the Egyptian long boats ruled the seas.

While Sesostris III and subsequent dynasties likely used early galleys to ferry soldiers and supplies to support military endeavours, the first recorded naval battle was the Battle of Artemisium in 2480 BCE. Xerxes, a Persian King, is reported to have assembled a fleet comprised 1,207 galleys, possibly carrying more than 200,000 men. While the battle against the Greek fleet was indecisive, it demonstrated the value of navies in supporting military operations (Mordal, 1959).

The age of navigation (between 1,000 and 3,000 years ago) saw the advent of the astrolabe (a device that calculates position using sun and star positions), compass and the use of more sophisticated boats based on the Egyptian longboat model. In addition to the Mediterranean nations, Arab, Chinese and Pacific Ocean cultures also exploited celestial navigation during this period, improving the reliability of the movement of goods.

The first large-scale use of the oceans was to explore and colonize other states. Most likely, the first colonizers to roam beyond the Mediterranean Sea were the Norse explorers of Northern Europe. In the 10th Century, they settled Greenland and parts of eastern North America. In addition, Polynesian explorers likely landed in the Hawaiian Islands in the 10th Century. In the 15th Century, guided by Papal Bull, Portugal and Spain began a protracted marine campaign to annex the Americas (see Chapter 3). The British and Dutch followed suit in the 17th Century, and the annexation of Africa occurred in the 19th and 20th Century by a number of European nations.

With colonization arose the need to transport goods and people to and from the colonies. The global shipping trade was born, complete with pirates, competition between shipping companies and the development of naval ships to protect trade routes. The first modern ship designed to carry bulk goods (e.g. coal, ore, grain) was built in 1852 (650 dead weight tonnes – dwt), followed in 1886 with the 3,000 dwt oil tanker named the 'Gluckauf' (IMO, 2012). With the advent of modern shipping came the establishment of the modern seaborne trade routes, which currently include the movement of:

- Coal from Australia, Southern Africa and North America to Europe and the Far East.
- Grain from North and South America to Asia, Africa and the Far East.
- Iron ore from South America and Australia to Europe and the Far East.

- Oil from the Middle East, West Africa, South America and the Caribbean to Europe, North America and Asia
- Containerized goods from the People's Republic of China, Japan and southeast Asia.

Source: IMO (2012)

This chapter presents a short overview of the operation of contemporary shipping. Specifically, it will discuss the configuration of the global shipping fleet, the processes by which shipping between states is conducted and the major international agreements that regulate shipping and the safety of seafarers.

An overview of contemporary shipping

The overall function of shipping has not changed over the millennia: Shipping moves goods from a region where they have a low value to a region where they have a high value (Branch and Roberts, 2014). Today, over 90 per cent of internationally traded goods are moved by shipping, and this percentage is higher for trade conducted between developing countries. In 2011, 90 per cent of the European Union's foreign trade and 40 per cent of trade between European Union member states was conducted by sea (IMO, 2012).

Shipping continues to be the most cost-effective method of moving goods over long distances. The continued globalization of trade combined with more efficient shipping technologies will likely result in the continued dominance of global trade by shipping. Currently the world's merchant shipping fleet consists of approximately 91,000 ships over 100 GT (gross tonnage is a measure of a ship's internal volume – see Box 9.1), owned by more than 250 companies representing approximately 1.8 billion dead weight tonnes (see Box 9.1) with an average ship age of 22 years (UNCTAD, 2017). The world's cargo fleet consists of more than 55,100 ships, representing more than 991 million GT with an average age of 22 years (IMO, 2012, Table 9.1). In 2010, shipping employed over 1.4 million seafarers from 150 states.

At present, 35 states control 95 per cent of the world's shipping fleet. Of these states, Greece, Japan, China, Germany, the United States and Singapore control approximately 50 per cent of the world's shipping fleet (by DWT), and, since 2000, China has led the world in the growth of shipping (Branch and Roberts, 2014). The most popular registries of ships (states in which ships are registered) are Panama, Liberia and the Marshal Islands, which registered 14,280 ships in 2016. This amounts to greater than 41 per cent of the world's shipping capacity (UNCTAD, 2017).

Box 9.1 Methods of categorizing ship size

Dead weight tonnage (dwt): The number of tons (2,240 lb) of cargo, stores and fuel a vessel can transport.

Cargo tonnage: The weight of cargo carried by the vessel, measured either in American short tons (2,000 lb), English (imperial) long tons (2,240 lb) or metric tonnes (1,000 kg). Note that one imperial ton is only slightly more than a metric tonne (1 tonne = 0.98 ton).

Gross tonnage (GT, G.T or gt): Volume of a vessel's enclosed spaces divided by 100. A GT is 100 cubic feet. Gross tonnage is not a measure of mass (weight) but volume.

Net tonnage (NT, N.T. or nt): Gross tonnage minus all areas of a ship not used to transport cargo. Also referred to as 'earning space'. Net tonnage is not a measure of mass (weight) but volume.

Table 9.1 World fleet size in 2016 by principal types of vessel

Vessel type	Thousands of dwt	Percentage of fleet
Bulk carriers	778,890	43.1
Oil tankers	503,343	27.9
General cargo ships	75,258	4.2
Container ships	244,274	13.5
Other types of ship	24,284	1.3
Liquefied gas carriers	54,469	3.0
Offshore supply	75,836	4.2
Chemical tankers	44,347	2.5
Ferries and passenger ships	5,950	0.3
Total	1,806, 650	100

Source: UNCTAD (2017)

Modern merchant shipping is performed by the following types of ships:

- **Container ships** carry containers that can be on- and off-loaded onto trucks and trains (referred to as 'intermodal freight transport' or 'fully cellular'). The size of container ships is based on how many 20-foot equivalent units (TEU) containers the ship can transport. Container ships have grown quickly, and the very largest container ships currently have capacities of more than 19,000 TEU, with several more rated at 18,000 TEU.
- **General cargo ships** include refrigerated cargo, specialized cargo, roll on-roll off (ro-ro – used to transport vehicles) and other types of special purpose ships.
- **Bulk carriers** transport raw materials including grain, ore and coal.
- **Tankers** transport crude oil, chemicals, cooking oils and other liquids. Some specialized tankers carry cargoes such as liquefied natural gas.
- **Ferries and cruise ships** transport passengers, cars and commercial vehicles.
- **Specialist ships** include supply vessels, tugs, ice breakers and research vessels.

In addition to classification of ships by purpose, ships are often categorized by size. Their categorization is according to their ability to transit the world's major canals and straights, as well as the types of ports they are able to service (Table 9.2).

Shipping can be further differentiated into the following discrete sectors operating under distinct commercial and regulatory regimes:

Passenger shipping services

The IMO defines a passenger ship as any vessel carrying or configured to carry more than 12 non-crew passengers. Passenger ships include ferries, ocean liners and cruise ships and are normally separated into two types: The **ferry fleet** consists of ships used for short voyages without accommodations that may also carry vehicles and cargo. The ferry fleet consists of conventional ferries and fast ferries. The latter includes monohull ships, catamarans, hydrofoils and hovercraft (Branch and Roberts, 2014). The **cruise fleet** consists of ships specifically designed for longer voyages. Currently, the global passenger fleet consists of 1,675 passenger-only vessels (including cruise ships) and 2,641 ships suitable for passengers and freight (ISL, 2016a). Worldwide, more

Table 9.2 Global ship classifications and sizes

Ship type	Applicability	Dead weight tonnes*	Maximum length (m)	Maximum draft (depth m)	Maximum air draft (height m)	Maximum beam (width m)
Handysize	Dry bulk carriers serving smaller ports	< 40,000	200	Unlimited	Unlimited	Unlimited
Handymax	Dry bulk carriers serving smaller ports	40,000–50,000	200	Unlimited	Unlimited	Unlimited
Supramax	Dry bulk carriers serving smaller ports	50,000 to 60,000	200	Unlimited	Unlimited	Unlimited
Suezmax	Maximum ship size to transit the Suez Canal	Unlimited but normally 120,000–150,000	Unlimited	20.1	68	50
Panamax	Maximum ship size to transit the Panama Canal	Unlimited but normally 50,000–80,000	290	12	58	32
Aframax	Oil tankers able to serve most ports. Named after 'Average Freight Rate Assessment'	< 120,000	Unlimited	Unlimited	Unlimited	44
Chinamax	Bulk carries able to serve most Chinese ports	< 400,000	360	24	Unlimited	65
Capesize	Bulk carriers and oil tankers too large for the Panama and Suez canals. Includes very large ore carriers (VLOC) and very large bulk carriers (VLBC) that service deepwater ports	150,000–400,000+	Unlimited	Unlimited	Unlimited	Unlimited
Malaccamax	Ships (bulk, oil or container ships) able to transit the Strait of Malacca	Unlimited	400	25	Unlimited	59
Very large crude carrier (VLCC)	Oil tankers for long-haul crude oil transport	200,000–349,999	Unlimited	Unlimited	Unlimited	60
Ultra large crude carrier (ULCC)	Oil tankers for long-haul crude oil transport	350,000+	Unlimited	Unlimited	Unlimited	Unlimited

*Sum of the weights of cargo, fuel, freshwater, ballast water, provisions, passengers and crew

Source: UNCTAD (2017) and IMO (2012)

than 2 billion passengers were transported in 2010, and the total number of cruise ship passengers was almost 21 million (IHS Fairplay, 2011). In 2011, more than 400 million passengers passed through European sea ports (UNCTAD, 2016).

The fastest growing segment of passenger shipping over the past two decades has been the cruise fleet. This segment consists of vessels that carry upwards of 3,000 passengers. The cruise market primarily serves the Mediterranean, Caribbean and Baltic Seas, the Atlantic Islands and North American and European rivers.

Passenger shipping is regulated through the international conventions (see below) and domestic law. In addition, the European Union regulates passenger shipping through a number of different legislative instruments including European Union directives.

Liner cargo services

Liner cargo services transport cargo for several shippers simultaneously and operate on regular schedules between ports ('liner trades'). This contrasts with bulk shipping (see below) where contracts are made with a single shipper on an as-needed basis. Vessels used in the liner trades include general cargo, container, refrigerated (reefer), roll-on/roll-off (ro-ro) and other specialized ships. The liner trade is increasingly multi-modal and comprised of large shipping companies that are able to guarantee delivery at specific times.

Currently, container shipping represents over 75 per cent of the liner trade by volume. Annually, half of the world's trade is shipped in over 15 million containers making over 230 million journeys. Over 700 million TEUs were shipped in 2014 with 5,224 container ships in operation in 2016 (ISL, 2016b; UNCTAD, 2017). More than 600 new container ships are on order.

Bulk cargo services

Bulk cargo services exploit a variety of ship configurations to convey their payloads around the planet. They carry petroleum products (e.g. crude oil, condensates and refined products), liquefied gasses (e.g. liquefied natural gas and liquid propane gas), liquid chemicals (e.g. cooking oil), foodstuffs (e.g. grains, sugar) or solids (e.g. ore, coal, sulphur, bauxite/alumina). In 2015, 11,289 bulk carriers were in service around the world, and 45 per cent of the world's crude oil production was shipped by sea (EMSA, 2016). This represents 1.8 billion tons, or about 70 per cent of the dwt shipped by bulk cargo services (IMO, 2012).

Bulk cargo ships are generally purpose-built for specific cargoes and bulk shipping companies often focus on specific cargoes (e.g. oil tankers). Ships participating in the bulk trades generally do not operate on scheduled services but are generally exclusively chartered/contracted by shippers on long-term contracts. This model incents competition between shipping companies but, compared to the liner trades, results in an older fleet. Bulk cargo services are not normally part of the liner conventions (see below). Instead, bulk shipping is regulated by national and international laws. As opposed to the liner trades where anti-competitive practices are a focus of regulation, bulk shipping focuses primarily on ship safety and environmental protection.

Tramp services

Tramp ships, or general traders, are cargo ships, normally with two to six holds, that carry a variety of cargos, including grains, ores, forest products and foodstuffs. Tramp services operate throughout the world but are prevalent in East and South Asia. These services are *ad hoc* in nature and do not adhere to fixed schedules. Rather, they anticipate and exploit niches and seasonal markets ('spot markets') that are often too uneconomical for liner and bulk cargo services. Tramp services are generally composed of smaller ships, are often family owned and respond quickly to market conditions and opportunities as they arise. However, many liner service shipping companies are acquiring tramp services to horizontally and vertically integrate their businesses. While tramp services are not as prevalent as they once were (outside of Asia), they continue to operate in densely populated regions such as the North American Great Lakes, the Baltic, the Mediterranean and other enclosed seas.

Introduction to marine transportation law and policy

The need to regulate shipping

Terrestrial transportation (truck, rail) is primarily focussed on moving goods within national boundaries or to-from adjacent states. In contrast, maritime shipping mostly occurs between states and often within ABNJ. Shipping is therefore inherently international in nature and requires cooperation amongst states in order to move goods efficiently and safely between ports. Shipping regulation is required for a number of reasons, including:

- Ensuring common environmental protection standards.
- Ensuring common safety standards for ship operations.
- Ensuring common crew training standards.
- Holding states and shippers accountable for accidents and negligence.
- Preventing and responding to piracy and maritime terrorism.
- Developing construction and design standards.
- Establishing navigation rules.
- Establishing communications protocols.
- Ensuring cargo liability.
- Ensuring safe and effective ship disposal.
- Meeting greenhouse gas reduction targets and adapting to climate change impacts.

International shipping is a unique economic activity where carriers from multiple states compete with one another to move goods or passengers between ports (often between states) at the lowest cost. States may promote or inhibit competition through subsidies, trade barriers, or transportation policies (Box 9.2). There are few economic incentives for shipping companies to invest in safety or environmental protection, especially if their competition fails to do so. Similarly, without shipping regulation or effective insurance regimes (see P&I clubs below), ship operators have no incentive to expend financial resources to clean up accidents, compensate shippers for lost cargoes or compensate others who may have been negatively affected by the ship operator. Compounding this situation is that most of the world's ships are registered in flag states (under flags of convenience – see Box 9.2) who may not have the desire or financial resources to enforce standards or make ship owners accountable for their actions. Lastly, responding to certain marine accidents (e.g. oil spills) is often out of financial reach for most ship owners. As such, regulation is needed to enable and compel those responsible for costly accidents and damages to be able to fund clean up and financially compensate affected third parties.

States also have a vested interest in shipping and in particular are concerned with ensuring:

- The safety of passengers, crews and cargos.
- The orderly movement of goods and people across waterways.
- Access to domestic and international shipping routes.
- Sufficient competition without compromising the ability for shipping companies, when necessary, to cooperate to provide scheduled, uninterrupted liner services.

Box 9.2 An overview of commercial shipping

Parties involved in commercial shipping

Carrier: The party transporting the goods or cargoes.

Coastal state: Any state with a marine coastline. To protect their interests, coastal states establish laws and regulations with which vessels must comply. The jurisdiction, duties and responsibilities of coastal states are set out in the LOSC.

Consignee: The party (agent, company or person) that receives goods once they reach their destination port (the 'receiver').

Consignor: The party (agent, company or person) sending consignment (the 'shipper' or 'merchant').

Classification societies: Provide technical and surveying services either on behalf of the ship's owner so that the ship can be issued a 'class certificate' (for insurance and operational purposes) or on behalf of flag states as a 'recognized organization'. There are 11 globally recognized classification societies currently in operation.

Flag state: Responsible for verifying that domestic (national) ships comply with national laws and international conventions to which the flag state is a party. These responsibilities are known as flag state control (FSC). Classification societies are often used to implement FSC.

International governmental organizations: Comité Maritime International, International Labour Organization and International Maritime Organization are discussed in the following sections.

Port state: Responsible for verifying compliance (including inspections and boarding) with international conventions for ships in their territorial waters. These responsibilities are known as port state control (PSC). PSC regimes are regional, enabling states to share verification responsibilities. The major PSC regimes include: Paris MoU (Europe and North Atlantic, 27 states), Tokyo MoU (Asia and the Pacific, 20 states), Indian Ocean MoU (Indian Ocean, 18 states), US Coast Guard, Vina del Mar Agreement (Latin America 15 states), Abuja MoU (West and Central Africa, 16 states), Black Sea MoU (Black Sea, 6 states) and Riyadh MoU (Persian Gulf, 6 states).

Protection and Indemnity (P&I) clubs: Co-operatives run by ship owners or operators to provide third-party marine insurance to their members. The 13 recognized P&I clubs are non-profit and insure for risks that the conventional insurance market refuses to cover.

Contracts of carriage

Bill of lading: Document that gives title (constructive possession) to the goods and the condition of goods at the time of loading. Usually prepared by the consignor, the bill of lading will state the date(s) goods were loaded and ports of loading and discharge. Since the bill of lading is transferrable, cargo may be sold many times during transit.

Charter party: A formal written contract between the carrier and cargo owner for the exclusive use of a ship for either a period of time ('time charter') or port-to-port voyage ('voyage charter'). In a 'demise' charter party (or bareboat charter), the charterer is responsible for providing the cargo and crew, whereas with a 'non-demise' charter party, the ship owner provides the crew and the charterer provides the cargo.

Waybill: A document outlining the terms and conditions of the contract of carriage as well as a receipt for the goods loaded aboard the carrying vessel. Under a waybill, the shipper owns the cargo until delivery. Unlike a bill of lading, a waybill is non-negotiable and not a document of title. Moreover, under a waybill a cargo cannot be bought or sold during transit under a waybill.

Types of shipping contracts

Free on board (fob) contract: The buyer pays for and arranges carriage. The seller's duty is to load the goods onto the ship.

Cost, insurance, freight (cif) contract: The seller pays for and arranges carriage as well as purchases an insurance policy on the goods, which will be assigned to the buyer.

Other policy-related shipping terminology

Cabotage: Most nations apply some elements of cabotage, where they reserve (protect) a portion of their domestic shipping trades for ships flying their national flag. States may require ships to be crewed or owned by domestic nationals, built at domestic shipyards or registered under the country's national flag.

Flags of convenience: Ships that are registered in states different than the states where the ship is owned/operated. Traditionally, flags of convenience have been used to reduce costs (avoid taxation, utilize lower wage crews) and avoid compliance with domestic regulations that may be more onerous.

Short Sea Shipping: Movement of cargo between any port on the Baltic, Black or Mediterranean seas.

Organizations that regulate marine transportation

Global shipping is regulated by over 150 conventions and agreements. In addition, the Organization for Economic Cooperation and Development (OECD) and World Trade Organization oversee economic competitiveness in shipping. While the key conventions are discussed later in this chapter, the following organizations oversee the regulation of the international shipping industry.

The LOSC and marine transportation

The LOSC (see Chapter 4) is not an organization per se, nor is it formally administered by the UN General Assembly. However, the Convention is a key regulatory instrument that informs the operations of the International Maritime Organization. The LOSC regulates marine transportation in two ways: First, the Convention enshrines a number of overarching shipping principles. Among these primary principles are the rights of innocent passage and transit passage; immunity of warships; freedom of the high seas and use of the high seas for peaceful purposes; rights of all states to participate in shipping; prohibitions on carrying slaves; the necessity of combating piracy; and the right of all states to fish.

Second, the LOSC provides the administrative framework for shipping. Within this framework lie the rights and duties of coastal and flag states by LOSC jurisdiction as well as the rights of states to determine the conditions for granting of their nationality to ships and standards for manning and training crews. The requirement that all flag states ensure that vessels flying their flag meet international shipping regulations also falls within this framework.

International Maritime Organization

Despite the importance of the LOSC, shipping is primarily regulated by the International Maritime Organization and the International Labour Organization. The United Nations established the International Maritime Organization (IMO) in 1958. Headquartered in London, England, the IMO is comprised of a 172-member Assembly (composed of states and three associate members), responsible for establishing the rules of international shipping. These rules address maritime safety, security and prevention of marine pollution from ships. The IMO is governed by a Council that is responsible for oversight of the organization. Between Assembly meetings the Council also performs all the functions of the Assembly except making recommendations to governments (including state parties) on maritime safety and pollution prevention.

The IMO's authority only applies to international shipping. Member states are expected to implement and enforce the 50 IMO conventions, protocols, as well as the codes and recommendations, which now exceed 800 in number. The work of the IMO, under direction from the Council, is undertaken by a number of committees and sub-committees. The current committees are:

- The *Maritime Safety Committee* is the senior IMO committee and responsible for all matters directly affecting marine safety. The Committee is responsible for nine sub-committees on: 1) safety of navigation; 2) radio communications and search and rescue; 3) standards of training and watchkeeping; 4) ship design and equipment; 5) fire protection; 6) stability, load lines and fishing vessel safety; 7) flag state implementation; 8) dangerous goods, solid cargoes and containers; and 9) bulk liquids and gases.
- The *Marine Environment Protection Committee* (MEPC) is responsible for the prevention and control of pollution from ships.
- The *Legal Committee* is responsible for all legal matters within the scope of the IMO.
- The *Technical Cooperation Committee* is responsible for providing technical assistance to member states and has a particular focus on ensuring less developed states are capable of meeting IMO conventions, protocols, codes and recommendations.
- The *Facilitation Committee* is responsible for simplifying the documentation and formalities in international shipping.

The draft conventions produced by these committees are submitted to conferences open to any member of the UN, including UN members that may not be IMO members. All governments, whether coastal, seafaring or landlocked, have the same status and participation in IMO discussions. Similar to other international conventions, the conference adopts a final text that is provided to member governments for ratification. The convention enters force once it is ratified by a set number of member states. Non-IMO members may accede to the convention. Each convention provides the conditions to be met prior to entering into force. For example, the International Convention for the Safety of Life at Sea (see below) required the ratification of at least 25 states whose merchant fleets comprised not less than 50 per cent of the world's tonnage. Conventions are binding on IMO member states while IMO codes and regulations are non-binding. Many states must pass or amend domestic laws to comply with IMO conventions (see Chapter 3).

Comité Maritime International

The Comité Maritime International (CMI), headquartered in Antwerp, Belgium, is a non-governmental organization established in 1897 to oversee the codification of maritime law, commercial practices and the promotion of national associations of maritime law. At present, the CMI is comprised of 51 national maritime law associations, including the IMO and IOPC Fund (see below).

Early work of the CMI focussed on regulations concerning collision and salvage. Among the 22 conventions produced between 1910 and 1969 are the 1952 Arrest of Sea-Going ships and 1968 Hague Rules (Table 3.2). Also, the CMI has assisted on the drafting of many IMO conventions, including the 1994 International Convention on Maritime Liens and Mortgages, the 1997 International Convention Relating to the Arrest of Sea-Going Ships and the 2007 Nairobi International Convention on the Removal of Wrecks. In addition, the CMI continues to be the responsible authority for international conventions of bills of lading and salvage (Table 3.2) (CMI, 2012).

International Labour Organization

The International Labour Organization (ILO) is responsible for establishing and overseeing international labour standards. Established in 1919 as part of the Treaty of Versailles (the accord that officially ended World War I), the ILO works to improve working conditions, worker safety and ensure equal pay for equal work between states. The ILO was established under the premise that labour peace is fundamental to national prosperity. The organization has four strategic objectives:

1 Promote and realize standards and fundamental principles and rights at work.
2 Create improved employment opportunities for women and men.
3 Enhance the coverage and effectiveness of social protection for all.
4 Strengthen tripartism and social dialogue.

Currently, the ILO consists of 187 member states and is headquartered in Geneva, Switzerland. The organization oversees 190 conventions and 190 recommendations that cover all aspects of employment. A key responsibility of the ILO is to oversee labour standards of the world's 1.5 million seafarers and 15 million fishers of various nationalities who crew ships that are often owned and flagged in different states. This is required because national labour codes may fail to protect seafarers and, in an effort to maximize profitability, some ships owners compromise worker safety and recruit low wage crews.

The ILO's primary contribution to maritime law is the 2006 Maritime Labour Convention, which is discussed below.

International law dealing with the ship

Key international agreements and conventions regulating the configuration and operation of ships are discussed below; however, four conventions are frequently referred to as the 'pillars' of maritime shipping law and are as follows:

1 International Convention for the Safety of Life at Sea (SOLAS).
2 International Convention on Standards of Training, Certification and Watchkeeping for Seafarers (STCW).
3 Maritime Labour Convention (MLC).

4   International Convention for the Prevention of Pollution from Ships, 1973 as modified by the Protocol of 1978 (MARPOL 73/78).

International Convention for the safety of life at Sea (SOLAS)

On 15 April 1912, the *Titanic* struck an iceberg and sank, leaving 1,503 casualties. The British Wreck Commissioner's Inquiry formed numerous conclusions from the tragedy. While the vessel was in full compliance with the United Kingdom *Merchant Shipping Acts* (1894–1906), they determined that these *Acts* were established for ships 13,000 tons or less. In the case of the 46,328-ton *Titanic*, this resulted in an insufficient number of lifeboats to accommodate her passengers. Since the iceberg was reported to the bridge, the Inquiry also concluded that the *Titanic's* speed was a factor in the accident. On top of this, the Inquiry noted that passenger lifeboat or safety drills were never conducted, nor were crews trained in the deployment and use of the safety equipment. Meanwhile, the ship's lack of a public address system and insufficient range of radio communications equipment were contributing factors (Titanic Inquiry, 2019).

Following the *Titanic* calamity, the first Safety of Life at Sea conference was convened in the United Kingdom in 1914 (IMO, 1974). Attended by 13 states, the conference produced regulations on emergency equipment (including lifeboats), emergency procedures and it prescribed new radio communication procedures. The 1914 SOLAS convention was updated in 1929, 1948 and 1960.

In 1974, in an attempt to maintain pace with the considerable number of amendments to the Convention, the Convention was rewritten completely and referred to as SOLAS 1974. The Convention was adopted 1 November 1974 by the International Conference on Safety of Life at Sea and entered into force on 25 May 1980. SOLAS was structured such that all future amendments would be implemented automatically unless the majority of members objected. Additional protocols were adopted in 1978 (International Conference on Tanker Safety and Pollution Prevention – the 1978 SOLAS Protocol – entered into force in 1981) and in 1988 (the International Conference on the Harmonized System of Survey and Certification – the 1988 SOLAS Protocol – entered into force in 2000) (IMO, 2004). The SOLAS 1974 convention has also been amended a further 58 times through resolutions approved by the IMO Maritime Safety Committee or by conferences of SOLAS's contracting governments. Furthermore, SOLAS 1974 encompasses over 30 mandatory codes and other instruments.

SOLAS 1974 contains 12 Chapters. They address:

- Chapter I: Definitions of types of ships and explanations of types of surveys and certificates.
- Chapter II-1: Structure, subdivision and stability, machinery and electrical installations related to ship construction.
- Chapter II-2: Fire protection, fire detection and fire extinction related to ship construction.
- Chapter III: Life-saving appliances and arrangements.
- Chapter IV: Radio communications.
- Chapter V: Safety of navigation.
- Chapter VI: Carriage of cargoes.
- Chapter VII: Carriage of dangerous goods.
- Chapter VIII: Nuclear ships.
- Chapter IX: Management for the safe operation of ships.
- Chapter X: Safety measures for high-speed craft.
- Chapter XI-1: Special measures to enhance maritime safety.
- Chapter XI-2: Special measures to enhance maritime security.
- Chapter XII: Additional safety measures for bulk carriers.

SOLAS applies to all ships and requires ship construction and equipping to conform to SOLAS on or after the dates specified in various regulations. SOLAS can be retroactive and may require modifications to existing ships to conform with new regulations. The SOLAS Conventions have been managed by the IMO since 1959.

Convention on the International Regulations for Preventing collisions at Sea

The Convention on the International Regulations for Preventing Collisions at Sea (COLREG) was adopted 20 October 1972 and entered into force 15 July 1977 (IMO, 1972). The Convention replaced the earlier 1960 International Regulations for Preventing Collisions at Sea, which had been based on a set of British rules formulated in 1862. The COLREG is best conceived as the 'rules of the road' for navigation and navigation safety and contains 38 Rules. The rules are divided into sections related to steering and sailing (Part B); lights and shapes (Part C); sound and light signals (Part D); and exemptions (Part D). Amendments to the Convention were adopted in 1981, 1987, 1989, 1993, 2001, 2007 and 2013. A short summary of the key Rules is as follows:

- Rule 1: The Convention applies to all vessels upon the high seas and in all waters connected to the high seas navigable by sea-going vessels. States reserve the right to retain and implement additional special rules in their inland waters or for specific operations so long as they are consistent with the Convention and do not compromise safety.
- Rule 2: Responsibility for complying with the Convention rests with vessel owners, masters and crew. Vessels may deviate from the Convention to avoid immediate danger.
- Rule 4: Rules apply in any condition of visibility.
- Rule 5: Every vessel shall at all times maintain a proper look-out by sight and hearing as well as by all available means appropriate in the prevailing circumstances and conditions.
- Rule 6: Vessels will be operated at speeds appropriate to the prevailing circumstances and conditions.
- Rule 7: Vessels will use all means to evaluate collision risks.
- Rule 8: Rules to avoid collisions.
- Rule 9: Rules for operation in narrow channels.
- Rule 10: Rules for traffic separation schemes.
- Rule 12: Rules for sailing vessels.
- Rule 13: Rules for overtaking.
- Rule 14: Rules for head-on situations.
- Rule 15: Rules for crossing situations.
- Rule 16: Rules for the 'give-way' vessel.
- Rule 17: Rules for the 'stand-on' vessel.
- Rule 18: Exceptions for Rules 9, 10 and 13.
- Rule 19: Conduct of vessels in restricted visibility.
- Rules 22–31: Lighting requirements in various circumstances.

International Convention on Load Lines

The International Convention on Load Lines (CLL 66) was adopted 5 April 1966 in London and entered into force 21 July 1968 (IMO, 1966). The purpose of the Convention is to prescribe loading limits on ships in terms of the amount of freeboard (draft) available to prevent overloading and stability issues. The Convention considers geographic location and season when setting freeboards (the height of a ship's deck above the water level) as well as the regulations for any access points through the hull below the water line. The Convention was amended by a Protocol in 1988 that entered into force 3 February 2000, and permits the Convention to be amended either by the IMO's Maritime Safety Committee or by a CoP. In 2003, the Convention was revised (entered into force 1 January 2005) to include additional regulations to improve the design and configuration of ships. Currently, the Convention has 40 signatories and was most recently amended in 2013.

International Convention on Arrest of Ships

The International Convention on Arrest of Ships (the 'ARREST' Convention) was adopted 12 March 1999 and entered into force 14 September 2011 (IMO, 1999). The Convention updated the 1952 Brussels Convention on the Arrest of Sea-Going Ships. Its purpose is to balance the interests of claimants seeking maritime claims against the owners and operators of ships. Specifically, the Convention outlines the conditions under which a ship may be arrested or released from arrest. Shipping is unique in international law in that a ship – rather than its owner – may be arrested. This is because tracing ownerships of ships can be very difficult.

The Convention applies to any ship within the jurisdiction of a contracting state and covers 22 types of marine claims. Ships may be arrested without proof of their liability but can only be arrested under the authority of a court of the state party in which the arrest is affected and only as a result of a marine claim (Article 2). Ships may be arrested only when owned or operated by the same individuals that owned or operated the ship at the time of the maritime claim (Article 3). Ships must be released when security (payment) has been received that is satisfactory to the state party in which the arrest was affected (Article 4.1). Courts will determine the amount of security if parties cannot agree but security must not be more than the value of the arrested ship (Article 4.2). Payment of the security does not acknowledge liability on behalf of the ship owner or operator (Article 4.3). Ships may be re-arrested if the security previously provided is inadequate or the owner or operator has not provided the security (Article 5). Furthermore, the arrest of multiple ships with the same owner or operator is permitted if the security provided for a marine claim is insufficient (Article 5). On the other hand, state courts may award damages for unjustified or wrongful arrest or in situations where the security was set improperly high (Article 6). Lastly, state courts have jurisdiction over these matters unless both parties agree either to submit the case to a court in another jurisdiction or send the case to arbitration (Article 7).

The Athens Convention relating to the Carriage of Passengers and their Luggage by Sea

The Athens Convention relating to the Carriage of Passengers and their Luggage by Sea (abbreviated as 'PAL') was adopted 13 December 1974 and entered into force 28 April 1987. The Convention's purpose is to harmonize regulations relating to passengers and their luggage and establish maximum liability limits for the carrier in the event that an incident was a result of the fault or neglect of the carrier. The Convention places the onus on the carrier to demonstrate it was not at fault.

The Convention is amended by three Protocols: The 1976 Protocol (entered into force 30 April 1989) substituted SDRs for the gold franc, the 1990 Protocol (not yet entered into force) increased the compensation limits and the 2002 Protocol introduced the principles of strict liability and compulsory insurance. The 2002 Protocol requires that the Convention and Protocol be adopted as a single instrument titled the Athens Convention relating to the Carriage of Passengers and their Luggage by Sea, 2002. It entered into force on 23 April 2014. The 2002 Convention increases liability limits from to 250,000 SDR per passenger, 2,250 SDR for cabin luggage and 12,700 SDR per vehicle.

International law dealing with the shipping company

International safety management code

The International Safety Management Code (ISMC) is not a stand-alone convention but an amendment to Chapter IX of SOLAS, adopted on 24 May 1994. The Code was enacted after previous voluntary guidelines agreed to in 1989, 1991 and 1993 were deemed to be of sufficient importance to make them mandatory (IMO, 2010). Its objectives are to ensure safety at sea, prevent human injury or loss of life and avoid damage to property and the environment. The Code became compulsory for all ships in 2002 and requires companies to implement a safety management system with the following components:

- A safety and environmental protection policy.
- Instructions and procedures to ensure safe operation of ships and protection of the environment in compliance with relevant international and flag state legislation.
- Authority and lines of communication between, and amongst, shore and shipboard personnel.
- Procedures for reporting accidents and non-conformities with the provisions of the Code.
- Procedures to prepare for and respond to emergency situations.
- Procedures for internal audits and management reviews.

United Nations Convention on Contracts for the International Carriage of Goods Wholly or Partly by Sea

The United Nations Convention on Contracts for the International Carriage of Goods Wholly or Partly by Sea (the 'Rotterdam Rules') was adopted by the UN General Assembly on 11 December 2008. The Rotterdam Rules replaced a number of previous conventions, including the 1924 International Convention for the Unification of Certain Rules of Law relating to Bills of Lading ('the Hague Rules' and the 'the Hague-Visby Rules') and the 1978 United Nations Convention on the Carriage of Goods by Sea ('the Hamburg Rules') (UNCITRAL, 2009). Sixteen states were initial signatories to the Convention while 24 are currently party to Convention. Currently, only three states (Congo, Spain, Togo) have ratified the convention.

The Rotterdam rules were negotiated over a 10-year period and coordinated by the United Nations Commission for International Trade Law (UNCITRAL). The Rotterdam rules address modern transport practices, ranging from containerization, door-to-door transport contracts and the use of electronic transport documents (UNCITRAL, 2009). In doing so, the Convention provides greater certainty for cargo owners. These assurances are provided through increases in the monetary limits on a carrier's liability (for loss, damage, or delay), removal of previous exceptions for unforeseen events (e.g. fire, navigation error) and clear articulation of responsibility and liability for all aspects of the transport process.

The Convention applies between the carrier and the consignee in any situation where the port of loading and port of discharge are in different states and the rules apply regardless of the vessel's nationality (Article 5). The general outline of the Convention is as follows:

- Chapter 1: General provisions and definitions.
- Chapter 2: Scope of the Convention and exclusions.
- Chapter 3: Electronic transport records and their application.
- Chapter 4: Obligations of the carrier.
- Chapter 5: Liability of the carrier for loss, damage or delay.
- Chapter 6: Additional provisions relating to particular stages of carriage.
- Chapter 7: Obligations of the shipper to the carrier.
- Chapter 8: Transport documents and electronic transport records.
- Chapter 9: Delivery of the goods.
- Chapter 10: Rights of the controlling party.
- Chapter 11: Transfer of rights.
- Chapter 12: Limits of liability.
- Chapter 13: Time for suit.
- Chapter 14: Jurisdiction.
- Chapter 15: Arbitration.
- Chapter 16: Validity of contractual terms.
- Chapter 17: Matters not governed by the Convention.
- Chapter 18: Final clauses.

UN Convention on a Code of Conduct for Liner Conferences

The liner trades utilize 'liner conferences' (also termed 'shipping conferences') to fashion formal or informal agreements between shipping companies. These conferences enable them to set freight rates or cooperate on the movement of goods between ports while avoiding antitrust scrutiny. Not coincidentally, liner conferences formed shortly after the opening of the Suez Canal (1869) – an engineering feat that halved the shipping time on the Asia-Europe trade route and that led swiftly to massive overcapacity. Since then, the objectives of liner conferences have been to avoid over-competition, to accommodate seasonal fluctuations in shipping demand and to provide economic stability to shippers (Munari, 2012). Liner conferences currently involve up to 40 separate shipping companies, and there are currently over 300 liner conferences in operation. While evidence that liner conferences create anti-competitive cartels is mixed (Clyde and Reitzes, 1998), most states view conferences in this light. Until recently, conferences were tolerated or exempted from most antitrust laws as they seemed to provide shippers with a stable access to their markets. This stability was perceived to have lucrative ramifications. Indeed, as far back as 1909, a British Royal Commission on Shipping Rings concluded that liner conferences were anti-competitive but, nonetheless, overall beneficial to the British economy (Liu, 2009).

The Convention on a Code of Conduct for Liner Conferences ('UN Liner Code') was signed in Geneva, 6 April 1974, and entered into force on 6 October 1983. The purpose of the Convention is to:

- Facilitate the orderly expansion of seaborne trade.
- Stimulate the development of regular and efficient liner services.
- Balance the interests between suppliers and users of liner shipping services.
- Foster communication and collaboration between shippers and liner services.

Many states, however, began to use the Convention to reserve a portion of the shipping trade for national shipping lines. Liner conferences thus became a means by which national shipping lines negotiated their involvement in international trade. The Convention was also used by less develop states – formerly excluded from participating in international shipping – to become involved in shipping. The Convention developed the 40:40:20 rule where 80 per cent of the shipping trade was reserved for national shipping lines of the source (40 per cent) and destination (40 per cent) states and 20 per cent made available for third-party shipping lines (Munari, 2012). Currently, there are 21 signatories and 76 parties to the Convention.

Since the adoption of the UN Liner Code, the influence of liner conventions and monopolistic shipping behaviour has waned. New entrants into the shipping business, relaxation of price fixing by shipping companies, growth of intermodal transportation and changes to competition laws in various nations have freed up the market. The Code has also enabled increased information on supply and demand trends in order to anticipate shipping needs. On top of this, monopolistic practices ebbed when the European Union, in 2006, repealed the exemption from the European Commission Treaty ban on restrictive business practices (Article 81) for liner conferences on routes to and from the European Union. The repealed exemption came into force in October 2008 (Hummels et al., 2009).

International law dealing with seafarer

On 25 October 1859, 195 ships were wrecked, and 685 souls perished in a single night in Northwest Europe. Casualties from the passenger ship *Royal Charter* were in excess of 450, and the storm is referred to eponymously as the *Royal Charter* Storm. By the end of the series of storms on 9 November 1859, 325 ships had been lost with 748 passengers and crew perishing. These losses triggered a global dialogue on how to improve marine safety and avoid such events in the future.

While merchant seafaring is considerably safer today when compared to several decades ago, it continues to be a high-risk occupation. Rates of homicide, suicide, illness and unexplained disappearances are higher for the 1.4 million seafarers than the general population. Furthermore, incidents are often not easy to investigate and prosecute given the international nature of shipping and the fact that incidents often occur in international waters. For example, the fatal accident rate for British merchant seafarers between 2003 and 2012 was 21 times that of the general British workforce. Overall, from 2000–2012, there were 4603 fatalities of which 65% of lives lost were on passenger ships (Roberts et al., 2014).

Potentially contributing to these risks is a global workforce from developing and east European states that is largely outsourced using a 'just in time' casual pool of labour. This workforce may have very different levels of training, certification and experience that may influence the safety of vessels and passengers. Currently, nine nations (Philippines, Russia, the Ukraine, China, India, Poland, Indonesia, Turkey and Myanmar) comprise two thirds of the international merchant crews. These crews are not employed by shipping companies but by international crewing agencies. Over the last decade, the frequency of ship accidents increased while the consequences of these accidents remained stable over this time.

While the IMO, through its various Conventions, regulates most aspects of vessel safety, the ILO is the primary regulator of the seafarer.

Maritime Labour Convention

The ILO is responsible for the 2006 Maritime Labour Convention (MLC), which consolidates more than 68 maritime labour standards adopted over 80 years (ILO, 2006). The MLC was adopted on 23 February 2006 in Geneva and entered into force on 20 August 2013. The purpose of the Convention, also called the 'seafarer's bill of rights', is to establish working conditions for seafarers and ensure fair competition for ship owners. The MLC applies to all public and private ships engaged in commercial activities except fishing vessels, traditional vessels (e.g. dhows and junks) and vessels used solely within a state's internal waters (Article 2). Like other international legal instruments, the MLC does not apply directly to seafarers or ship owners/operators. Instead, it requires state parties to the Convention to pass domestic legislation or enact regulations consistent with the Convention. As such, the MLC provides considerable flexibility to states to develop new or use existing standards that are equivalent to the requirements in the Convention.

The Convention consists of Articles and Regulations that set out the basic principles and responsibilities of state parties to the Convention. Under each Regulation, the Code contains both mandatory standards (Part A) and voluntary (Part B) guidelines. Key Articles embodied within the MLC are:

- Article III: Parties to the Convention will: permit seafarers the right to collective bargaining and union association; eliminate forced labour; abolish child labour and eliminate discrimination.
- Article IV: Seafarers have the right to a safe work environment, medical services, decent living conditions and fair terms of employment.
- Article V: States must ensure that, for ships flying their flag, they carry evidence that they are in compliance with MLC. Flag states must also undertake inspections and oversee seafarer recruitment in their jurisdictions. Port states may inspect ships of another flag when they are in port of a port state.

The Regulations are separated into the following five Titles under which Parts A and B of the Code are supplied:

Title 1: Minimum requirements for seafarers to work on a ship

- Regulation 1.1: Minimum age.
- Regulation 1.2: Medical certificate.

Title 2: Conditions of employment

- Regulation 2.1: Seafarers' employment agreements.
- Regulation 2.2: Wages.
- Regulation 2.3: Hours of work and hours of rest.
- Regulation 2.4: Entitlement to leave.
- Regulation 2.5: Repatriation.
- Regulation 2.6: Seafarer compensation for the ship's loss or foundering.
- Regulation 2.7: Manning levels.
- Regulation 2.8: Career and skill development and opportunities for seafarers' employment.

Title 3: Accommodation, recreational facilities, food and catering

- Regulation 3.1: Accommodation and recreational facilities.
- Regulation 3.2: Food and catering.

Title 4: Health protection, medical care, welfare and social security protection

- Regulation 4.1: Medical care on board ship and ashore.
- Regulation 4.2: Ship owners' liability.

- Regulation 4.3: Health and safety protection and accident prevention.
- Regulation 4.4: Access to shore-based welfare facilities.
- Regulation 4.5: Social security.

Title 5: Compliance and enforcement

- Regulation 5.1: Flag State responsibilities.
- Regulation 5.2: Port State responsibilities.
- Regulation 5.3: Labour-supplying responsibilities.

International Convention on Standards of Training, Certification and Watchkeeping for Seafarers (STCW)

The International Convention on Standards of Training, Certification and Watchkeeping for Seafarers (STCW) was adopted on 7 July 1978 and entered force on 28 April 1984. The Convention responded to many state-specific regulations on training and certification that caused problems when ships transited foreign waters or put into foreign ports. The STCW establishes the basic requirements and standards for masters (captains), officers and watch personnel on merchant ships (IMO, 1978). The overall format of the convention is as follows:

- Chapter I: General provisions.
- Chapter II: Master and deck department.
- Chapter III: Engine department.
- Chapter IV: Radiocommunication and radio personnel.
- Chapter V: Special training requirements for personnel on certain types of ships.
- Chapter VI: Emergency, occupational safety, medical care and survival functions.
- Chapter VII: Alternative certification.
- Chapter VIII: Watchkeeping.

The Convention was amended in 1995 to provide clarification on such phrases as 'to the satisfaction of the Administration'; move the technical regulations into a new STCW Code that can be amended without convening a full conference; and require parties to the Convention to provide evidence of complaints – the first incidence of IMO involvement in compliance. There are currently 155 parties to the Convention.

The STCW Code has mandatory (Part A) and optional (Part B) sections where Part B provides advice to assist parties with complying with the Convention. Both the STCW Convention and Code were further amended on 25 June 2010 and entered force on 1 January 2012. Referred to as the 'Manila Amendments', they updated the standards necessary to comply with the Convention.

The STCW does not apply to ships less than 200 GT or ships operating only in internal waters. The Convention does apply to ships registered in non-party states when they visit port states that have ratified the Convention.

Maritime Security

Maritime security encompasses piracy, terrorism and criminal activities related to the safety of ships, crews, passengers and cargoes when underway as well as in port. Shipping is a unique endeavour with respect to safety as illegal acts are often perpetrated in international waters

(ABNJ). These ships are often registered in flag states different from their ownership state and crewed by citizens of neither state. As such, maritime security regulation requires a degree of international collaboration far in excess of most other modes of transportation. Furthermore, modern terrorism utilizes not just the ship as a means to transport weapons, for the ship itself can be used as a weapon, particularly if the ship is carrying dangerous maritime cargoes (DMCs), which include hydrocarbons and chemicals (Nincic, 2005).

Addressing maritime security is important: Research has shown that, for example, without international action against Somali pirates operating in the Gulf of Aden, shipping along the Far East–Europe route would decline by 30 per cent (Lu et al., 2010). However, the costs of maritime security are prohibitively high and leaves the industry vulnerable. As an example, currently less than 1 per cent of the 230 million container voyages are inspected (Broder, 2004).

International Ship and Port Facility Security Code

The International Ship and Port Facility Security Code (ISPS) was adopted on 13 December 2002, partly in response to the 11 September 2001 terrorist attacks on the USA (IMO, 2003). The ISPS applies to all passenger ships, cargo ships greater than 500 GT, mobile offshore drilling units and ports serving such ships engaged on international voyages. Broadly, the Code:

- Establishes cooperative mechanisms between governments, shipping and port industries to detect security threats and take preventative measures to protect ships and ports involved in international trade.
- Establishes roles and responsibilities between governments, shipping and port industries with respect to maritime security.
- Ensures the exchange of maritime security information.
- Provides a methodology for security assessments in order to develop ship and port security plans able to react to changing security levels.

The Code entered into force 1 July 2004 in the form of the following amendments to SOLAS:

- Chapter V Safety of Navigation: Regulation 19 was amended to require upgraded navigation equipment and to ensure that ships equipped with automated information systems use the system at all times.
- Chapter XI-I Special Measures to Enhance Maritime Safety: Regulation 3 was amended to specify how a ship's identification number is to be marked on the vessel.
- Chapter XI-I Special Measures to Enhance Maritime Safety: Regulation 5 was amended to require ships to carry a Continuous Synopsis Record, which is an on-board document record of the history of the vessel with respect to information including the vessel's previous names, ownership, classification societies and administrations/organizations issuing the ISM Code Safety Management Certificate.
- Chapter XI-II Special Measures to Enhance Maritime Safety: Regulation 1 provides updated definitions.
- Chapter XI-II Special Measures to Enhance Maritime Safety: Regulation 2 provides the types of ships and ports (discussed above) to which the Code applies, and exempts warships or ships owned by a contracting government in non-commercial service.
- Chapter XI-II Special Measures to Enhance Maritime Safety: Regulation 3 commits states to convey security status information to ships flying their flag and their ports.
- Chapter XI-II Special Measures to Enhance Maritime Safety: Regulations 4 and 10 commit states to comply with part A of the Code and to take into account part B.

- Chapter XI-II Special Measures to Enhance Maritime Safety: Regulation 5 requires a ship's Master to be able to identify who is responsible for hiring crew and determining the business of the ship.
- Chapter XI-II Special Measures to Enhance Maritime Safety: Regulation 6 requires ships to carry a security alert system.
- Chapter XI-II Special Measures to Enhance Maritime Safety: Regulation 7 commits states to communicate security threat levels to vessels operating in their waters.
- Chapter XI-II Special Measures to Enhance Maritime Safety: Regulation 8 provides a ship's Master authority over a ship's security and allows the Master, in the interests of security, to override decisions/policies of the ship owners.
- Chapter XI-II Special Measures to Enhance Maritime Safety: Regulation 9 specifies the information that port states may require of vessels intending to enter its ports. If information is not produced the port state may inspect the ship or deny entry into its ports.
- Chapter XI-II Special Measures to Enhance Maritime Safety: Regulations 11 and 12 permit contracting states to enter into bi-lateral/multi-lateral security agreements and develop their own security standards so long as they are equivalent to the Code.

The Convention for the Suppression of Unlawful Acts against the Safety of Maritime Navigation and Protocol for the Suppression of Unlawful Acts against the Safety of Fixed Platforms Located on the Continental Shelf

The Convention for the Suppression of Unlawful Acts against the Safety of Maritime Navigation (SUA) and Protocol for the Suppression of Unlawful Acts against the Safety of Fixed Platforms Located on the Continental Shelf was adopted on 10 March 1988 and entered into force 1 March 1992 (IMO, 1988). The Convention is currently ratified by 156 states. It was partially a result of the growing awareness of maritime security issues but ultimately convened after the Italian cruise ship *Achille Lauro* was high-jacked sailing from Alexandria to Port Said, Egypt. While piracy and hot pursuit are addressed in the LOSC, the Convention is silent on the international response to maritime terrorism.

The IMO requested the Maritime Safety Committee develop measures to address unlawful acts against civilian shipping. The primary purpose of the Convention is to require state parties to the Convention to extradite or prosecute persons committing unlawful acts against ships. Key Articles of the Convention include:

- Article 2: The Convention does not apply to warships and government ships operated for non-commercial purposes, nor does it apply to ships withdrawn from navigation or ships removed from water for repair.
- Article 4: The Convention applies to ships operating or planning to operate outside of their flag state's territorial waters, or when an alleged offender is found in the territory of a state party.
- Articles 6–12: Responsibilities of state parties (including ship's masters) in the apprehension, extradition, conviction and punishment of those committing unlawful acts against ships and fixed platforms.
- Article 13: Commits state parties to share information on marine security and cooperate to prevent marine terrorism.

The Convention was amended on 14 October 2005 via a Protocol (2005 SUA Convention and 2005 SUA Fixed Platforms Protocol) (IMO, 2005a, 2005b, 2005c). The Protocol was prompted by increased attention to maritime security following the 11 September 2001 terrorist attacks on the United States as well as the increased security threats as a result of the growth in containerized shipping (cargo ownership becomes more difficult to trace when shippers use liner trades to move cargo in containers). Furthermore, the rise of suicide bombers was never contemplated in the original SUA. Also required in the amendments was the need to expand the categories of 'unlawful acts' and to authorize boarding of foreign flagged ships when on the high seas in a manner that respected freedom of navigation and right of innocent passage. The 2005 SUA Convention is significantly amended from the original Convention and fundamentally expands the definition of unlawful acts. It permits, with the permission of the flag state, a state party to the Convention to board a ship on the high seas if there is reasonable suspicion that individual(s) aboard a ship have been or are expected to contravene the Convention (IMO, 2003).

Piracy

The first documented acts of piracy were in Mesopotamia around 2000 BCE. Since this time, piracy has evolved from 'privateering', where states supplemented their navies through issuing letters of marque to private operators to modern piracy driven by lawlessness, weak governance and lack of opportunity.

In 2016, 221 incidents of piracy were recorded worldwide. This is a significant decrease from 2011 where 544 incidents were recorded with 50 captured vessels and 569 seafarers taken hostage (IMB, 2017). The decreases are largely thought to be a result of changes in vessel routeing and additional security measures for ships transiting the Horn of Africa Region in combination with broader nation building and international engagement as opposed to criminal enforcement (Paige, 2017). While piracy is declining in the Somalia region, incidents remain high in the South China Sea (armed robbery) and West Africa (theft of oil cargos). Piracy, however, may be underreported by up to 50 per cent since ship owners may want to avoid dealings with the authorities, avoid increases to insurance rates or conceal actions or inactions by their crews (Chalk, 2008). Moreover, the globalization of the shipping industry can make it difficult to identify the victims of piracy. For example, on 4 April 2010, Somali pirates captured the MV *Sambo Dream*, which was registered in the Marshal Islands, owned by a South Korean company and crewed predominantly by Filipino nationals (Coggins, 2012).

The LOSC addresses piracy in Articles 100–107 and 110 and provides the overall legal foundation for addressing piracy and related incidents such as armed robbery. With respect to piracy, perhaps the most important aspect of the LOSC is that the Convention defines piracy in Article 101 as an action that can only occur on the high seas or areas outside of national jurisdiction and encompasses 'any acts of illegal violence or detention, or act of depredation, committed for private ends by the crew or the passengers of a private ship or a private aircraft'. Acts of piracy in territorial waters are considered domestic crimes under the LOSC. Furthermore, under the LOSC, piracy requires one ship (or aircraft) to seize another, and therefore the actions of a ship's crew or passengers against their ship cannot be considered piracy. Another key component of the LOSC regarding piracy includes the ability of states to seize pirate ships and ships under the control of pirates on the high seas, arrest persons on board and seize cargo (Article 105). A ship of another state may only be seized by a warship or government ship (Article 107) while a state may board a ship believed to be engaged in piracy (Article 110).

The problem with the LOSC's definition of piracy is that acts of piracy often occur in territorial waters of jurisdictions without effective control of their seas (e.g. Somalia, West Africa). As such, a more modern definition of piracy is 'an act of boarding or attempting to board any ship with the apparent intent to commit theft or any other crime and with the apparent intent or capability to use force in the furtherance of that act'. Piracy is often differentiated from terrorism by the fact that piracy is generally undertaken for financial gain while terrorism is conducted to further political or ideological causes (Nincic, 2005).

Since 2000, over 30 reports on piracy and related activities have been submitted to the UN General Assembly, resulting in multiple resolutions on piracy. In addition, between 2008 and 2017, the UN Security Council passed 16 specific resolutions on piracy adjacent to the coast of Somalia and the Gulf of Guinea. Lastly, the IMO is the lead international agency on combating piracy and passed the Code of Practice for the Investigation of Crimes of Piracy and Armed Robbery Against Ships on 2 December 2009 (A.1025(26)) in order to facilitate the investigation of piracy and robbery against ships (IMO, 2009a). The Code urges states to adopt national legislation to prosecute offences of piracy and armed robbery (Article 3.1) and to implement the provisions of the LOSC, the Convention for the Suppression of Unlawful Acts Against the Safety of Navigation and the Protocol for the Suppression of Unlawful Acts Against the Safety of Fixed Platforms Located on the Continental Shelf (Article 3.2). The Code implores masters to report piracy and armed robbery attempts and encourages coastal/port states ensure investigations do not unduly delay ships (Article 3.3). States are also requested to cooperate with each other on investigations (Article 3.4), train investigators (Article 4) and follow established procedures regarding investigation (Article 5) and initial response (Article 6) (IMO, 2009a).

In addition to the IMO Code of Practice, the IMO adopted the Code of Conduct concerning the Repression of Piracy and Armed Robbery against Ships in the Western Indian Ocean and the Gulf of Aden on 29 January 2009 (the 'Djibouti Code of Conduct'). Currently, 20 of 21 eligible states have signed the Code and agree to cooperate on reducing piracy in the Western Indian Ocean. The Jeddah Amendment to the Djibouti Code of Conduct 2017 was adopted 12 January 2017 by 18 Gulf States. The 'Revised Code of Conduct' calls on signatories to cooperate to the fullest possible extent to repress transnational organized crime in the maritime domain, maritime terrorism, illegal, unregulated and unreported (IUU) fishing and other illegal activities at sea (IMO, 2017).

Other maritime security measures

Proliferation Security Initiative

The United States launched the Proliferation Security Initiative (PSI) with the purpose of halting the transport of weapons of mass destruction (WMD), their constituent materials or their delivery systems. The PSI allows states to develop bilateral agreements to board each other's vessels in ABNJ if believed to be involved in fostering terrorism or illegal activities. The seizure of WMD and related cargo in ABNJ if found is also authorized by the PSI (Nincic, 2005). The principles of the PSI include:

- Interdicting the transfer of WMD or their components to states and non-state actors believed to be seeking WMD.
- Exchanging information between states.
- Strengthening domestic laws to facilitate interdiction.
- Taking specific interdiction actions to halt the proliferation of WMD.

The PSI is not an international agreement or convention. Rather than becoming signatories, states either endorse or participate in the PSI. The PSI has no secretariat, no statistics on success or compliance, and, as of 2015, 105 states are participating in the PSI (US State Department, 2003).

Container Security Initiative

The Container Security Initiative (CSI) was established by the United States in 2002. The CSI ensures that high risk containers are identified and inspected prior to loading on US-bound ships. The program stations United States nationals at 58 ports in North America, Europe, Asia, Africa, the Middle East and Latin and Central America. Since the establishment of the initiative, Canada and Japan have signed bilateral agreements to share information and station their nationals at US ports to monitor exports to their nations (Haveman et al., 2007). The European Union is examining the CSI for use in European trade.

Stowaways

Stowaways are individuals or groups of people who hide (or are hidden) on ships (or in their cargos) to flee one state for another for economic, political or religious reasons. Addressing the issue of stowaways encompasses aspects of basic human rights, immigration and refugee policies and efforts to combat organized crime. Stowaways are relevant to this section because international agreements and conventions related to them generally focus on preventing access to ports and ships rather than addressing the underlying causes (e.g. inequity and oppression) that cause people to flee their jurisdiction. In 2014, the IMO reported 61 stowaway cases involving 120 stowaways found on all types of commercial ships (IMO, 2015). This is a decrease from 253 cases involving 721 stowaways in 2010 (IMO, 2011a). Currently, nearly 50 per cent of the world's stowaways are from West Africa. Stowaways interfere with shipping as they must be processed by the relevant authorities in port and cared for by a vessel's crew while underway.

The overarching international law related to stowaways is the United Nations Declaration of Human Rights. The 1948 Declaration confers upon stowaways the right to life, and freedom from torture, slavery, discrimination and degradation. The 1957 International Convention Relating to Stowaways (the 'Brussels Convention'), adopted 10 October 1957, attempted to establish common procedures to address stowaways; however, the Convention has not yet entered into force (UNHCR, 1957).

The next official attempt to address stowaways was the Convention on Facilitation of International Maritime Traffic (the 'FAL'), adopted 9 April 1965, and entered into force 5 March 1967 (IMO, 1965). The Convention set the formalities, documentary requirements and procedures to be used on the arrival, stay and departure of a ship. It was amended in 2002 (adopted 10 January 2002, entered into force 1 May 2003) to add new standards and recommended practices to grapple with issues surrounding stowaways.

Other efforts to address stowaways include:

- The 1997 IMO Guidelines on the Allocation of Responsibilities to seek the Successful Resolution of Stowaway Cases (A.871(20)) (IMO, 1997).
- The 2002 SOLAS amendments and adoption of the ISPS Code (see above) to prevent unauthorized access to ships and port facilities (IMO, 2003).
- The 2011 IMO Revised Guidelines on the Prevention of Access by Stowaways and the Allocation of Responsibilities to Seek the Successful Resolution of Stowaway Cases (FAL.11(37)) (IMO, 2011b).

Shipping and the environment

Shipping has many potential negative consequences to the marine environment, to adjacent terrestrial environments and to the atmosphere. Nearly all commercial ships in operation today are powered by hydrocarbons, which contribute to atmospheric pollution and greenhouse gasses.

Ships may intentionally (dumping) or unintentionally (accidents, anti-fouling hull paint) pollute (e.g. oil, garbage, cargoes, emissions) the ocean, land or atmosphere. Besides suffering detrimental effects of pollution, marine mammals are affected by shipping noise and ship strikes. Light pollution from ships may alter navigational patterns by marine birds. Ships are vectors for invasive or alien species that may reside in ballast water, on a ship's hull or within a ship's piping (see Chapter 6). Recycling or disposal of retired ships often releases toxic substances into the environment.

This section outlines the primary global institutions and agreements that govern shipping's impacts on the marine environment. This is not a comprehensive examination but, instead, provides the reader with an understanding of how environmental impacts from shipping are addressed, both in terms of preventing or minimizing environmental impacts and ensuring adequate compensation to those affected by the environmental externalities of shipping. This section begins with the International Convention for the Prevention of Pollution from Ships (MARPOL 73/78). It continues with the introduction of a number of other marine protection-related conventions.

Note that Chapter 6 addresses the non-shipping related impacts to marine environments from human activities.

International Convention for the Prevention of Pollution from Ships (MARPOL)

Early conventions to protect the marine environment included the 1921 Conference on the Protection of the Area from Pollution by Oil and the 1954 International Convention for the Prevention of Pollution of the Sea by Oil. Termed the 'Oil Pollution Convention' (OPC), it prohibited the deliberate discharge of oil or oily mixtures from all sea-going vessels. The Conventions were amended in 1969 to prohibit the discharge of oil cargo within 50 nmi of the nearest land and to regulate the proportion of oil and ballast water discharges. The OPC was amended again in 1971 to provide additional protection to Australia's Great Barrier Reef.

The OPC Convention was eventually replaced by the International Convention for the Prevention of Pollution from Ships, which is now the primary convention that governs the protection of the marine environment from shipping. The Convention addresses both the operational and accidental causes of pollution in solid, liquid and airborne forms. The pollutants covered by the Convention include oil, chemicals, harmful substances in packaged form, sewage and garbage. The Convention, often referred to as 'MARPOL 73/78', is a combination of the 1973 and 1978 Treaties and a 1997 Protocol.

The Convention contains six Annexes. Annexes I and II are mandatory for those states that have ratified the Convention while Annexes III, IV, V and VI are optional:

- ANNEX I: Regulations for the Prevention of Pollution by Oil (1983). Reinforces the 1954 Oil Pollution Convention with additional provisions for minimizing oil discharges when loading/unloading, prohibiting the use of cargo oil tanks for carrying ballast water and structural requirements to prevent spills in the event of collisions or stranding.
- Annex II: Regulations for the Control of Pollution by Noxious Liquid Substances in Bulk (1967). Identifies four categories (X, Y, Z, other) of liquid substances transported in bulk ranked by relative hazard to marine resources or human health. Ships constructed after 1 July 1986 must comply with different safety codes depending on which category of liquid substances they are constructed to transport.
- Annex III: Regulations for the Prevention of Pollution by Harmful Substances Carried at Sea in Packaged Form (1992). Outlines eight regulations related to the labelling, packing, stowing, documenting and transport of harmful substances. Revised in 2010.

- Annex IV: Regulations for the Prevention of Pollution by Sewage from Ships (2003). Regulates the discharge of sewage into the sea from ships and specifies ship- and land-based sewage management facilities. Does not apply to the high seas.
- Annex V: Regulations for the Control of Pollution by Garbage from Ships (1998). Prohibits the disposal of plastics into the sea and limits other types of garbage that can be introduced into the marine environment including food, domestic and operational waste. The Annex also requires port states to provide reception facilities to accept and process ship-borne garbage.
- Annex VI: Regulations for the Prevention of Air Pollution from Ships (1997). Regulates the harmful emissions from ships and, in particular, diesel emissions (sulphur oxide SOx and nitrogen oxide NOx). Establishes emissions thresholds in Sulphur Emission Control Areas (see below). Regulations on reducing the release of greenhouse gasses from shipping are currently under development.

MARPOL 73/78 introduced the concept of 'special areas' for Annexes I, II, IV, V and VI. For certain environments of concern ('special areas'), more stringent regulations were applied. Special areas in Annex I prevent oil discharges of any kind and currently include the Mediterranean Sea (1973), the Black Sea (1973), the Baltic Sea (1973), the Red Sea (1973), the Persian Gulf area (1973, 1987, 2004), the Antarctic (1992), Northwest European (1997) and Southern South African (2006) waters. Annex II prohibits the release of noxious liquid substances in Antarctic areas (1992). Annex IV outlines special provisions for the release of sewage into the Baltic Sea (1992) and Annex V further restricts the dumping of garbage from ships into the Baltic (1973), Black (1973), Mediterranean (1973), North (1989) and Red (1973) Seas as well as the Antarctic (1992), the 'Gulfs' area (1973) and wider Caribbean (1991). Annex VI establishes Emission Control Areas in the Baltic (1997) and North (2005) Seas, North America (2010) and United States Caribbean Sea (2011).

MARPOL 73/78 has no direct enforcement provisions but does permit any flag state or party to the Convention to enforce its provisions. Under both the Convention and the LOSC, primary responsibility for enforcing the Conventions rests with the flag state. Generally, the flag state's duties are to issue international certificates (e.g. International Oil Pollution Prevention Certificate) and to carry out periodic inspections of vessels to ensure that they comply with international laws. Coastal and port states have no direct obligation under the LOSC and MARPOL 73/78 to enforce environmental protection aspects of the Conventions, yet many coastal and port states are active in ensuring the compliance and enforcement of international and domestic laws. Furthermore, port states maintain the right to inspect and detain vessels that either fail to produce valid international certificates or are believed to be in contravention of MARPOL 73/78 and the LOSC.

It has long been recognized in international law that a ship's captain (master) can be criminally liable for contravention of international laws (including MARPOL 73/78) and can also be liable for the actions of those under his/her command.

Regulation of marine oil spills and oil spill response and countermeasures

Approximately half of the world's oil production (approximately 1.6 billion tonnes) is transported by sea. Much of this transport is over long distances in VLCC and ULCC class tankers (Table 9.2) that pass through straits and transit along coastlines. This means of transport has resulted in many serious accidents and spills over the past several decades (Table 9.3). In addition, a significant amount of oil is transported short distances by seabed pipelines from

Table 9.3 Largest marine oil spills from ships, pipelines and oil production platforms

Ship/platform	Estimated spilled (max tonnes)	Year	Location of spill
Gulf War oil spill	820,000	1991	Persian Gulf
Deepwater Horizon	627,000	2010	Gulf of Mexico
Ixtoc I	480,000	1980	Gulf of Mexico
Atlantic Empress	287,000	1979	Trinidad and Tobago
Nowruz Field Platform	260,000	1983	Persian Gulf
ABT Summer	260,000	1991	Angola
Castillo de Bellver	252,000	1983	South Africa
Amoco Cadiz	227,000	1983	France
MT Haven	144,000	1991	Italy
Odyssey	132,000	1988	Eastern Canada
Torrey Canyon	119,000	1967	United Kingdom
Sea Star	115,000	1972	Gulf of Oman
Exxon Valdez	104,000	1989	Alaska

Source: UNCTAD (2012)

production wells to offshore or onshore facilities. While the average number of major oil spills per year has dropped from 25 in the 1970s to 3 today, spills continue to impact fisheries, tourism and coastal economic activities (UNCTAD, 2012).

While ship owners, ship registries (flag states), classification societies, port states and cargo owners all have responsibilities to avoid oil spills into the marine environment, the IMO is the primary international agency responsible for establishing the regulatory framework related to the prevention, clean up and compensation for oil spills from ships. Although MARPOL 73/78 is the primary international instrument used to prevent the introduction of pollution into the marine environment from shipping, the mandated standards for the construction, equipment and operation of vessels under SOLAS may reduce the risk of oil pollution. The IMO also administers a number of oil pollution specific conventions, which are discussed below.

International Convention relating to Intervention on the High Seas in Cases of Oil Pollution Casualties

The International Convention relating to Intervention on the High Seas in Cases of Oil Pollution Casualties ('Intervention Convention') was adopted on 29 November 1969 and entered into force on 6 May 1975 (IMO, 1969a). The Convention currently has 89 contracting states representing 75.1 per cent of the world tonnage (UNCTAD, 2012). It was developed following the March 1967 loss of the tanker *Torrey Canyon*, which ran aground off the western coast of Cornwall, England, enroute from Kuwait to Milford Haven in Wales. The tanker spilled 108,000 tonnes of oil and extensively damaged 190km of the western coast of southern England.

The Convention enables coastal states to take necessary actions in their territorial waters or the high seas to prevent, mitigate or eliminate oil pollution that would otherwise be illegal under international law (Article I). The Convention applies to all ships except warships, non-commercial government vessels and coastal installations and defines 'oil' as crude oil, fuel oil, diesel oil and lubricating oil (Article II). However, the coastal state must first consult with the flag state prior to initiating intervention (Article III).

In 1973, the Intervention Convention was modified by a Protocol relating to Intervention on the High Seas in Cases of Marine Pollution by Substances other than Oil. The Protocol entered force 30 March 1983 and amends the Convention by adding a number of non-oil substances listed in the Annex to Protocol. Among non-oil substances are noxious substances, liquefied gasses and radioactive substances. Currently the Protocol has been ratified by 54 contracting states representing 50.4 per cent of the world's tonnage. Amendments to the Protocol by the IMO's MEPC to add further substances to the Annex occurred in 1991 (MEPC.49(31)), 1996 (MEPC.72(38)), 2002 (MEPC.100(48)), and 2007 (MEPC.165(56)).

International Convention on Civil Liability for Oil Pollution Damage

The International Convention on Civil Liability for Oil Pollution Damage (also referred to as the 'Civil Liability Convention' or CLC 1969) was adopted 29 November 1969 and entered into force 19 June 1975 (IMO, 1969b). The Convention was amended by Protocols in 1976 (CLC 1976) and 1992 (CLC 1992) with the CLC 1992 consolidating the 1969 text and 1976 Protocol. The CLC 1992 has effectively replaced CLC 1969 and was adopted 17 November 1992 and entered into force 30 May 1996 (IMO, 1992). As of 2017, the CLC 1992 has 136 contracting parties representing almost 98 per cent of the world's fleet.

Maritime law has a long tradition of operating under the principle of limitation of liability (financial caps). The purpose of the CLC 1992 is to ensure that adequate compensation is available to persons who suffer damage caused by pollution resulting from the escape or intentional discharge of oil from ships. Claimants may be governments, private bodies (e.g. tourism operators) or individuals so long as the incident occurs within the jurisdiction of a state party to the Convention. The Convention also sets out the procedures for determining liability and appropriate compensation and creates a system of compulsory liability insurance. Key Articles of the Convention include:

- Article I: The Convention only applies to tankers carrying oil as a bulk cargo (fuel and lubricants are excluded).
- Article II: The Convention only applies to the territory (i.e. coastal land areas), territorial sea and EEZ of a contracting state.
- Article III: The ship owner is not liable if a spill is a result of hostilities, exceptional natural phenomena, intentional acts by third parties, or failure of (or incorrect) navigation aids. Only the ship owner may be held liable under CLC 1992.
- Article V: Liability limits are set at 4,510,000 SDR (as defined by the International Monetary Fund, approximately $US 7 million in 2019) for ships not exceeding 5,000 units of tonnage. Each additional unit of tonnage adds 631 units of account (approximately $US 1,000) up to a maximum of 89,770,000 units of account (approximately $US 133 million).
- Article VII: Owners of ships registered in a contracting state carrying more than 2,000 tons of oil as bulk cargo must have insurance or guarantees to the liability limits outlined in Article V.
- Article XI: The Convention does not apply to warships or government ships used for non-commercial purposes.

Source: IPIECA (2016)

Marine transportation and safety policy

International Convention on the Establishment of an International Fund for Compensation for Oil Pollution Damage

The International Convention on the Establishment of an International Fund for Compensation for Oil Pollution Damage ('1971 Fund Convention') was adopted 18 December 1971 and entered into force 16 October 1978. The original Convention was superseded by the 1992 Protocol to amend the 1971 International Convention on the Establishment of an International Fund for Compensation for Oil Pollution Damage ('1992 Fund Protocol'), adopted 27 November 1992 and entered into force 30 May 1996 (IMO, 1992). The 1992 Fund Protocol has 114 parties as of 2017 (Figure 9.1).

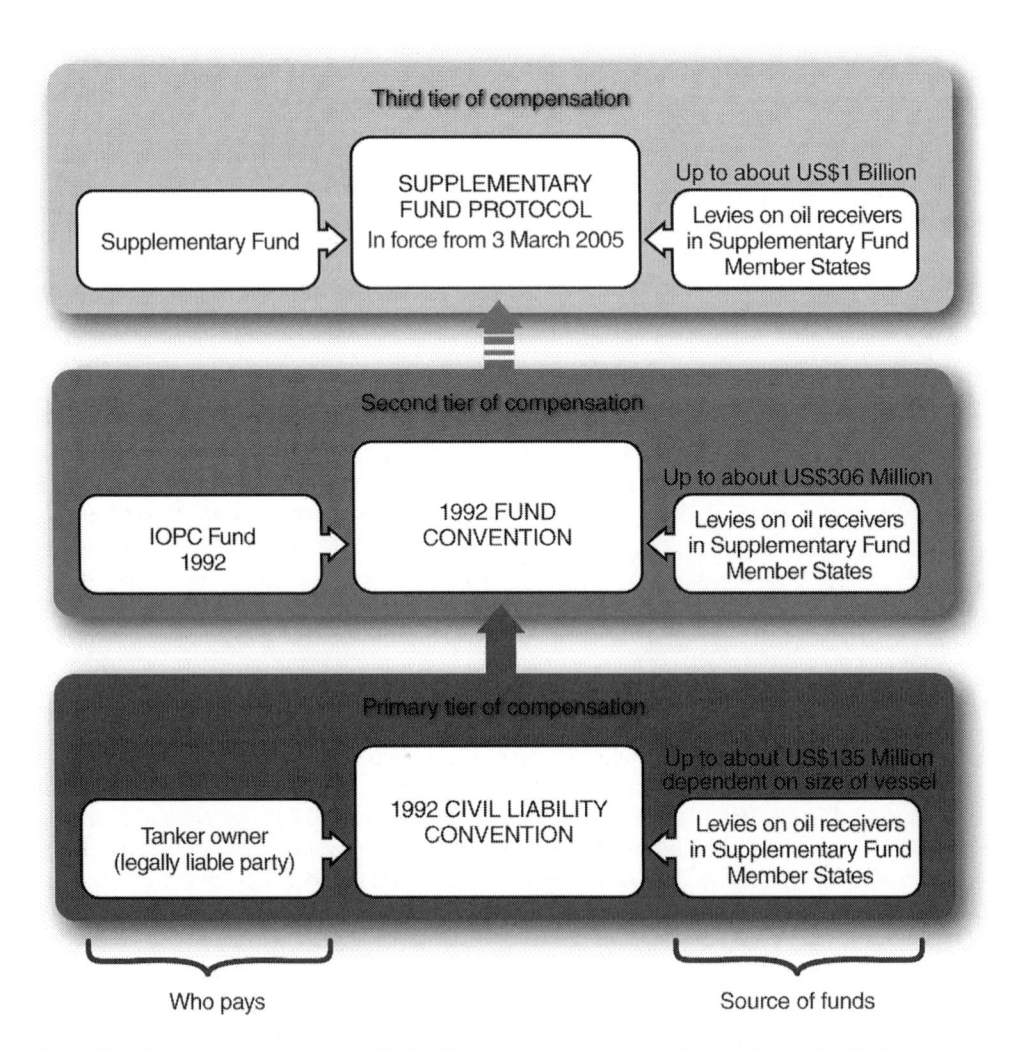

Figure 9.1 Compensation levels established by international conventions to fund oil spill clean up

Source: Reproduced with permission from IPIECA
www.oilspillresponseproject.org/wp-content/uploads/2017/01/Compensation_2016.pdf

The purpose of the Fund Conventions is to provide compensation for pollution damage in instances where the CLC 1969/1992 is inadequate. This transpires when the ship owner can demonstrate an exemption under Article III of the CLC 1992 or the ship owner lacks the financial resources (or insurance) to pay the full compensation required under the CLC 1992. The 1971 Fund Convention created the International Oil Pollution Compensation Fund (the 'IOPC Funds') through a levy on both shippers and cargo owners.

The IOPC Funds, based in London, are used to pay damages over and above the limits of the CLC 1992. Three funds – which have different member states and amounts of compensation – are managed by the IOPC Funds: the 1971 Fund, the 1992 Fund and the Supplementary Fund. Compensation maximums in the 1971 Fund were deemed by many members to be insufficient, and, as such, the 1992 Fund and Supplementary Fund were established. The 1971 Fund currently has no members and ceased to be in force on 24 May 2002. Nevertheless, it continues to exist to address compensation claims for older incidents.

As of November 2003, the liability limits of the 1992 Fund are 203 million SDR (US$ 312 million). The 1992 Fund is financed by a levy on any person (e.g. business, government) operating within a state party to the Convention who has received more than 150,000 tonnes of crude oil and heavy fuel oil in one calendar year (IOPC, 2018).

In 2005, the Supplementary Fund was established with a maximum liability limit of 750 million SDR (US$ 1.15 billion). It is financed in a similar fashion to the 1992 Fund with the exception that, for contribution purposes, each state will be levied for a minimum of one million tonnes of contributing oil. Thirty-one states are party to the Supplementary Fund.

International Convention on Oil Pollution Preparedness, Response, and Co-operation

The International Convention on Oil Pollution Preparedness, Response, and Co-operation (OPRC) was adopted 30 November 1990 and entered into force 13 May 1995 (IMO, 1990). Article 3 requires ships to carry onboard an oil pollution emergency plan that may be inspected when a ship is in the port of a state party to the Convention. Ships must report the discharge of oil to the nearest coastal state while oil handing facilities must report to the appropriate national authority (Article 4). Each state party to the Convention must establish a national organization with responsibility for oil spill preparedness and response. This authority must develop national contingency plans for oil spills, maintain a minimum level of pre-positioned oil spill clean-up equipment, coordinate oil spill response exercises and share information amongst various parties and governments (Article 6). Parties must also cultivate research and development related to oil spill preparedness and response (Article 8). In addition, they must cooperate with other jurisdictions to provide technical support, training and technology transfer (Article 9).

The Protocol on Preparedness, Response and Co-operation to Pollution Incidents by Hazardous and Noxious Substances (OPRC-HNS Protocol) was adopted 15 March 2000 and entered into force 14 June 2007. The Protocol mirrors the OPRC Convention but is augmented by regulations concerning discharge, release/emission of hazardous or noxious substances that threaten the marine environment, coastline, or other interest of a state (Article 2).

Emissions and climate change

While shipping is an efficient means of moving goods over short and long distances, the burning of fossil fuels for propulsion results in emissions of sulphur dioxide (SO_2), nitrogen oxides (NOx), volatile organic compounds (VOCs), particular matter (PM), carbon dioxide (CO_2) and other greenhouse gasses (Miola et al., 2011). It is estimated that globally shipping contributed about

14 per cent of the world's total SO_2 and NO_x emissions, resulting in 60,000 annual cardiopulmonary and lung-cancer deaths worldwide, partially as a result of using low-grade marine fuels that contain up to 3,500 times more sulphur than road diesel (Wan et al., 2016). Passenger and cruise ships must also produce power when in ports (when shore power is unavailable), which result in increases in air pollution adjacent to the port.

The global shipping industry emits significant amounts of greenhouse gasses. The IMO estimated that in 2014 shipping was responsible for one billion tonnes of CO_2 emissions, or about 3.1 per cent of global emissions (Wan et al., 2016). If the global shipping industry were a country, it would be the sixth largest emitter of CO_2 after the US, China, Russia, India and Japan. Without additional regulations on CO_2 emissions or the adoption of more efficient shipping technologies, shipping emissions are forecasted to increase by a factor of 2 to 3 by 2050. On the other hand, if technical and operational improvements are implemented, they could reduce CO_2 emissions by 25 per cent to 75 per cent below current levels.

The IMO and other organizations have been active in reducing ship emissions. The IMO has set a GHG emissions reduction target of 15 per cent by 2018 and a much more ambitious target of 50 per cent 2008 levels by 2050. Furthermore, parts of Europe and western North America are Emission Control Areas where low-sulphur fuels must be used when steaming in territorial waters.

The IMO has introduced new measures, which entered into force on 1 January 2013, that apply to the entire global international shipping industry and are the first example of global industry-wide CO_2 reduction measures. The Energy Efficiency Design Index (EEDI) for new ships mandates energy efficiency targets of 10 per cent improvement for ships built between 2015–2019, 15 per cent or 20 per cent for 2020–2024 (depending on the type of ship) and 30 per cent for ships delivered after 2024. The Ship Energy Efficiency Management Plan (SEEMP), required for all new and existing ships, establishes best practices for fuel efficiency and voluntary guidelines for measuring CO_2 emissions.

There are a number of ways to reduce ship emissions through either the design or operation of a vessel. Broadly, ships may improve energy efficiency (productivity), use renewable energy sources (e.g. wind, solar, biofuels), use low carbon fuels (e.g. natural gas) or reduce emissions through chemical conversion, captures and storage. Additional operational strategies include improving ship routeing (including weather routeing) and reducing vessel speeds ('slow steaming'). Between 60–70 per cent of the costs of running are energy-related; therefore, there is an incentive on shippers to improve efficiency.

Between 2007 and 2012, the shift to 'slow steaming' resulted in improved fuel economy of the global shipping fleet and reduced emissions. In an ironic turn of events, as the Arctic Ocean warms and sea ice retreats northward, commercial shipping may begin to use the northwest passage (Canada) and northern sea route (Russia) as alternatives to the Panama and Suez Canals. The northern sea route is 43 per cent shorter from Rotterdam (Europe) to Yokohama (Asia), thus fuel consumption is expected to be reduced by up to 37 per cent on this route.

The 1992 UNFCCC commits state parties to the Convention to develop national policies for limiting greenhouse gas emissions (see Chapter 7). The 1997 Kyoto Protocol requires 37 developed states to reduce their greenhouse gas emissions to 1990 levels. However, the Kyoto Protocol acknowledges that emissions from international shipping are difficult to ascribe to particular states and, as such, tasks the IMO with addressing greenhouse gas reductions for the global shipping industry.

In July 2011, the IMO's MEPC adopted mandatory energy efficiency measures for international shipping that were added as a new Chapter 4 to MARPOL 73/78 Annex VI. The measures, which entered into force on 1 January 2013, apply to the entire global international shipping industry and are the first example of global industry-wide CO_2 reduction measures.

While the IMO MEPC has considered a number of market-based mechanisms (incentives) for reducing ship emissions, no such market mechanisms are yet in place. However, market mechanisms will likely be incorporated into international shipping in the future, and, if the IMO is unable to broker a global consensus, the European Union is likely to bring into force an emissions trading program for shipping (Miola et al., 2011).

Green shipping practices

Green shipping practices (GSPs) is a broad term that captures efforts by shipping companies to reduce waste and to conserve resources. Opportunities to 'green' shipping are numerous: reducing atmospheric pollution (e.g. NOx, SOx, CO_2) from ship propulsion systems, more efficient routeing and reducing shipboard waste are important measures. Two other practices to help make shipping greener are the reduction of 'vapour emissions' or the minimization of cargo loss (e.g. hydrocarbons) to the atmosphere while in transit and tracking improvements related to the environmental impacts of shipping (Mair, 1995). Green shipping is driven by regulation (including the Emissions Control Areas outlined in MARPOL 73/78 Annex VI and the Kyoto Protocol), consumer demand for products shipped in a sustainable manner and the need to improve efficiencies (e.g. reduce fuel costs).

Industry led green shipping efforts include the International Chamber of Shipping and International Shipping Federation (www.shippingandco2.org) and Green Award program, which certifies ships according to a number of environmental management and safety criteria. Green Award certified ships received reduced port dues at more than 20 ports worldwide (www.greenaward.org).

Ship recycling

Ship recycling, also termed 'ship breaking', is the process of disassembling and disposing of ships at the end of their life. It is estimated that approximately 1,000 ships per year are recycled, while more than 40,000 ships over the next 30 years will need to be recycled (Mikelis, 2007). Currently, 98 per cent of the world's recycled ships (by tonnage) are recycled in Bangladesh,

---

Box 9.3 Der Blaue Engel ('the blue angel') eco-label and shipping

In 2002, Germany included ships in their Der Blaue Engel ('the blue angel') environmental labelling ('eco-labelling') program. The program sets 10 mandatory and 20 optional standards to reduce air emissions and harmful discharges into the marine environment. Standards include environmental protection management, staff management, ship design and equipment, collision protection and leakage stability, redundant systems, hull stress monitoring, emergency towing systems, SOx emissions, particle emissions, emissions from cooling and refrigerating devices, waste disposal, incineration of waste, wastewater, bilge water, antifouling, ballast water and firefighting foams. Ships of any flag may be certified, but the program excludes warships, high-speed passenger vessels and fishing vessels. Since inception, many new vessels have been constructed and operated to the Der Blaue Engel standards.

Source: Der Blaue Engel (2015)

China, India, Pakistan and Turkey. Around the globe, the industry provides considerable local employment and economic benefits. Nearly every part of a recycled ship can be reused. On the downside, recycling practices are often hazardous to both workers, and the environment since ships – and particularly older ships – often contain hazardous materials including asbestos, polychlorinated biphenols (PCBs), ozone depleting substances, anti-fouling compounds, heavy metals and radioactive substances.

To minimize the impact of these hazards, the 1989 Basel Convention on the Control of Transboundary Movements of Hazardous Wastes and their Disposal (Chapter 6) prohibits the transboundary dumping or transfer of hazardous wastes. Not until 2004 did the Convention contemplate the ship recycling industry as a vector for the movement of hazardous waste. It was in this year that the CoP invited the IMO to establish regulations to ensure ship recycling conformed to the Basel Convention. Exploiting their experience with their 2003 voluntary Guidelines on Ship Recycling, the IMO agreed to lead the development of a new convention for ship recycling in 2005.

The Convention on the Safe and Environmentally Sound Recycling of Ships (the 'Hong Kong' Convention) was adopted 19 May 2009 (IMO, 2009b). Its purpose is to ensure that the disposal and recycling of ships is executed in a way that protects human health and the environment. In addition, the Convention includes provisions that apply to the construction and operation of ships in order to improve public and environmental health throughout a ship's life cycle. The Convention builds upon existing conventions related to hazardous substances, worker safety and past work of the IMO on ship recycling. It includes 25 regulations grouped into the following categories:

- Requirements for ships (regulations 4–14).
- Requirements for ship recycling facilities (regulations 15–23).
- Reporting requirements (regulations 24–25).

Discussion questions

- What are the most effective ways to make international shipping safer for seafarers and the environment?
- What changes to international shipping are likely to occur in the face of climate change?
- How can the individual consumer make informed choices to limit his/her footprint on the world ocean for purchased goods that are shipped internationally?

Further reading

Bellwood, P. (1997) 'Ancient seafarers', *Archaeology*, vol 50, no 2.
Branch, A. E. and Roberts, M. (2014) *Branch's Elements of Shipping*, Routledge, London.
Broder, J. M. (2004) 'At ports cargo backlog raises security questions', *New York Times*, 27 July.
Chalk, P. (2008) *The Maritime Dimension of International Security: Terrorism, Piracy and Challenges for the United States*, RAND Corporation, Santa Monica, CA.
Clyde, P. and Reitzes, J. D. (1998) 'Market power and collusion in the ocean shipping industry: Is a bigger cartel a better cartel?', *Economic Inquiry*, vol 36, no 2, pp. 292–304.
Coggins, B. L. (2012) 'Global patterns of maritime piracy, 2000–2009: Introducing a new dataset', *Journal of Peace Research*, vol 49, no 4, pp. 605–617.

Comité Maritime International (CMI) (2012), www.comitemaritime.org, accessed 04 January 2019.
Der Blaue Engel (2015) 'Environment-conscious ship operation RAL-UZ 110', www.blauer-engel.de/en/products_brands/search_products/produkttyp.php?id=506, 04 January 2019.
European Marine Safety Agency (2016) 'Equasis Statistics: The world merchant fleet in 2015', www.emsa.europa.eu/emsa-documents/latest/download/4429/472/23.html, accessed 20 June 2017.
Gibbons, A. (1998) 'Ancient island tools suggest *homo erectus* was a seafarer', *Science*, vol 279, no 5357, pp. 1635–1637.
Haveman, J. D., Shatz, H. J., Jennings, E. M. and Wright, G. C. (2007) 'The container security initiative and ocean container threats', *Journal of Homeland Security and Emergency Management*, vol 4, no 1, pp. 1–19.
Hummels, D., Lugovskyy, V. and Skiba, A. (2009) 'The trade reducing effects of market power in international shipping', *Journal of Development Economics*, vol 89, no 1, pp. 84–97.
IHS Fairplay (2011) *2011 World Fleet Statistics*, IHS Fairplay, Englewood, CO.
ILO (2006) *Maritime Labour Convention*, Geneva, Switzerland, www.ilo.org/wcmsp5/groups/public/@ed_norm/@normes/documents/normativeinstrument/wcms_090250.pdf, accessed 04 January 2019.
IMO (1965) *Convention on Facilitation of International Maritime Traffic*, London, treaties.un.org/doc/Publication/UNTS/Volume%20591/volume-591-I-8564-English.pdf, accessed 04 January 2019.
IMO (1966) *International Convention on Load Lines*, London, www.riigiteataja.ee/aktilisa/2160/1201/3001/Conv_on_Load_Lines.pdf, accessed 04 January 2019.
IMO (1969a) *International Convention Relating to Intervention on the High Seas in Cases of Oil Pollution Casualties*, London, treaties.un.org/doc/Publication/UNTS/Volume%20970/volume-970-I-14049-English.pdf, accessed 04 January 2019.
IMO (1969b) *International Convention on Civil Liability for Oil Pollution Damage*, London, treaties.un.org/doc/Publication/UNTS/Volume%20973/volume-973-I-14097-English.pdf, accessed 04 January 2019.
IMO (1972) *Convention on the International Regulations for Preventing Collisions at Sea*, IMO, London, treaties.un.org/doc/Publication/UNTS/Volume%201050/volume-1050-I-15824-English.pdf, accessed 04 January 2019.
IMO (1974) *International Convention for the Safety of Life at Sea*, IMO, London, treaties.un.org/doc/Publication/UNTS/Volume%201184/volume-1184-I-18961-English.pdf, accessed 04 January 2019.
IMO (1978) *International Convention on Standards of Training, Certification and Watchkeeping for Seafarers*, IMO, London, www.imo.org/en/OurWork/humanelement/pages/stcw-f-convention.aspx, accessed 04 January 2019.
IMO (1988) *Convention for the Suppression of Unlawful Acts against the Safety of Maritime Navigation*, IMO, London, treaties.un.org/doc/db/Terrorism/Conv8-english.pdf, accessed 04 January 2019.
IMO (1990) *International Convention on Oil Pollution Preparedness, Response and Co-Operation*, IMO, London, treaties.un.org/doc/Publication/UNTS/Volume%201891/volume-1891-I-32194-English.pdf, accessed 04 January 2019.
IMO (1992) *1992 Protocol to Amend the 1969 International Convention on Civil Liability for Oil Pollution Damage*, IMO, London, www.imo.org/en/About/Conventions/ListOfConventions/Pages/International-Convention-on-Civil-Liability-for-Oil-Pollution-Damage-%28CLC%29.aspx, accessed 04 January 2019.
IMO (1997) *Resolution A.871(20), Guidelines on the Allocation of Responsibilities to Seek the Successful Resolution of Stowaway Cases*, FAL.2/Circ.43, Annex; Resolution A.871(20), IMO, London, www.unhcr.org/refworld/docid/3ae6b31db.html, accessed 04 January 2019.
IMO (1999) *International Convention on Arrest of Ships*, IMO, London, treaties.un.org/doc/Publication/MTDSG/Volume%20II/Chapter%20XII/XII-8.en.pdf, accessed 04 January 2019.
IMO (2003) *International Ship and Port Facility Security Code*, IMO, London, www.portofantwerp.com/sites/portofantwerp/files/ISPS_code_en.pdf, accessed 04 January 2019.
IMO (2004) *International Convention for the Safety of Life at Sea*, IMO, London, library.arcticportal.org/1696/1/SOLAS_consolidated_edition2004.pdf
IMO (2005a) *Convention for the Suppression of Unlawful Acts against the Safety of Maritime Navigation*, IMO, London, www.unodc.org/tldb/pdf/Convention&Protocol%20Maritime%20Navigation%20EN.pdf, accessed 04 January 2019.

IMO (2005b) *2005 Protocol to the 1988 Protocol for the Suppression of Unlawful Acts against the Safety of Fixed Platforms Located on the Continental Shelf*, IMO, London, www.imo.org/en/About/Conventions/ListOfConventions/Pages/SUA-Treaties.aspx, accessed 04 January 2019.

IMO (2005c) *2005 Protocol to the 1988 Convention for the Suppression of Unlawful Acts against the Safety of Maritime Navigation*, IMO, London, www.imo.org/en/About/Conventions/ListOfConventions/Pages/SUA-Treaties.aspx, accessed 04 January 2019.

IMO (2009a) *Code of Practice for the Investigation of Crimes of Piracy and Armed Robbery Against Ships*, IMO, London, www.imo.org/OurWork/Security/PiracyArmedRobbery/Guidance/Documents/A.1025.pdf, accessed 04 January 2019.

IMO (2009b) *The Hong Kong International Convention for the Safe and Environmentally Sound Recycling of Ships*, IMO, London, ec.europa.eu/environment/waste/ships/pdf/Convention.pdf, accessed 04 January 2019.

IMO (2010) *International Safety Management Code*, IMO, London, www.imo.org/ourwork/humanelement/safetymanagement/documents/ismcode_4march2010_.pdf, accessed 04 January 2019.

IMO (2011a) *Reports on Stowaway Incidents: Annual Statistics for the Year 2010*, FAL.2/Circ.121, IMO, London, www.imo.org/blast/blastDataHelper.asp?data_id=30513&filename=121.pdf, accessed 04 January 2019.

IMO (2011b) *Revised Guidelines on the Prevention of access by Stowaways and the Allocation of Responsibilities to Seek the Successful Resolution of Stowaway Cases*, FAL.11(37), IMOO, London, www.imo.org/ourwork/facilitation/stowaways/documents/resolution%2011(37)_revised%20guidelines%20on%20the%20prevention%20of%20access%20by%20stowaways%20and%20the%20allocation%20of%20responsibilities.pdf, accessed 04 January 2019.

IMO (2012) *International Shipping Facts and Figures Information Resources on Trade, Safety, Security, Environment*, IMO, London, imo.libguides.com/MaritimeFactsandFigures, accessed 04 January 2019.

IMO (2015) *Reports on Stowaway Incidents: Annual Statistics for the Year 2014*, FAL.2/Circ.129, IMO, London, www.maritimesecurity.org/IMO/IMO_report_on_stowaway_incidents_2014.pdf, accessed 09 July 2017.

IMO (2017) *Revised Code of Conduct Concerning the Repression of Piracy, Armed Robbery Against Ships, and Illicit Maritime Activity in the Western Indian Ocean and the Gulf of Aden Area*, IMO, London, www.imo.org/en/OurWork/Security/PIU/Documents/DCOC%20Jeddah%20Amendment%20English.pdf, accessed 20 June 2017.

Institute of Shipping Economics and Logistics (2016a) *Shipping Statistics and Market Review*, vol 60, nos 1–2, www.isl.org/sites/default/files/sites/news/news/2016-04-14-01/Web-Short-Comment_SSMR_60-1-2.pdf, accessed 20 June 2017.

Institute of Shipping Economics and Logistics (2016b) *Shipping Statistics and Market Review*, vol 60, nos 5–6, www.isl.org/sites/default/files/sites/news/news/2016-04-14-01/Web-Short-Comment_SSMR_60-5-6.pdf, accessed 20 June 2017.

International Maritime Bureau (IMB) (2017) *Maritime Piracy and Armed Robbery Reaches 22-Year Low, Says IMB report*, Piracy Reporting Centre, London, https://iccwbo.org/media-wall/news-speeches/maritime-piracy-armed-robbery-reaches-22-year-low-says-imb-report/, accessed 04 January 2019.

International Oil Pollution Compensation Funds (2018) *The International Regime for Compensation for Oil Pollution Damage: Explanatory Note*, www.iopcfunds.org/fileadmin/IOPC_Upload/Downloads/English/explanatory_note.pdf, accessed 04 January 2019.

International Petroleum Industry Environmental Conservation Association (2016) 'Economic assessment and compensation for marine oil releases: Good practice guidelines for incident management and emergency response personnel', www.oilspillresponseproject.org/wp-content/uploads/2017/01/Compensation_2016.pdf, accessed 06 March 2019.

Liu, H. (2009) *Liner Conferences in Competition Law: A Comparative Analysis of European and Chinese Law*, Springer, Berlin.

Lu, C.-S., Chang, C.-C., Hsu, Y.-H. and Prakash, M. (2010) 'Introduction to the special issue on maritime security', *Maritime Policy & Management*, vol 37, no 7, pp. 663–665.

Mair, H. (1995) 'Green shipping', *Marine Pollution Bulletin*, vol 30, no 6, p. 360.

Mikelis, N. E. (2007) 'A statistical overview of ship recycling', International Symposium on Maritime Safety, Security & Environmental Protection, Athens, September.

Miola, A., Marra, M. and Ciuffo, B. (2011) 'Designing a climate change policy for the international maritime transport sector: Market-based measures and technological options for global and regional policy actions', *Energy Policy*, vol 39, pp. 5490–5498.

Mordal, J. (1959) *Twenty-Five Centuries of Sea Warfare*, Abbey Library, London.

Munari, F. (2012) 'Competition in liner shipping', in J. Basedow, M. Jürgen and R. Wolfrum (eds.), *The Hamburg Lectures on Maritime Affairs 2009 & 2010*, Springer-Verlag, Berlin.

Nincic, D. J. (2005) 'The challenge of maritime terrorism: Threat identification, WMD and regime response', *Journal of Strategic Studies*, vol 28, no 4, pp. 619–644.

Paige, T. P. (2017) 'The impact and effectiveness of UNCLOS on counter-piracy operations', *Journal of Conflict and Security Law*, vol 22, no 1, pp. 97–123.

Roberts, S. E., Nielsen, D., Kotlowski, A. and Jaremin, B. (2014) 'Fatal accidents and injuries among merchant seafarers worldwide', *Occupational Medicine*, vol 64, pp. 259–266.

Titanic Inquiry (2019), www.titanicinquiry.org, accessed 04 January 2019.

UNCITRAL (2009) *United Nations Convention on Contracts for the International Carriage of Goods Wholly or Partly by Sea*, London, www.uncitral.org/uncitral/en/uncitral_texts/transport_goods/2008rotterdam_rules.html, accessed 04 January 2019.

UNCTAD (2012) *Liability and Compensation for Ship-Source Oil Pollution: An Overview of the International Legal Framework for Oil Pollution Damage from Tankers*, New York, unctad.org/en/PublicationsLibrary/dtltlb20114_en.pdf, accessed 04 January 2019.

UNCTAD (2017) *Review of Maritime Transport 2016: The Long-Term Growth Prospects for Seaborne Trade and Maritime Business*, New York, http://unctad.org/en/PublicationsLibrary/rmt2016_en.pdf, accessed 21 June 2017.

UNHCR (1957) *International Convention Relating to Stowaways*, www.unhcr.org/refworld/docid/3ae6b3a80.html, accessed 04 January 2019.

United States Department of State (2003) *Proliferation Security Initiative*, www.state.gov/t/isn/c10390.htm, accessed 04 January 2019.

Wan, Z., Zhu, M., Chen, S. and Perline, D. (2016) 'Three steps to a green shipping industry', *Nature*, vol 530, pp. 275–277.

Chapter 10

International law and policy of the Polar oceans

Governing the Southern and Arctic Oceans

Key topics

- The Southern Ocean is defined by the Antarctic Convergence, which is found between 50–60 degrees south latitude except in the Western Indian Ocean where the Convergence extends to 45 degrees south latitude.
- The boundary that separates Antarctic management from the management of oceans elsewhere is 60 degrees south latitude (the 'Antarctic maritime area').
- The primary conventions governing the Antarctic are the International Whaling Convention (IWC), the Antarctic Treaty (AT), the LOSC, and the Protocol on Environmental Protection to the Antarctic Treaty.
- The primary conventions that apply only to the Southern Ocean are as follows: The Convention for the Conservation of Antarctic Seals (CCAS), the Convention on the Conservation of Antarctic Marine Living Resources (CCAMLR) and the Agreement on the Conservation of Albatrosses and Petrels (ACAP) under the Convention on the Protection of Migratory Species of Wild Animals (CMS).
- The Arctic is defined as areas north of the Arctic Circle (66.5 degrees north latitude) or areas where the average temperature in the warmest month (July) is below 10°C.
- The Arctic is bordered by five coastal states (Canada, Denmark, Norway, Russia and the United States), and three non-coastal states (Finland, Iceland and Sweden). In addition, the 'near-Arctic' states (China, European Union, Japan and Korea) have interests in the Arctic but no territory above the Arctic Circle.
- No specific international treaty regime governs the Arctic, and the Arctic is not specifically referenced in the LOSC.
- The Arctic Council (founded by the eight Arctic coastal and non-coastal states) promotes cooperation, coordination and interaction among member states and Indigenous communities on common Arctic issues including sustainable development and environmental protection.
- The Arctic Council has no independent legal powers.

Introduction

The Polar Regions – comprising the Antarctic and Arctic – are characterized by their remoteness, inhospitable climate, distinctive biological systems and, until recently, perceived limited economic utility. Yet, while similar in climate, the Antarctic and Arctic Regions differ from each other in almost all other respects, including governance, patterns of human use and oceanographic structure.

For the Antarctic Ocean (termed the 'Southern Ocean') and its continental shelf, no new sovereignty claims are permitted, and historical ones have been put into 'abeyance' (put on hold). Most human activities are attentively regulated in the region under international agreements. As such, the Southern Ocean is a global 'commons', available for use by any states or individual with the means to operate in its inhospitable environs.

Whereas Antarctica is a continent surrounded by an ocean, the Arctic is an ocean surrounded by continents, on which five states have near-exclusive jurisdiction. This is a result of the extent of the Arctic continental shelves, and, as such, the Arctic has few elements of a global commons. Furthermore, with climate change engendering the retreat of sea ice, no ocean region is changing faster. Indeed, from many perspectives, the Arctic is a 'new' ocean that functionally did not exist until recent ice retreat opened what was once a mostly inaccessible area. These transformations are focussing global attention on the potential for resource development (particularly hydrocarbons) and the use of the Arctic for shipping.

This chapter introduces how the Antarctic and Arctic environments are governed and also explores some of the challenges of managing the Polar Regions. A more detailed summary on Antarctic governance can be found in Joyner (1998) and Stokke and Vidas (1996). Emmerson (2010), Byers (2009) and Osherenko and Young (2005) are excellent resources on the management of the Arctic.

Introduction to the Antarctic and Southern Ocean

The Southern Ocean comprises approximately 35–36 million square kilometres, about 15 per cent of the world ocean surface. The northern boundary of the Southern Ocean is oceanographically defined by the Antarctic Convergence, which is a 30–50km wide zone where the warmer, oligotrophic (nutrient poor – see Chapter 2) waters of the Atlantic, Indian and Pacific Oceans merge with nutrient and biologically rich Antarctic waters. This convergence creates a zone of high biological productivity (Figure 10.1). The Antarctic Convergence is between 50–60 degrees south latitude except in the Western Indian Ocean where the Convergence extends to 45 degrees south latitude.

For the purposes of Antarctic administration, two boundaries are commonly employed. The Antarctic Treaty System (discussed below) is generally applicable to all areas south of 60 degrees south latitude (the 'Antarctic maritime area'). This latitude, therefore, is the boundary that separates Antarctic management from the management of oceans elsewhere (Vidas, 2000). However, the 1980 Convention on the Conservation of Antarctic Marine Living Resources (see below) is defined by the Antarctic Convergence.

The Antarctic continent is 14 million square kilometres (about twice the size of Australia). It was first discovered in the 1820s by British, American and Russian fleets. By the end of the century, Belgium, Germany, the United Kingdom and Sweden were staging overwintering expeditions. Norway's Roald Amundsen reached the South Pole in December 1911, and exploration activities continued throughout the early part of the 20th Century. Between 1943 and 1980, seven nations – Argentina, Australia, Chile, France, New Zealand, Norway and the United Kingdom – made territorial claims to Antarctica. Despite these declarations, Antarctica is the only continent that is free from permanent human non-research-related settlement and use.

Throughout much of human history, the Southern Ocean was too inhospitable to support human activities. In the 18th Century, however, humans began hunting in the Southern Ocean. Populations of fur seals (> two million harvested), whales (> 500,000 harvested), elephant seals (~one million harvested) and penguins (unknown harvest) were all driven to near commercial

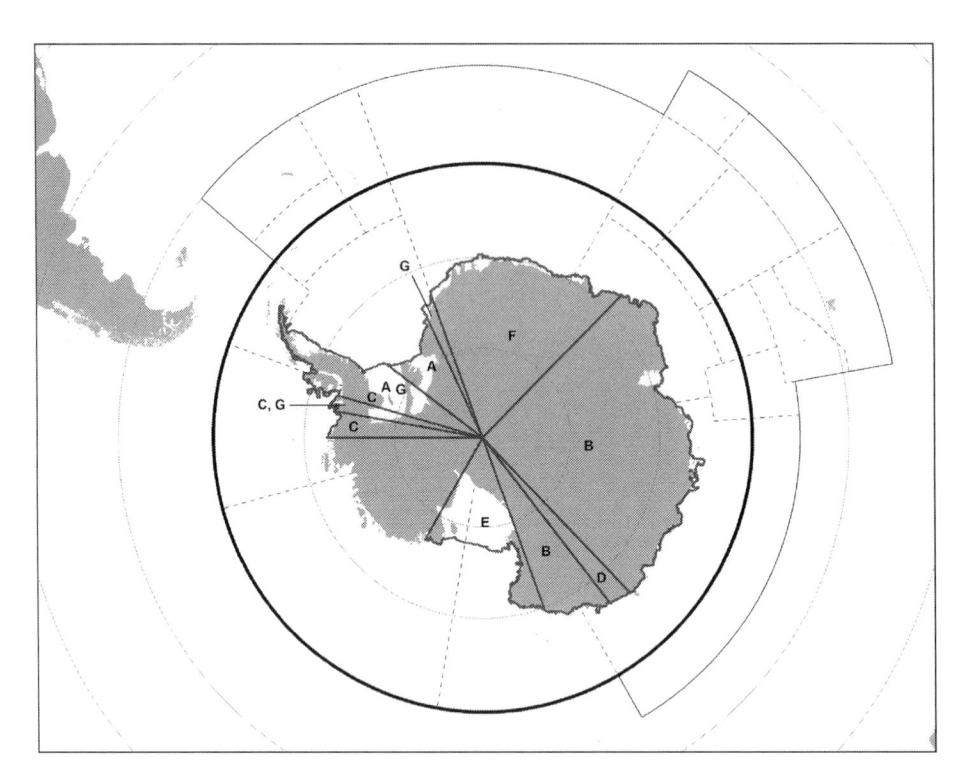

Figure 10.1 Overview of the Antarctic maritime area (Southern Ocean) defined as areas south of 60 degrees south latitude (thick line). The Antarctic Convergence is found between 50–60 degrees south latitude except in the Western Indian Ocean where the Convergence extends to 45 degrees south latitude (thin line). The Antarctic Convergence is used to define the northern boundary of the Commission for the Conservation of Antarctic Marine Living Resources (subareas shown as dashed lines). Sovereignty claims (in abeyance) are as follows: A) Argentina, B) Australia, C) Chile, D) France, E) New Zealand, F) Norway, G) Britain.

extinction (Kock, 2007). In the 20th Century, commercial fishing interests moved into the area. First, cod icefishes (*Notothenia spp*) were targeted using trawling methods; then, in the 1980s, a longline fishery for Patagonian toothfish (*Dissostichus eleginoides*) emerged. In the process, bycatches caused declines of albatrosses and petrels. In the 1970s, a fishery developed for krill (*Euphausia superba*), which is a food staple for many fish, birds and marine mammals. This raised concerns that a large, unregulated low trophic level fishery may have ecological repercussions throughout the Southern Ocean.

The threats of sovereignty claims, commercial interests and environmental damage led to a suite of international agreements negotiated over a period of several decades beginning in the 1940s. The most important international agreements governing the Southern Ocean are (the year of entry into force in parentheses):

- The International Whaling Convention (IWC – see Chapter 6) (1946).
- The Antarctic Treaty (AT – see below) (1961).

- The United Nations Convention on the Law of the Sea (LOSC – see Chapter 2) (1994).
- The Protocol on Environmental Protection to the Antarctic Treaty (1998).

In addition, many other treaties and conventions (e.g. marine pollution, fishing, climate change) apply to the Southern Ocean area and are discussed in other chapters. Specific conventions that apply only to the Southern Ocean are as follows (the year of entry into force in parentheses):

- The Convention for the Conservation of Antarctic Seals (CCAS) (1978).
- The Convention on the Conservation of Antarctic Marine Living Resources (CCAMLR) (1982).
- The Agreement on the Conservation of Albatrosses and Petrels (ACAP) (2004) under the Convention on the Protection of Migratory Species of Wild Animals (CMS).

The Antarctic Treaty and Antarctic Treaty System

The Antarctic Treaty regulates the management and governance of all land and ice shelf areas south of 60 degrees south latitude. The Treaty and its related agreements are generally termed the Antarctic Treaty System (ATS). Negotiations were concluded over a six-week period, and the Treaty was signed on 1 December 1959. It entered into force on 23 June 1961 after ratification by the 12 parties (Argentina, Australia, Belgium, Chile, France, Japan, New Zealand, Norway, South Africa, UK, USA and USSR) (ATS, 1959). The primary intent of the Treaty is to ensure 'in the interests of all mankind that Antarctica shall continue forever to be used exclusively for peaceful purposes and shall not become the scene or object of international discord'. The Treaty currently has 12 signatory parties plus 41 other parties that have since acceded, representing 65 per cent of the world's population. It is administered by the Antarctic Treaty Secretariat in Buenos Aires, Argentina.

The specific Treaty articles are outlined in Table 10.1; however, the primary focus of the Treaty is to prohibit military activity, ban nuclear explosions and the disposal of nuclear waste, promote scientific cooperation and prevent additional territorial claims.

With respect to the Southern Ocean, only Argentina, Australia and Chile have the ability under the LOSC to claim marine jurisdiction south of 60 degrees south latitude. To date, however, these states have not done so and continue to work within the ATS. However, Australia and Argentina have submitted claims to parts of the Antarctic's extended continental shelf. Japan is prevented from claiming Antarctic areas by its post-war Peace Treaty.

The ATS is administered annually through the Antarctic Treaty Consultative Meeting (ATCM) where all parties (53 to date) to the Treaty have the right to attend (Box 10.1). However, only the 29 Consultative Parties (CPs) have the right to participate in decision-making. Consultative Parties include the 12 original signatories to the Treaty as well as 17 additional states that have carried out significant scientific investigations in the Antarctic region. The Netherlands is the only CP without a scientific base in Antarctica. Conversely, Pakistan established a summer station on the continent but was not granted CP status. The remaining parties, termed Non-Consultative Parties, are those states who have acceded to the AT but without sufficient scientific activities on the continent to warrant CP status.

Although unable to participate in resolutions, observers and invited experts are permitted to attend ATCM meetings. Currently, the three observers are the Scientific Committee on Antarctic Research (SCAR), the Commission for the Conservation of Antarctic Marine Living Resources (CCAMLR) and the Council of Managers of National Antarctic Programs (COMNAP – see Box 10.2). Recurrent invited experts include the Antarctic and Southern Ocean Coalition (ASOC), the International Association of Antarctica Tour Operators (IAATO) and Coalition of Legal Toothfish Operators (COLTO) (Dudeney and Walton, 2012).

Table 10.1 Articles of the Antarctic Treaty

- Article I: Antarctica to be used for peaceful purposes and no state may establish or present a military presence.
- Article II: Antarctica is free to all for scientific investigation.
- Article III: States will cooperate on scientific investigation.
- Article IV: No state may claim new or additional sovereignty in the Antarctic region.
- Article V: Nuclear tests and the disposal of nuclear wastes are prohibited.
- Article VI: The Treaty applies to terrestrial and ice shelf areas south of 60 degrees south latitude.
- Article VII: Contracting parties to the Treaty may designate inspectors who may inspect any part of Antarctica at any time.
- Article VIII: Anywhere within Antarctica, observers discussed in Article VII shall be subject to the jurisdiction of the Contracting Party of which they are nationals.
- Article IX: Contracting Parties agree to meet to discuss the Treaty after the Treaty enters into force.
- Article X: Contracting Parties may not defeat the purpose and intent of the Treaty.
- Article XI: Disputes will be resolved through negotiation, inquiry, mediation, conciliation, arbitration, judicial settlement or other peaceful means of their own choice. If a settlement cannot be reached the dispute may be referred to the International Court of Justice.
- Article XII: The Treaty may be modified by unanimous agreement of the Contracting Parties and enter into force when ratified by the national governments of all the Contracting Parties. The Treaty expires 30 years from the date of entry into force.
- Article XII: Non-party states may enter the Treaty by accession and are bound by the Treaty once it is received by the depositary government.
- Article XIV: The Treaty will be deposited in the archives of the Government of the United States of America.

Box 10.1 List of parties to the Antarctic Treaty and their status

State	Entry into force	Consultative status	Environment Protocol[1]	CCAS[2]	CCAMLR[2]
Argentina	23 Jun 1961	23 Jun 1961	14 Jan 1998	X	X
Australia	23 Jun 1961	23 Jun 1961	14 Jan 1998	X	X
Austria	25 Aug 1987				
Belarus	27 Dec 2006		15 Aug 2008		
Belgium	23 Jun 1961	23 Jun 1961	14 Jan 1998	X	X
Brazil	16 May 1975	27 Sep 1983	14 Jan 1998	X	X
Bulgaria	11 Sep 1978	05 Jun 1998	21 May 1998		X
Canada	04 May 1988		13 Dec 2003	X	X
Chile	23 Jun 1961	23 Jun 1961	14 Jan 1998	X	X
China	08 Jun 1983	07 Oct 1985	14 Jan 1998		X
Colombia	31 Jan 1989				
Cuba	16 Aug 1984				
Czech Republic	14 Jun 1962		24 Sep 2004		
Denmark	20 May 1965				

State	Entry into force	Consultative status	Environment Protocol[1]	CCAS[2]	CCAMLR[2]
Ecuador	15 Sep 1987	19 Nov 1990	14 Jan 1998		
Estonia	17 May 2001				
Finland	15 May 1984	20 Oct 1989	14 Jan 1998		X
France	23 Jun 1961	23 Jun 1961	14 Jan 1998	X	X
Germany	05 Feb 1979	03 Mar 1981	14 Jan 1998	X	X
Greece	08 Jan 1987		14 Jan 1998		X
Guatemala	31 Jul 1991				
Hungary	27 Jan 1984				
India	19 Aug 1983	12 Sep 1983	14 Jan 1998		X
Italy	18 Mar 1981	05 Oct 1987	14 Jan 1998	X	X
Japan	23 Jun 1961	23 Jun 1961	14 Jan 1998	X	X
Korea (DPRK)	21 Jan 1987				
Korea (ROK)	28 Nov 1986	09 Oct 1989	14 Jan 1998		X
Malaysia	31 Oct 2011		14 Sep 2016		
Monaco	30 May 2008		31 Jul 2009		
Netherlands	30 Mar 1967	19 Nov 1990	14 Jan 1998		X
New Zealand	23 Jun 1961	23 Jun 1961	14 Jan 1998		X
Norway	23 Jun 1961	23 Jun 1961	14 Jan 1998	X	X
Pakistan	01 Mar 2012		31 Mar 2012		X
Papua New Guinea	16 Mar 1981				
Peru	10 Apr 1981	09 Oct 1989	14 Jan 1998		X
Poland	23 Jun 1961	29 Jul 1977	14 Jan 1998	X	X
Portugal	29 Jan 2010		10 Oct 2014		
Romania	15 Sep 1971		05 Mar 2003		
Russian Federation	23 Jun 1961	23 Jun 1961	14 Jan 1998	X	X
Slovak Republic	01 Jan 1993				
South Africa	23 Jun 1961	23 Jun 1961	14 Jan 1998	X	X
Spain	31 Mar 1982	21 Sep 1988	14 Jan 1998		X
Sweden	24 Apr 1984	21 Sep 1988	14 Jan 1998		X
Switzerland	15 Nov 1990		01 Jun 2017		
Turkey	24 Jan 1996		27 Oct 2017		
Ukraine	28 Oct 1992	04 Jun 2004	24 Jun 2001		X
United Kingdom	23 Jun 1961	23 Jun 1961	14 Jan 1998	X	X
United States	23 Jun 1961	23 Jun 1961	14 Jan 1998	X	X
Uruguay	11 Jan 1980	07 Oct 1985	14 Jan 1998		X
Venezuela	24 Mar 1999		31 Aug 2014		

Source: www.ats.aq/devAS/ats_parties.aspx?lang=e

1 Protocol on Environmental Protection to the Antarctic Treaty (1991)
2 Convention for the Conservation of Antarctic Seals (1978)
3 Convention on the Conservation of Antarctic Marine Living Resources (1982)

> **Box 10.2 Organizations formed to advise the Antarctic Treaty System**
>
> The **Scientific Committee on Antarctic Research** (SCAR) is an independent, interdisciplinary scientific body established to provide scientific advice to the ATCMs and other organizations on issues of science and conservation affecting the management of Antarctica and the Southern Ocean. It was established in 1958 and currently has 37 member states. SCAR establishes standing scientific groups and committees who address a wide range of scientific topics related to the life sciences, geosciences and physical sciences.
>
> The **Council of Managers of National Antarctic Programs** (COMNAP) was established in 1988 to foster the coordination of scientific research by the 28 consultative parties. COMNAP's purpose is to 'develop and promote best practice in managing the support of scientific research in Antarctica'. The organization's goals are to:
>
> - Serve as a forum to develop practices that improve effectiveness of activities in an environmentally responsible manner.
> - Facilitate and promote international partnerships.
> - Provide opportunities and systems for information exchange.
> - Provide the ATS with objective, technical and non-political advice.
>
> The **Committee for Environmental Protection** (CEP) was convened in 1998 to provide advice with respect to the implementation of the Protocol on Environmental Protection to the Antarctic Treaty. The CEP meets annually in conjunction with the ATCM and provides specific advice on environmental impact assessment, environmental emergencies, the establishment and operation of protected areas, scientific information management and environmental monitoring.
>
> Source: www.comnap.aq, www.cep.aq, www.scar.org

Decision-making at the ATCM is by consensus. Discussion topics are presented in two forms: Working Papers, which require debate and action, and Information Papers, which do not. Working Papers may be brought forward by CCAMLR, SCAR, COMNAP and the Committee for Environmental Protection (CEP – see below). Other invited experts/groups may only introduce Information Papers. During each ATCM, there is also a meeting of the Committee of Environmental Protection.

The ATS has been lauded as one of the most successful international treaty arrangements. It has managed over 10 per cent of the earth's surface both peacefully and in the interests of scientific research. Not only has the ATS avoided disputes over sovereignty between the seven claimant states, but it has adroitly balanced the interests of the non-claimant states and indeed attracted new states to become signatories to the ATS, even though the primary activity under the ATS is science. Despite continued pressure to allow exploitation and development, the ATS has to date prohibited the exploration (and exploitation) for mineral resources. All of this has been accomplished during five decades of profound political, technological and economic change (Triggs, 2009).

Nonetheless, prickly issues in the Southern Ocean continue: the threat of overfishing (both legal and illegal), whaling, ozone depletion, pollution from inside and outside the region, climate change (regional warming, ocean acidification, changes in sea ice distribution), tourism, bioprospecting and human security (Chown et al., 2012). Additionally, there has been considerable debate on whether to establish large marine protected areas (MPAs). While the CLCS (see Chapter 4) has not yet been asked to review continental shelf claims for claimant states, there is a legal risk of these states requesting the CLCS to do so. On top of this, 30 states currently operate research bases on the continent, and there are concerns that some could use the area to advance their military objectives (Anon, 2012). With 200 organizations representing 27 states, complications are inevitable while undertaking commercial-based research in the AT region.

Other serious challenges are emerging. Disputes between claimant and non-claiming states over access to offshore hydrocarbon resources, ownership of intellectual property from bioprospecting, and conflicts between whaling and non-whaling nations all endanger the fragile co-existence in the Southern Ocean (Joyner, 2009).

Many blame the Treaty itself for the myriad of conflicts. In contrast to other international treaties, it has been criticized for its ambiguous and vague language as well as weak compliance and enforcement mechanisms. Others argue the opposite, contending that the absence of definitive provisions in the Treaty and associated agreements have, perhaps inadvertently, created a political climate where claimant states feel comfortable entering into agreements without jeopardizing their claims.

A recent study by Liggett et al. (2017) concluded there were four future scenarios for the Antarctica and the Southern Ocean. The first is a 'collaborative-conservationist' scenario where signatories to the AT continue adhere to and build on current governance that prioritizes environmental management informed by scientific research. The second is a 'collaborative-exploitive' scenario where parties continue to respect the AT/ATS but with a focus on the development of marine resources that could include mining, bioprospecting and aquaculture in addition to increased utilization of fisheries. The third is in 'individualistic-conservationist' scenario where interest and investment in the AT/ATS wanes but state actors continue to manage for environmental values first and foremost. Lastly, an 'individualistic-exploitive' scenario would result in a drift away from AT/ATS with states adopting positions that the Southern Ocean should be exploited for economic gain.

The Protocol on Environmental Protection to the Antarctic Treaty

The Protocol on Environmental Protection to the Antarctic Treaty (the 'Environmental Protocol' or 'Madrid Protocol') was signed on 4 October 1991 (ATS, 1991). It entered into force on 14 January 1998. The Protocol has been ratified by 37 states and signed by a further 16 states. The Protocol designates Antarctica as a 'natural reserve, devoted to peace and science' (Article 2). Article 3 commits to the protection of Antarctica's environmental values and its vital role for scientific pursuits. The Article then sets out the principles to avoid or mitigate human impacts on Antarctic systems.

The Protocol is clear that it neither modifies nor amends the Treaty and that nothing in the Protocol shall relieve any party from their rights and obligations under the ATS. The Protocol also commits signatories to cooperate on the planning and conduct of activities in the Antarctic, including completing environmental assessments and evaluating cumulative impacts. Importantly, except for scientific research, the Protocol prohibits any activity related to mineral resources. Lastly, the Protocol establishes the Committee for Environmental Protection (CEP) to provide advice with respect to the implementation of the protocol.

The Environmental Protocol contains the following six Annexes:

- Annex I: Guidelines and processes for conducting environmental impact assessments in the Antarctic region.

- Annex II: Instructions regarding permitting for the removal or use of Antarctic flora and fauna for scientific purposes and rules on the introduction or use of non-native species in Antarctica.
- Annex III: Guidelines on the storage, removal and disposal of wastes.
- Annex IV: Guidelines on the prevention of marine pollution, including emergency preparedness and the disposal of garbage and sewage.
- Annex V: Instructions for the establishment of Antarctic Specially Protected Areas and Antarctic Specially Managed Areas.
- Annex VI: Prevention of environmental emergencies and setting liability limits.

Convention for the Conservation of Antarctic Seals

For almost three centuries, hunting for pelts has decimated various marine mammal species. By the 1820s, populations of southern fur seals (*Arctocephalus* sp.) were believed to be commercially extinct. Close to 95 per cent of the fur seal population breeds on South Georgia Island, making them particularly vulnerable to over-exploitation. The decline of the fur seal combined with interest in the commercial harvest of other Antarctic seal species resulted in early prohibitions on sealing on South Georgia Island. These were followed by the 1964 Agreed Measures for the Conservation of Antarctic Fauna and Flora, which were endorsed to at the third ATCM.

The Convention for the Conservation of Antarctic Seals (the 'Antarctic Seals agreement') was signed on 1 June 1972 and entered into force on 11 March 1978 (BAS, 1972). There are 17 parties to the Convention. These include Argentina, Australia, Belgium, Brazil, Canada, Chile, France, Germany, Italy, Japan, New Zealand (signed but not ratified), Norway, Poland, Russia, South Africa, the United Kingdom and the United States. The Convention applies to all marine waters south of 60 degrees south latitude and the following species:

- Southern elephant seal *Mirounga leonine*.
- Leopard seal *Hydrurga leptonyx*.
- Weddell seal *Leptonychotes weddelli*.
- Crabeater seal *Lobodon carcinophagus*.
- Ross seal *Ommatophoca rossi*.
- Southern fur seals *Arctocephalus* sp.

The Convention commits signatories to establish a permitting system for the harvest of species covered by the Convention, to share scientific and harvest information with other contracting parties and to conform to harvest regulations outlined in the Annex. These regulations address protected species, harvest levels by species, closed seasons, seal reserves and sealing methods.

Conference on the Conservation of Antarctic Marine Living Resources

The 1977 ATCM invoked Article IX of the ATS and held a number of meetings that, on 1 August 1980, led to the adoption of the Convention on the Conservation of Antarctic Marine Living Resources (CCAMLR) (ATS, 1980). With 31 signatories, the Convention entered into force on 7 April 1982. The overall objective is to conserve the marine life of the Southern Ocean while allowing 'rational use' of its resources. In particular, and unique to international marine law in 1982, CCAMLR used Article II of the Convention as a basis to manage the Southern Ocean using both 'precautionary' and 'ecosystem' approaches (discussed in Chapter 4). It has now successfully exploited these approaches for three decades (Constable, 2011). CCAMLR specifically excludes whales and seals, which are managed through the ICRW (Chapter 6) and Convention for the Conservation of Antarctic Seals (see above). CCAMLR is complimentary to and cooperates with the Annex II (Conservation of Antarctic Fauna and Flora) of the Protocol on Environmental Protection to the Antarctic Treaty. The Convention currently has 25 member states and an additional 11 acceding states.

Key Articles of CCAMLR include:

- Article I: Outlines the CCAMLR area (Figure 10.1) and specifies that the Convention applies to apply to all living marine organisms in the CCAMLR area, including birds.
- Article II: Defines the principle of 'conservation' for the Convention and outlines how harvesting and associated activities are to be conducted in accordance to the definition of 'conservation'.
- Article III: Commits parties to the Convention to also respect the Antarctic Treaty (AT) whether or not they are signatories.
- Article IV: Asserts that no state may claim new or additional sovereignty in the Antarctic region.
- Article V: Parties to the Convention recognize their obligation to the living resources provisions in the AT.
- Article VI: The Convention will not override the ICRW and the Convention for the Conservation of Antarctic Seas.
- Article VII: A Commission for the Conservation of Antarctic Marine Living Resources will be established.
- Article VIII: The Commission discussed in Article VII will have the authority to operate in all CCAMLR areas.
- Article XIV: The Scientific Committee for the Conservation of Antarctic Marine Living Resources will be established.

The CCAMLR Commission, headquartered in Hobart, Tasmania, Australia, oversees the implementation of the Convention. The Commission is comprised of 25 members who are involved in fishing and/or scientific research in the Southern Ocean. The primary duties of the Commission are to establish catch levels for harvested species and minimize impacts to non-target species. Compliance and enforcement of harvest levels are the responsibility of Commission members. The Commission meets annually to review reports on the members' activities, analyse compliance with Commission decisions, study and adopt regulatory measures and discuss the financing and administration of the Commission.

In addition to the Commission, sub-committees on science (Scientific Committee – SC-CAMLR), ecosystem monitoring (Ecosystem Monitoring and Management – WG-EMM), fish stock assessment (Fish Stock Assessment – WG-FSA), finance and administration (Standing Committee on Administration and Finance – SCAF) and implementation and compliance (Standing Committee on Implementation and Compliance – SCIC) were established.

CCAMLR is widely viewed as successful in its efforts to balance environmental protection with sustainable use. At the request of the Commission, CCAMLR underwent a performance review in 2008. The review suggested that CCAMLR devote attention to invasive species issues, vessel safety standards, marine pollution, krill management and bycatch issues. The review also recommended that CCAMLR be more proactive with respect to the designation of marine protected areas (see Chapter 12), controlling fishing effort and revisiting mechanisms to resolve disputes. Since then, CCAMLR has made progress on all these issues, but the establishment of MPAs has been fraught with political differences in opinion (generally between the fishing and non-fishing states) as to their definition, size, location and permitted activities.

Introduction to the Arctic

The Arctic comprises 30 million square kilometres, or one sixth of the earth's surface. The region is generally defined as areas north of the Arctic Circle (66.5 degrees north latitude), which is the approximate limit of the midnight sun and the polar night. The Arctic is also defined as the areas where the average temperature in the warmest month (July) is below 10°C (Figure 10.2). There are eight Arctic states (Canada, United States, Russia, Norway, Sweden, Finland, Denmark [including Greenland and Faroe Islands] and Iceland), five of which border the Arctic Ocean.

The Arctic Ocean is a relatively shallow entrained sea, primarily covered by ice, and exhibits the lowest salinities of any ocean as a result of freshwater inputs. Due to the remote location and harsh environment, there is still much unknown about the region. Recent technological advances have allowed a growth in knowledge of the region's physiographic, oceanographic and ecological characteristics. In addition, traditional knowledge stemming from the regions' vast Indigenous populations has emerged as an important source of information to better understand Arctic phenomena.

Humans have been active in Arctic areas for tens of thousands of years. Approximately four million people living in eight states inhabit the Arctic, of which Indigenous populations number approximately 300,000 (ACIA, 2004). Indigenous peoples, numbering in the hundreds of thousands, have domesticated reindeer and continue to rely on the ocean for whales, walrus, seals, fish and polar bears.

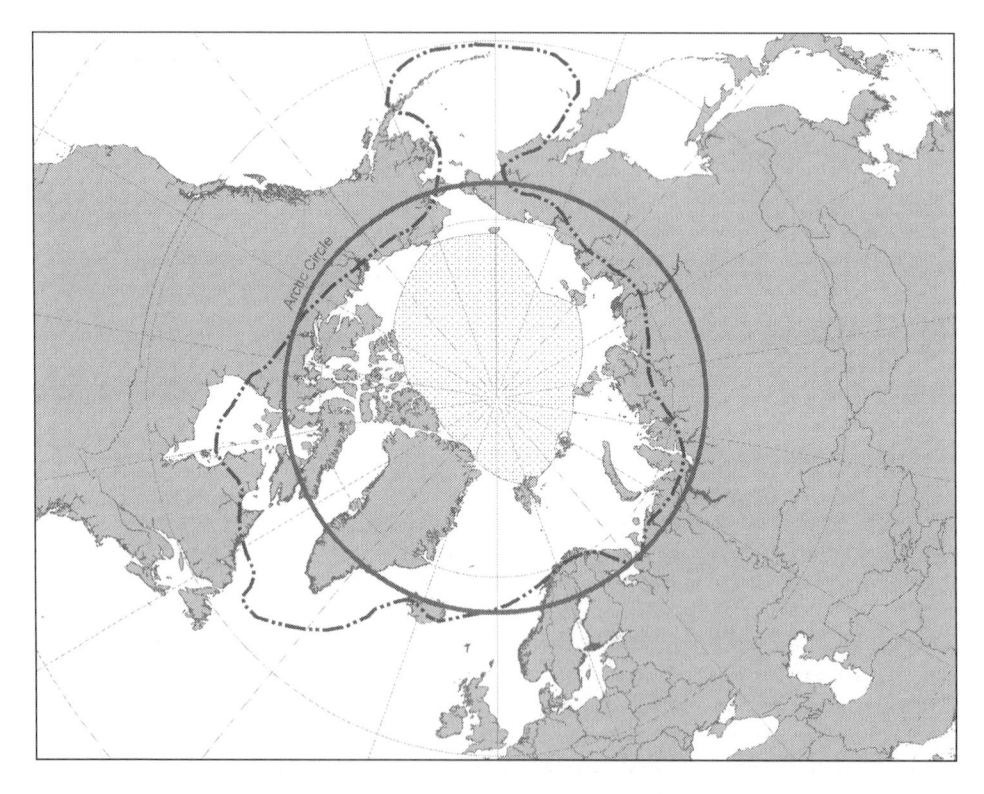

Figure 10.2 Overview of Arctic circumpolar area (north of the Arctic Circle (66° 33′N) – solid grey line), the 10°C July isotherm and historical summer extent of permanent sea ice (shaded area)

With some notable differences, the Arctic Ocean experiences similar environmental impacts from human activities as other ocean areas. Oceanic and atmospheric advection of pollutants from more southern areas is common in the Arctic. In particular, PCBs, persistent organic pollutants, heavy metals and radionuclides accumulate through the Arctic food web (Kallenborn et al., 2011). As part of this web, Indigenous Arctic peoples, whose traditional diets consume marine and terrestrial mammals, exhibit some of the highest contaminant levels on earth.

Arctic sovereignty and politics

Historically, the Arctic has been occupied by Indigenous peoples for tens of thousands of years followed by sojourns by western explorers, missionaries and scientists. With the establishment of the state of Alaska in 1959 and the advent of the Cold War, nuclear submarines – able to transit undersea ice – made the Arctic an important military interest and led to decades of submarine activity. Similarly, the threat of missiles and bombers using Great Circle routes to reach North America and the former USSR resulted in the establishment of distant early warning radar installations that, to this day, continue to be a source of pollution. More recently, melting sea ice has provoked an interest in resource development and the use of the northern sea passages for trade. This has caused both Canada and Russia to declare the Northwest Passage and Northern Sea Route, respectively, as internal waters under the LOSC. As early as the 15th Century, traders pondered the use of the Arctic for shipping between Europe and Asia. A trip between London and Yokohama through the Northwest Passage is 15,700km and 13,841km through the Northern Sea Route (Northeast Passage). These distances are significantly shorter than routes through the Suez (21,200km) or Panama Canals (23,300km) (Lasserre and Pelletier, 2011) (Figure 10.3). Only approximately 20 per cent of the Arctic Ocean is considered as areas beyond national jurisdiction (ABNJ, Chapters 3 and 4).

Indigenous populations are governed by capitals much further south. As such, many Arctic residents feel removed and isolated from decisions made in their territories. The converse can be said of their fellow countrymen to the south. While they rarely visit their northern climes, they often exhibit nationalist sentiments towards the Arctic and feel a strong kinship with Arctic regions. The differentiated northern/southern sensibilities of ownership towards Arctic environments have been growing more polarized in recent years as offshore hydrocarbons are discovered and the potential for the commercialization of northern shipping routes is realized.

Arctic governance

Unlike Antarctica, Arctic governance is relatively recent. It was not until the end of the Cold War in the early 1990s that Arctic states began to work together. The interests of the five Arctic coastal states are reasonably well known (Table 10.2). With the exception of the 1973 Agreement on the Conservation of Polar Bears, no specific international treaty regime governs the Arctic. Within the Arctic region, continental shelves within the EEZ, as per the LOSC, are governed by domestic law. All other marine regions are considered ABNJ, so the LOSC is the de facto primary international instrument governing the Arctic along with the Rovaniemi and Nuuk Declarations in 1991 and 1993, respectively, that broadly commit Arctic states to the protection of the environment. The Arctic is not specifically addressed in the LOSC. The only reference within the Convention related to the Arctic is Article 234, providing that coastal states have the right to adopt and enforce non-discriminatory laws and regulations for the prevention, reduction and control of marine pollution from vessels in ice-covered areas.

Article 76 of the LOSC permits states to claim sovereignty over areas beyond 200nmi provided applications are approved by the CLCS (see Chapter 4). Arctic states that have applied

Table 10.2 Summary of national Arctic interests

Nation	Arctic Interests
Canada	Sovereignty and responsible development of the Northwest Passage
	Defining Arctic waters as 'internal waters' under the LOSC to better regulate shipping
	Oil and gas exploration and development
	Tourism
United States	Does not acknowledge Canadian sovereignty claims
	Sovereignty and responsible development of the Northwest Passage
	Oil and gas exploration and development
Russia	Development of the Northern Sea Route for transit
	Oil and gas exploration and development
Sweden/Denmark/Norway	Strive for regulation of Arctic environmental pollution while maintaining protection for Arctic communities
	Responsible development of the Northern Sea Route
Indigenous Peoples	Concern that changes in ecosystem structure due to sea ice melting and increase coastal development could have detrimental effects on community structure and traditional lifestyles
	Want responsible economic development and social progress
	Fear that increased oil pollution due to expansion of Arctic shipping, oil and gas development, could lead to ecological destruction and health risks for local populations
	Want Indigenous rights and title recognized by states and international bodies

Source: Modified from Pietri et al. (2008)

to the CLCS to increase their territorial waters beyond 200nmi include Russia (2001), Norway (2006), Iceland and Denmark (2009) and Canada (2010). If all five Arctic states are successful in their claims, approximately half of the Arctic will be managed as territorial seas and therefore will be under national jurisdiction.

Despite the absence of a formalized Arctic treaty system, the Arctic states have cooperated among themselves for many decades to manage fisheries, undertake joint search and rescue exercises, establish scientific programs and collaborate on boundary identification. In 1987, the president of the Soviet Union, Mikhail Gorbachev, called for Arctic states to cooperate on arms control and resource development. This demand soon prompted the International Arctic Science Committee in 1990, the Arctic Environmental Protection Strategy in 1991 and the Arctic Council in 1996 (Young, 2011).

Despite these beneficial actions, certain government and non-government organizations maintain that existing cooperative and jurisdictional arrangements are insufficient to address the scope, scale and pace of climate change impacts in the region and the increase in economic activity. In particular, warming is expected to shift the distribution and composition of biological communities. This may contribute additional management challenges for commercially important fish stocks, including the movement of stocks across national borders and calls to increase fishery quotas. Currently, there is no pan-Arctic RFMO to manage rapidly changing fish stocks, and there have been a number of calls to establish more binding and comprehensive international governance mechanisms to address Arctic issues (Pietri et al., 2008). In 2018, as proposed by the European Union, Canada, the People's Republic of China, Denmark (in respect of Greenland and the Faroe Islands), Iceland, Japan, the Republic of Korea, Norway, the Russian Federation and the United States signed (not yet ratified) an *Agreement to prevent unregulated high seas Fisheries in the Central Arctic Ocean*, effecting a moratorium until scientists confirm that it can be done sustainably and until the Parties agree on mechanisms to ensure the sustainability of fish stocks.

The absence of an Arctic legal framework has led the European Parliament (2008) and World Wildlife Fund to propose the development of an Arctic Treaty (European Parliament, 2008; Koivurova and Molenaar, 2010). However, to date the coastal Arctic states oppose the development of an Arctic treaty and have expressed their position in the Ilulisaat Declaration (Ilulisaat Declaration, 2008).

Arctic Council

Signed on 19 September 1996 in Ottawa and referred to as the Ottawa Declaration, the Arctic Council emerged from earlier efforts by the eight Arctic nations (Canada, Denmark – including Greenland and the Faroe Islands – Finland, Iceland, Norway, Russian Federation, Sweden and the United States) to develop the 1989 Arctic Environmental Protection Strategy (AEPS) that was formally adopted on 14 June 1991 (Arctic Council, 1996). The overall purpose of the Council (Article 1.a) is to provide a means for promoting cooperation, coordination and interaction among member states and Indigenous communities on common Arctic issues including sustainable development and environmental protection. Other duties of the Council are to oversee AEPS programs (Article 1.b); supervise a sustainable development program (Article 1.c); and disseminate information, encourage education and promote interest in Arctic-related issues (Article 1.d). The Council has no independent legal power, avoids military matters and is a forum for discussion only (Jeffers, 2010). The chairmanship of the Arctic Council rotates between Arctic nations every two years. Since its inception, each Arctic state has chaired the council once. Canada was chair from May 2013–2015, followed by the United States 2015–2017 and Finland 2017–2019.

The following six Indigenous organizations are permanent participants on the Arctic Council:

1 Arctic Athabaskan Council (AAC).
2 Aleut International Association (AIA).
3 Gwich'in Council International (GGI).
4 Inuit Circumpolar Council (ICC).
5 Russian Arctic Indigenous Peoples of the North (RAIPON).
6 Saami Council (SC).

Observers are permitted on the Council if they are endorsed by the Council. Observer status is open to non-Arctic states, inter-governmental organizations and non-governmental organizations. In May 2013, the Arctic Council added six new non-Arctic state observers: China, India, Italy, Japan, Singapore and South Korea. These are in addition to France, Germany, The Netherlands, Poland, Spain and the United Kingdom. Thirteen inter-governmental organizations (e.g., the International Union for the Conservation of Nature – IUCN) and 13 non-governmental organizations (e.g., the World Wildlife Fund) have received observer status. The European Union has requested observer status on the Council, but as of February 2019 it has not yet been approved.

The Council directs the activities of the following working groups comprised of government representatives and researchers:

- Arctic Contaminants Action Program (ACAP).
- Arctic Monitoring and Assessment Programme (AMAP).
- Conservation of Arctic Flora and Fauna (CAFF).
- Emergency Prevention, Preparedness and Response (EPPR).
- Protection of the Arctic Marine Environment (PAME).
- Sustainable Development Working Group (SDWG).

Additionally, the Council has established a number of Task Forces that address time-limited issues. Examples of Task Forces include the Task Force on Ecosystem-Based Management, Circumpolar Business Forum, Black Carbon, Arctic Marine Oil Pollution Preparedness and Response, and Search and Rescue.

Since the Ottawa Declaration, the Council has met eight times. A summary of the key issues and provisions is provided in Table 10.3.

Table 10.3 Declarations of the Arctic Council and key provisions

Declaration	Signed	Key provisions
Iqaluit Declaration	17 September 1998	Approval of new Permanent and Observer Participants in the Arctic Council.
		Establish the Sustainable Development Program and Working Group.
		Announce the University of the Arctic.
Barrow Declaration	13 October 2000	Approval of new Permanent and Observer Participants in the Arctic Council.
		Approval of new proposals to conduct a Survey of Living Conditions in the Arctic.
		Endorse and adopt the Arctic Climate Impact Assessment.
Inari Declaration	10 October 2002	Approval of new Permanent and Observer Participants in the Arctic Council.
		Expansion of the work of the Council.
		Continued focus on human conditions in the Arctic.
		Partnering with other organizations.
Reykjavik Declaration	24 November 2004	Approval of new Permanent and Observer Participants in the Arctic Council.
		Request for Council working groups to review Arctic contaminants, conduct a marine shipping assessment and assess oil and gas and acidification issues.
Salekhard Declaration	26 October 2006	Approval of new Permanent and Observer Participants in the Arctic Council.
		Request to continue climate research.
		Request working groups to research the behaviour of oil and other hazardous substances in Arctic waters.
Tromso Declaration	29 April 2009	Approval of new Permanent and Observer Participants in the Arctic Council.
		Urge early implementation actions on methane and other short-lived climate forcers.

Declaration	Signed	Key provisions
		Urge Council members to strengthen their work on adaptation.
		Encourage cooperation with the International Maritime Organization on measures to reduce shipping impacts.
		Decide whether to hold deputy minister-level Council meetings.
Nuuk Declaration	12 May 2011	Announce the Agreement on Cooperation in Aeronautical and Maritime Search and Rescue in the Arctic as the first binding agreement under the Arctic Council.
		Establishment of the Arctic Council Secretariat.
		Establish Ecosystem-Based Management task force.
Kiruna Declaration	15 May 2013	Announce the second binding Agreement on Oil Spill Preparedness and Response.
		Establish Circumpolar Business Forum task force, Black Carbon task force, Scientific Cooperation task force.
Iqaluit Declaration	24 April 2015	Establish an Arctic Economic Council.
		Address health and mental health issues.
		Establish a telecommunications infrastructure experts group.
		Implement the Framework for Action on Enhanced Black Carbon and Methane Emissions.
		Implementing the Framework Plan for Cooperation on Prevention of Oil Pollution from Petroleum and Maritime Activities in the Marine Areas of the Arctic.
		Approve the Arctic Marine Strategic Plan.
Fairbanks Declaration	11 May 2017	Provide expertise on Arctic shipping matters, including at the International
		Maritime Organization.
		Continue efforts to study the effects of ocean acidification, and marine debris.
		Continue efforts to improve internet connectivity.

Hydrocarbons and transport have galvanized interest in the area. Almost half of the Arctic marine area is comprised of continental shelves, the largest percentage of any ocean. These shelves are estimated to contain 40–83 billion barrels of oil (13 per cent of the world's undiscovered reserves) and 770–1,547 trillion cubic feet of natural gas (30 per cent of the world's undiscovered reserves) at 95 and 50 per cent probabilities, respectively. In addition, up to 44 billion barrels of natural gas liquids are predicted (Bird et al., 2008; Gautier et al., 2009). As sea ice continues to recede, these resources will become more accessible to development.

Despite economic interests in the Arctic, rapid industrial expansion in the region is unlikely. For offshore hydrocarbon exploration and development to occur, many remaining boundary disputes will need to be resolved, and advances in technology to address ice risks (for platforms, ships and pipelines) will need to be developed. Additionally, a social license will need to be obtained from Indigenous Arctic peoples and citizens further south. For shipping, the northern routes may be decades away from significant commercial use. Arctic shipping requires lower average vessel speeds in order to circumvent ice, thus reducing the competitive advantage over

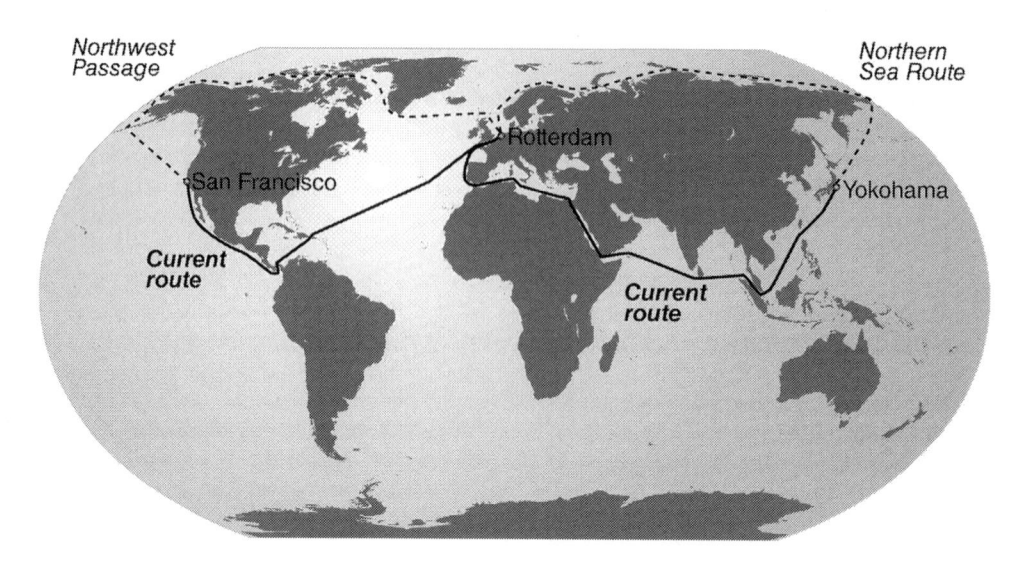

Figure 10.3 Northern Sea Route (A) and Northwest Passage (B)

the Panama and Suez routes. In addition, shipping can only operate in the near ice-free summer (at least for the immediate future). There are other disincentives as well: Vessels will require specialized navigation equipment and training, revised cargo management practices and technologies to prevent cargo freezing will be required and the northern routes will most likely require higher insurance rates (Stokke, 2011).

Overall, the Arctic is a politically stable and peaceful region. Sovereignty claims are mature and generally accepted, states work well together and towards common purposes, and the remaining boundary disputes between states are – in contrast to some other parts of the world – proceeding towards resolution. Yet, challenges remain. The Arctic Ocean as a global commons, continues to shrink as seabed mapping technologies lengthen the boundaries of the Arctic states' extended continental-shelf claims (as allowed under the LOSC). Indeed, this expansion has developed to a point where Russia and Canada are expected to require a maritime boundary treaty (Baker and Byers, 2012). With the exception of a few high latitude areas, it is quite conceivable that in the future most of the accessible Arctic will be within the jurisdiction of one or more Arctic coastal states.

China, while currently lacking an Arctic policy, considers itself an Arctic stakeholder and has lobbied the international community to prevent Arctic coastal states from securing claims to their extended continental shelves (see Box 10.3). The state has expressed a strong interest in several aspects of the Arctic: the effects of changes in Arctic climate on China, securing access to the Northern Sea Route to move goods to Europe and beyond, and participating in Arctic fisheries and natural resource development.

Climate change and political implications in the Arctic

Relative to other areas of the planet, the effects of climate change have a disproportionate impact on Arctic environments. While the reasons for the Arctic's accelerated rate of warming

> **Box 10.3 China and the Polar oceans**
>
> China has been very active in the Polar oceans in recent years. China's 13th Five Year Plan (2016–2020) commits the state to increased participation in 'strategic frontiers' defined as cyberspace, the deep ocean, the Polar oceans and space. By 2014, the third largest source of tourists to Antarctica were Chinese, and Chinese state media frequently reports on the exploits of its Polar 'science heroes'.
>
> China's largest shipping company (COSCO) began using the Northern Sea Route in 2017 on a scheduled basis, and the icebreaker Xuelong (Snow Dragon) participates in annual expeditions that include circumnavigating the Arctic Ocean. A second icebreaker, the Xuelong 2, will be completed in 2019. China's Belt and Road Initiative – a 2013 marine- and land-based transportation initiative to link East Asia with South Asia, Africa and European markets – includes the 'Polar Silk Road', a partnership with Russia to develop the Northern Sea Route to provide safe passages for energy imports and trade with Russia and Europe. While not geographically an Arctic state, China obtained observer status on the Arctic Council in 2013 by recognizing the sovereign rights of the Arctic states and acknowledging that international law, and the LOSC in particular, governs the Arctic.
>
> China has also become invested in Antarctic governance and has constructed four research stations. Unlike the Arctic where China recognizes current sovereignty claims of the Arctic states bordering the Arctic Ocean, China opposes the territorial claims of Australia and other Antarctic claimants and is dissatisfied with the Antarctic Treaty System and its precautionary approach to developing the area's resources.
>
> Overall, China's approach to the Polar oceans mirrors the state's approach to international relations elsewhere. China's use of non-coercive diplomacy and financial investment are the cornerstones of its 'soft power' approach where the objective is to improve China's reputation as a peaceful partner while opening new markets and achieving national security interests.

are not fully understood, increases in its average temperature are almost twice that of the rest of the world. This can be attributed in part to sea ice melt, reducing regional albedo and increasing ocean heat absorption. Annual temperatures have increased by an average of 1°C over northwest Russia, east Greenland and Scandinavia over the past 50 years, yet they have cooled by 1°C over Iceland and the North Atlantic Ocean over the same period. In addition, average atmospheric temperatures in the Arctic have increased up to 3°C over land and are projected to increase between 6–10°C over the Arctic Ocean by 2090 (ACIA, 2004).

As Arctic environments warm, sea ice and glaciers melt and thin. This adverse circumstance contributes to sea level rise and decreases in the earth's albedo, the amount of incoming solar radiation reflected back into space. Since the Arctic is an important regulator of the entire earth's temperature, malignant changes on a global scale are also on the horizon. Over the past 30 years, Arctic sea ice has decreased in area by approximately 30 per cent, and modelling suggests that the Arctic may be ice free by 2030 (Jones, 2011). Other Arctic-specific impacts of climate change include increased methane emissions with the melting of permafrost, reductions in atmospheric ozone in the Arctic and human settlements being jeopardized with escalating coastal erosion.

Climate change is expected to have significant economic, social and environmental impacts. Along with coastal erosion, the continued retreat of sea ice and increased ocean temperatures will affect the distribution and abundance of commercially and socially important fisheries and marine mammal species (Huntington, 2009). In particular, polar bear populations are expected to be imperilled by climate-induced habitat change (Laidre et al., 2008). Increasing resource development – and oil and gas exploration and development in particular – while creating economic opportunities will increase risks of accidents and spills (Meek, 2011).

> **Discussion questions**
>
> - Among Arctic states, Canada and Russia most strenuously object to non-Arctic states becoming observers on the Arctic Council. What might be their rationale?
> - Describe the key differences between Antarctic and Arctic governance and the advantages and disadvantages of each.
> - Consider the implications to the global economy if both the Northern Sea Route and Northwest Passage become navigable for commercial vessels. What might be the effects upon the coastal states?
> - In addition to the issues in the above questions, what are some other implications of retreating sea ice?

Further reading

ACIA (2004) *Arctic Climate Impact Assessment*, Cambridge University Press, Cambridge.
Anonymous (2012) 'Antarctic treaty is cold comfort', *Nature*, vol 481, p. 237.
Antarctic Treaty Secretariat (1959) *Antarctic Treaty*, Buenos Aires, www.ats.aq/documents/ats/treaty_original.pdf, accessed 02 September 2018.
Antarctic Treaty Secretariat (1980) *Convention on the Conservation of Antarctic Marine Living Resources*, Buenos Aires, www.ats.aq/documents/ats/ccamlr_e.pdf, accessed 02 September 2018.
Antarctic Treaty Secretariat (1991) *Protocol on Environmental Protection to the Antarctic Treaty*, Buenos Aires, www.ats.aq/documents/recatt/Att006_e.pdf, accessed 02 September 2018.
Arctic Council (1996) *Declaration on the Establishment of the Arctic Council*, www.arctic-council.org/index.php/about/documents/category/5-declarations?download=13:ottawa-declaration, accessed 02 September 2018.
Baker, J. S. and Byers, M. (2012) 'Crossed lines: The curious case of the Beaufort Sea maritime boundary dispute', *Ocean Development and International Law*, vol 43, no 1, pp. 70–95.
Bird, K. J., Charpentier, R. R., Gautier, D. L., Houseknecht, D. W., Klett, T. R., Pitman, J. K., Moore, T. E., Schenk, C. J., Tennyson, M. E. and Wandrey, C. J. (2008) 'Circum-arctic resource appraisal; estimates of undiscovered oil and gas north of the Arctic Circle', *U.S. Geological Survey Fact Sheet 2008–3049*, pubs.usgs.gov/fs/2008/3049/, accessed 02 September 2018.
British Antarctic Survey (1972) *Convention for the Conservation of Antarctic Seals*, Cambridge, UK, www.bas.ac.uk/about/antarctica/the-antarctic-treaty/convention-for-the-conservation-of-antarctic-seals-1972/, accessed 02 September 2018.
Byers, M. (2009) *Who Owns the Arctic?* Douglas & McIntyre, Toronto, ON.
Chown, S. L., Lee, J. E., Hughes, K. A., Barnes, J., Barrett, P. J., Bergstrom, D. M., Convey, P., Cowan, D. A., Crosbie, K., Dyer, G., Frenot, Y., Grant, S. M., Herr, D., Kennicutt, M. C., Lamers, M., Murray, A., Possingham, H. P., Reid, K., Riddle, M. J., Ryan, P. G., Sanson, L., Shaw, J. D., Sparrow, M. D., Summerhayes, C., Terauds, A. and Wall, D. H. (2012) 'Challenges to the future conservation of the Antarctic', *Science*, vol 337, pp. 158–159.

Constable, A. (2011) 'Lessons from CCAMLR on the implementation of the ecosystem approach to managing fisheries lessons from CCAMLR on EBFM', *Fish and Fisheries*, vol 12, no 2, pp. 138–151.

Dudeney, J. R. and Walton, W. H. (2012) 'Leadership in politics and science within the Antarctic Treaty', *Polar Research*, vol 31, pp. 1–9.

Emmerson, C. (2010) *The Future History of the Arctic*, Perseus Books, Jackson, TN.

European Parliament (2008) *Resolution of 9 October 2008 on Arctic governance*, www.arcticgoverance.org, accessed 04 January 2019.

Gautier, D. L., Bird, K. J., Charpentier, R. R., Grantz, A., Houseknecht, D. W., Klett, T. R., Moore, T. E., Pitman, J. K., Schenk, C. J., Schuenemeyer, J. H., Sørensen, K., Tennyson, M. E., Valin, Z. C. and Wandrey, C. J. (2009) 'Assessment of undiscovered oil and gas in the Arctic', *Science*, vol 324, no 5931, pp. 1175–1179.

Huntington, H. (2009) 'A preliminary assessment of threats to arctic marine mammals and their conservation in the coming decades', *Marine Policy*, vol 33, no 1, pp. 77–82.

Ilulisaat Declaration (2008) *Declaration of 28 May 2008 Adopted by the Representatives of Canada, Denmark, Norway, the Russian Federation, and the United States*, www.arcticgoverance.org, accessed 04 January 2019.

Jeffers, J. (2010) 'Climate change and the Arctic: Adapting to changes in fisheries stocks and governance regimes', *Ecology Law Quarterly*, vol 37, no 3, pp. 917–977.

Jones, N. (2011) 'Towards an ice-free Arctic', *Nature Climate Change*, vol 1, p. 381.

Joyner. C. C. (1998) *Governing the Frozen Commons: The Antarctic Regime and Environmental Protection*, University of South Carolina Press, Columbia, SC.

Joyner, C. C. (2009) 'Potential challenges to the Antarctic Treaty', in P. A. Berkman, M. A. Lang, D. W. H. Walton and O. R. Young (eds.), *Science Diplomacy: Antarctica, Science and the Governance of International Spaces*, Smithsonian Institution Scholarly Press, Washington, DC, www.atsummit50.org/media/book-15.pdf, accessed 02 September 2018.

Kallenborn, R., Borgå, K., Christensen, J. H., Dowdall, M., Evenset, A., Odland, J. Ø., Ruus, A., Aspmo Pfaffhuber, K., Pawlak, J. and Reiersen, L.-O. (2011) *Combined Effects of Selected Pollutants and Climate Change in the Arctic Environment*, Arctic Monitoring and Assessment Programme, Oslo.

Kock, K.-H. (2007) 'Antarctic marine living resources: Exploitation and its management in the Southern Ocean', *Antarctic Science*, vol 19, no 2, pp. 231–238.

Koivurova, R. and Molenaar, E. J. (2010) *International Governance and the Regulation of the Marine Arctic: A Proposal for a Legally Binding Agreement*, WWF International Arctic Programme, Oslo.

Laidre, K. L., Stirling, I., Lowry, L. F., Wiig, Ø., Heide-Jørgensen, M. P. and Ferguson, S. H. (2008) 'Quantifying the sensitivity of Arctic marine mammals to climate-induced habitat change', *Ecological Applications*, vol 18, Supplement, pp. S97–S125.

Lasserre, F. and Pelletier, S. (2011) 'Polar super seaways? Maritime transport in the Arctic: An analysis of shipowners' intentions', *Journal of Transport Geography*, vol 19, pp. 1464–1473.

Liggett, D., Gilbert, N. and Morgan, F. (2017) 'Is it all going south? Four future scenarios for Antarctica', *Polar Record*, vol 53, no 5, pp. 459–478.

Meek, C. L. (2011) 'Adaptive governance and the human dimensions of marine mammal management: Implications for policy in a changing north', *Marine Policy*, vol 35, no 4, pp. 466–476.

Osherenko, G. and Young, O. R. (2005) *The Age of the Arctic: Hot Conflicts and Cold Realities*, Cambridge University Press, Cambridge, UK.

Pietri, D., Soule, A. B., Kershnera, J., Solesa, P. and Sullivana, M. (2008) 'The Arctic shipping and environmental management agreement: A regime for marine pollution', *Coastal Management*, vol 36, no 5, pp. 508–523.

Stokke, O. S. (2011) 'Environmental security in the Arctic: The case for multilevel governance', *International Journal*, vol 66, pp. 835–848.

Stokke, O. S. and Vidas, D. (1996) *Governing the Antarctic: The Effectiveness and Legitimacy of the Antarctic Treaty System*, Cambridge University Press, Cambridge, UK.

Triggs, G. (2009) 'The Antarctic Treaty System: A model of legal creativity and cooperation', in P. A. Berkman, M. A. Lang, D. W. H. Walton and O. R. Young (eds.), *Science Diplomacy: Antarctica, Science and the Governance of International Spaces*, Smithsonian Institution Scholarly Press, Washington, DC, www.atsummit50.org/media/book-8.pdf, accessed 02 September 2018.

Vidas, D. (2000) 'Emerging law of the sea issues in the Antarctic maritime area: A heritage for the new century?', *Ocean Development and International Law*, vol 31, nos 1–2, pp. 197–222.

Young, O. R. (2011) 'If an Arctic Ocean treaty is not the solution, what is the alternative?', *Polar Record*, vol 47, no 4, pp. 327–334.

Chapter 11

International law and policy related to offshore energy and mining

Renewable and non-renewable resource development on continental shelves and in the Area

> **Chapter summary**
> - One third of the world's petroleum is produced from offshore fields.
> - It is estimated that the deep-sea areas outside of continental shelves could contain two thirds of the world's economically accessible mineral reserves mineral wealth.
> - Under the UN Law of the Sea Convention (LOSC), continental shelves and their corresponding mineral and petroleum resources fall under the sovereignty of coastal states and therefore are managed under domestic legal and regulatory regimes.
> - The 'Area' is defined by the LOSC as the seabed and ocean floor and subsoil beyond the limits of national jurisdiction. The mineral resources of the Area (which includes petroleum and geological minerals) are available to all states to be used for the benefit of humankind.
> - All states, regardless of geographic location (e.g. land-locked) and state of economic development, have equal rights to participation in the Area. Economic benefits derived from the Area will be shared among states party to LOSC, using equitable sharing criteria, taking into account the interests and needs of developing States, particularly the least developed and land-locked.

Introduction

This chapter explores the different non-biological resources that are extracted – or anticipated to be extracted – from offshore locations, including the continental shelf and deep-sea. These realms are treated differently legally. The 'continental shelf', as defined under the United Nations Law of the Sea Convention (LOSC), falls under the jurisdiction of coastal states, whereas most of the deep-sea lies beyond. As noted in Chapter 4, the legal continental shelf is not limited to the geological continental shelf and can include ridges as well as some deep-sea areas under state sovereignty.

The nearshore marine environment has long been utilized as a source for materials and more recently, energy. For centuries, beaches and nearshore areas were mined for gold, tin, diamonds, sand and gravel (Scott, 2007). As mining and shipping technologies improved, exploration and development radiated seaward onto the continental shelves. Offshore oil drilling was pioneered in the 1930s in the Gulf of Mexico while the first offshore windfarm was established in Denmark in the 1990s. Currently, about one third of the world's petroleum is produced from offshore reserves. As sea ice retreats, and as drilling technology continues to go progressively deeper and further offshore, the role of the marine environment in meeting the world's petroleum requirements is likely to remain significant, albeit controversial in some places owing to environmental sensitivities (e.g. in the Arctic) and political commitments to reduce greenhouse gas emissions. At the same time, renewable wind, wave and tidal energy can be expected to be further developed; thus, the role of the offshore environment in meeting global energy needs, both renewable and non-renewable, will continue to increase.

Since the 1960s, mining the mineral-rich deep-sea has been discussed and anticipated. However, scarcities on land that were predicted to drive interest into this new frontier have failed to materialize, due to a combination of increased metal recycling, refinements in industrial processes, improved mining technology and a steady stream of new terrestrial discoveries. This left commodity prices insufficiently high to make deep-sea mining economically attractive. The rapid industrialization of China recently altered this dynamic. Driven in part by Chinese government stimulus projects initiated after the 2008–09 economic crisis, mineral prices quickly recovered and continued to rise (until around 2013) such that commercial deep-sea mineral production was again seen as attractive to some investors looking for alternatives to the (distrusted) stock market. In the 10 years since then, the International Seabed Authority (ISA) approved 20 *exploration* contracts for 'the Area' beyond national jurisdictions (ABNJ), thereby more than trebling the global total to 29 (as of January 2019). The first deepwater offshore mineral *exploitation* had been expected to occur within two licenses awarded to Nautilus Minerals in 2011 by the Government of Papua New Guinea (PNG). These two hydrothermal vent sites are rich in copper, gold and silver, amongst other minerals. However, the subsequent decline in global commodity prices beginning in 2014, combined with internal financial and contractual difficulties, have left the future of Nautilus' operations unclear and the promise of deep-seabed mining more generally as yet still unanswered (Anon, 2018).

Differences between continental shelves and deep ocean areas with respect to energy and mineral development

Marine energy and mineral development can be separated into activities that occur on the continental shelf versus activities that occur in the deep ocean. To date, offshore hydrocarbons such as oil, natural gas and coal have been exploited on, or adjacent to, geological continental shelves within national jurisdictions. Similarly, continental shelves contain aggregates (sand and gravel) transported from terrestrial environments that may also contain metals (e.g. gold, tin) or diamonds that can be mined using placer operations. The shallow depths of continental shelves also provide two other opportunities to harvest energy: anchoring wind energy projects to the seabed and generating electricity by harnessing wave and water movement (tides and currents) amplified by shallow depths and geographical constraints.

Under the LOSC, the legal continental shelf – and its corresponding mineral and energy resources – fall under the sovereignty and control of the coastal state. These areas are therefore managed under domestic legal and regulatory regimes similarly to their terrestrial counterparts. (There is one exception, concerning the outer continental shelf, which is discussed further below.) It should be noted that legal continental shelves are subject to more jurisdictional disputes than terrestrial borders.

Deep-sea environments contain significant mineral resources that plausibly could become within economic reach this century. Proponents of deep-seabed mining suggest minerals found in the deep-sea will be increasingly in demand for use in electronics and batteries, as well as renewable energy, such as solar panels and wind generators. However, these minerals are not unique to the deep-sea, and one study suggests that terrestrial-based mining could readily meet projected demand for the foreseeable future (Teske et al., 2016), casting doubt on these claims. Rare earth elements, despite their name, are actually moderately abundant throughout the world on land but are very costly to process in an environmentally safe manner; therefore, seabed mining for rare earths (which are not especially abundant) is unlikely to be economical in the foreseeable future, though some countries without terrestrial reserves may wish to mine them for reasons of national security.

The 'commodity cycle', though currently low, could again rebound; therefore, the economic viability of deep-seabed mining (DSM) is being watched carefully by some countries. Island states in the Pacific, with limited terrestrial resources, view DSM as a potential source of natural resource revenue, though most remain cautious about potential environmental impacts. The Cook Islands, Kiribati, Nauru and Tonga have entered into exploration agreements with foreign-owned DSM companies or their local subsidiaries.

While conventional hydrocarbons (i.e. oil, gas and coal) have not historically been explored or exploited in the deep-sea, frozen methane hydrates have been found to be widely distributed where temperature and pressure regimes are favourable. These hydrates could represent a source of energy security for those states with few conventional hydrocarbons.

Unlike continental shelves, the deep ocean beyond national jurisdiction is not currently used for energy production. However, there have been proposals and small projects within national jurisdictions to use the temperature gradient between surface and deeper ocean water to either directly cool buildings, when close enough to shore, or to generate electricity. One engineering company has already successfully installed commercial air conditioning from cool deepwater sources in three places, with four more ongoing (www.makai.com/sea-water-air-conditioning/).

The LOSC and energy and mineral development

In many respects, the LOSC owes its existence to uncertainties regarding ownership of deep-sea mineral resources. Before negotiation of LOSC, deep-sea environments were truly examples of *res nullis* (belonging to no one – Chapter 3). Interest in seabed mining in the 1960s raised concerns that deep-sea areas would be developed in the absence of an international regulatory framework and without due consideration of the economic and environmental concerns of developing states (some of which had large terrestrial mines) and states adjacent to potential mineral development. After a now-famous speech in 1967 by Ambassador Arvid Pardo of Malta to the UN General Assembly (UNGA) extolling the potential of deep-seabed mining to supply the world's needs, the UNGA created a committee to look into the exploitation of the deep seabed, which resulted in the passing of a resolution in 1970 that the exploration of the Area and the exploitation of its resources shall be carried out for the 'benefit of mankind as a whole' (UN, 1968). This was followed by further resolutions placing a moratorium on DSM and prohibiting the recognition of any claims to the seabed until an international regulatory regime was in place. Because expectations for DSM were high, its international regulation was a major driver behind the third set of LOSC negotiations, from 1973 to 1982, resulting in the Convention itself. However, Part XI, which concerns DSM, still remained contentious with pro-mining countries refusing to ratify. Therefore, it was partially re-negotiated through the 1994 Part XI Agreement, resulting in the countries engaged in DSM exploration ratifying the treaty, save the United States, which signed but to date has not ratified (Brown, 2018).

The LOSC confers on the coastal state the rights to explore, exploit, conserve and manage the resources in the water column on and under the seabed within the state's territorial waters and exclusive economic zone (EEZ; Article 56). For the legal continental shelf, which may in some circumstances extend beyond a coastal state's EEZ, the state has rights to the mineral resources as well as the living sedentary species (Article 77). This includes the exclusive right to authorize and regulate exploration and exploitation activities, such as geological surveys, drilling and extraction.

Regarding profits produced from the exploitation of any part of the legal continental shelf that extends beyond a coastal state's EEZ (the 'extended continental shelf'), coastal states are required to make payments to the International Seabed Authority (see Chapter 4). Though it has not yet occurred, the revenue is to be equitably redistributed to other states, taking into account the needs of less developed states (Article 82). Planned petroleum production on the extended continental shelf off Canada's east coast of may become the first implementation of this LOSC provision.

Under the LOSC, the seabed and ocean floor and subsoil beyond the limits of national jurisdiction are termed 'the Area' (see Chapter 4) and comprise approximately 260 million square kilometres, or about three times the area of all marine areas under state control. Non-biological resources, including petroleum, are all called 'minerals' (Article 133), regardless of whether they are metallic in nature. The Area does not apply to the water column or airspace above the seabed outside of continental shelves (Article 135).

Under the LOSC, the Area is open to all for the benefit of mankind and is to be used exclusively for peaceful purposes. States are required to cooperate to achieve these ends (Articles 136 and 138). The LOSC provides for the economic development of the Area under the condition that its development contributes to the world economy and growth in international trade (Article 150). Specifically, all states, regardless of geographic location (e.g. landlocked) or state of economic development, have equal rights to access and gainful participation (Article 141). Economic benefits derived from the Area will be shared among all states (Article 139).

No state is permitted to claim territory or the natural resources within the Area. However, mineral resources from the Area may be explored and exploited but only in accordance with LOSC and the International Seabed Authority (Article 137). In addition, archaeological artefacts or discoveries of historical significance must be managed for the benefit of humanity, recognizing the importance of the state or culture of origin (Article 149). States bear the responsibility for actions of their nationals (individuals, corporations or government) and are liable for their actions (Article 139).

Management of marine energy and mineral resources on continental shelves

The following sections outline the various types of mining and energy activities that take place on continental shelf areas (i.e. within national jurisdiction) and how they are managed.

Aggregate sand and gravel deposits

Aggregate deposits are non-mineral (mostly sand and gravel) resources formed by the erosion of terrestrial areas and their subsequent deposition in the marine environment. Mined in shallow marine areas in depths up to 50m, aggregates (terrestrial and marine) are the world's largest mined commodity by tonnage. They are most commonly used for construction purposes, although marine sands are also used for beach nourishment. Many states, including Europe, Japan and the United States, currently mine marine aggregates. Nearshore aggregate mining is expected to increase as existing terrestrial sand and gravel resources become depleted and new deposits become more difficult to mine due to public opposition and conflicts with other land-based activities (Morgan, 2012).

Although countries mining marine aggregates usually have domestic legislation in place, there are very few international agreements that address their broader management. The Convention for the Protection of the Marine Environment of the Northeast Atlantic 1992 (OSPAR Convention) was convened to eliminate pollution and protect the marine environment. In 2003, an Agreement on Sand and Gravel Extraction was adopted that required OSPAR signatories to adhere to the ICES *Guidelines for the Management of Marine Sediment Extraction*, which established a set of procedures for environmentally responsible marine aggregate development (Radzevičius et al., 2010). In addition, under the Barcelona Convention (see Chapter 6), the Protocol for the Protection of the Mediterranean Sea against Pollution Resulting from Exploration and Exploitation of the Continental Shelf and the Seabed and its Subsoil (adopted in 1994 but not yet in force) provides guidance on marine aggregate extraction. The European Code of Conduct for Coastal Zones (www.eucc.net) was developed by the Coastal Union and adopted by the Council of Europe Ministers in 1991. The Code provides detailed recommendations to minimize the effects of marine aggregate extraction on coastal processes, seasonal biological events and water quality (Radzevičius et al., 2010).

Marine placer deposits

Marine placer deposits are formed by the deposition and transport of eroded terrestrial materials into marine environments. Certain placer deposits may concentrate heavy minerals (e.g. gold, silver, tin) or diamonds in certain locations. These deposits, termed 'secondary occurrences', are not produced at the mining site but transported there by fluvial (river) processes. They may be further sorted or concentrated by tidal or wave action.

Diamonds are transported by fluvial processes into shallow nearshore sediments where they can be excavated in waters up to 200m deep. For example, marine diamonds have been mined adjacent to the Namibian coast since 1989 and now exceeds its terrestrial production (Garnett, 2002).

Conventional offshore oil and gas

Marine environments currently produce about one third of the world's oil and more than a quarter of the gas, with potential for significant expansion (Sandrea and Sandrea, 2007). Globally, 500 billion barrels of oil have been discovered in offshore areas, of which 200 billion barrels have been extracted. It is estimated that another further 300 billion barrels may yet be discovered in offshore areas (Sandrea and Sandrea, 2007). The shallow marine (< 400 metres) and deep marine (> 400 metres) environments are estimated to contain 36 per cent and 11 per cent of the global distribution of oil-bearing reservoir rocks, respectively. Hydrocarbon-bearing rocks are found in the deep marine as a result of avalanches that move sediments into the deep ocean. Other deep-sea hydrocarbon reserves can be found through drilling but remain poorly known or understood.

Early efforts to develop offshore hydrocarbons were undertaken in the 1920s and 1930s in Azerbaijan, the Gulf of Mexico and Venezuela. However, it was not until the 1950s that offshore oil and gas extraction became economically and technically feasible in the Gulf of Mexico. In the 1970s, the first deepwater (400–1,500 metres water depth) wells were completed. Ultra deepwater (> 1,500 metres depth) wells are currently producing in Brazil (1,500 metres) and in the Gulf of Mexico (2,700 metres) (Sandrea and Sandrea, 2007). Current investment in offshore exploration and development is approximately 20 per cent of all oil and gas investment. Between 1999 and 2009, over half of new oil and gas discoveries were found in offshore areas. Offshore production, in order of total production, is concentrated in the Persian Gulf/Middle East, North Sea, West Africa, the Gulf of Mexico (US and Mexico), Asia/Australasia, Brazil, China, Caspian Sea and Russian Arctic. About 17,000 platforms are operating, and another

400 production facilities are constructed annually. Of the new offshore exploration, over half is centred in Southeast Asia (www.ifpenergiesnouvelles.com).

With respect to the management and regulation of offshore oil and gas resources, individual states issue exploration and development leases on their seabed areas and prescribe the operational framework under which activities are conducted. The exception is the European Union that – through its Hydrocarbons Licensing Directive (94/22/EC) – has a common approach for the prospecting, exploration and production of offshore activities. Moreover, the European Commission is contemplating establishing common safety standards for all aspects of offshore hydrocarbon exploration and development.

Methane hydrates

Methane hydrates are produced by the microbial and thermochemical alteration of organic materials. Consisting of molecules of natural gas (predominately methane) trapped in ocean sediments by a combination of low temperature and high pressure, they are found in in cold water (0–4 C) at depths greater than about 400 m, depending on the location, usually on the margins and slopes of continental shelves. One cubic metre of the most common hydrate is equivalent to 160–180 cubic metres of methane at atmospheric pressure and temperature. The world's reserves could be many times greater than conventional natural gas reserves, and indeed could contain more carbon than known conventional hydrocarbon reserves, though there is considerable uncertainty in the estimates (Antrim, 2005; Morgan, 2012; Collett et al., 2013).

Methane hydrates can be recovered by increasing the temperature or releasing the pressure within the deposit. This dissociates the gas from its bonds with water, whereupon the gas can be piped to the surface. Given the engineering challenges, there are currently no commercial methane hydrate recovery operations. Furthermore, low natural gas prices have meant that development of gas hydrates has generally not been a priority. However, for countries with few conventional petroleum reserves such as Japan (the largest importer of liquid natural gas in the world), India and Korea, gas hydrates in their maritime jurisdiction represent potentially increased national energy security. Research into their extraction continues.

Marine renewable energy production

'Marine renewables' is a broad term encompassing all types of energy (generally electricity) production generated from renewable sources involving the ocean. Marine renewables include the use of wind, waves, currents, or temperature gradients between bottom and surface waters to generate electricity, as well as deepwater-sourced cooling.

Offshore energy production using wind-driven turbines ('wind farming') to generate electricity is the most mature of the marine renewable industries. While wind has been used for millennia to drive pumps and produce mechanical energy, the earliest commercial wind turbine installation for the purposes of generating electricity was installed in 1980. The first offshore turbine was installed in 1991 in Denmark. Although it adds another level of complexity to construction and maintenance, locating wind turbines in offshore environments reduces public opposition and capitalizes on much more consistent and favourable wind conditions (Lozano-Minguez, 2011).

Total offshore wind production at the end of 2017 was 18.4 gigawatts (GW), or sufficient energy to power 20 million homes. To illustrate the growth in the industry, in 2011 world production was 4 GW. The United Kingdom is the largest producer (6.8 GW), followed by Germany (5.4 GW), China (2.8 GW) and Denmark (1.23 GW) (gwec.net/wp-content/uploads/2018/04/offshore.pdf). Globally, offshore wind production is expected to reach 75

GW by 2020, predominantly centred in Europe and Asia. The average annual growth rate of the industry is expected to be greater than 80 per cent until 2016 and approximately 15.6 per cent thereafter (BTM, 2013).

Wave energy is harvested by harnessing the kinetic energy produced by floating structures as waves move them up and down on the ocean surface. The first wave energy demonstration projects were initiated in the 1970s, and while no large-scale wave energy installations are in operation, many states, including Australia, Chile and the UK, have significant opportunities to develop wave energy projects (Bahaj, 2011). Relative to offshore wind energy production, wave energy has fewer environmental impacts (e.g. bird strikes) and is less visible (as buoys have a low visibility profile), and waves can be predicted accurately up to a week in advance, which allows for less reliance on alternative power sources (Dunnett and Wallace, 2009). Because the moving parts are located in the ocean, however, there are greater technological challenges to be overcome with regard to material fatigue, natural 'fouling' (the growth of marine life on the devices) and storm events.

Tidal energy production uses water movement generated by tidal streams/marine currents to turn impellers to produce electricity. While tidal energy production is in its infancy, the approach is appealing as, unlike wind and wave energy, tidal stream velocity is predictable to the hour years in advance. Furthermore, generation sites can be located in constrained topography that increases the speed of water movement (Bahaj, 2011). Certain geographies such as northern Scotland are well-situated for tidal energy production, and, in 2016, the first large-scale tidal energy system (6 megawatts (MW)) was installed near Inverness. Previously, tidal energy has been harnessed through the filling and emptying of dammed embayments, driving turbines in both directions. However, most current research is looking at underwater turbines *in situ*, anchored to the seabed.

Phosphate deposits

Phosphorous (P) is a necessary element to support life and is a limiting nutrient in most biological systems (including marine environments). It is also a key ingredient in modern agriculture and, as such, is a commercially valuable resource that cannot be substituted by any other element or compound. Eighty per cent of mined phosphate – predominantly from China, Morocco and the United States – is used in the production of fertilizer and the remainder in other uses such as detergents.

Phosphorus accumulates in the marine environment from a number of sources but primarily from terrestrial erosion and the deposition of material by fluvial processes. Phosphorous transported into the ocean is consumed by biota and later deposited and buried in ocean sediments. Bacteria and enzymes then dissolve the organic matter, which releases the element into ocean sediments (Delaney, 1998). The process is anything but rapid: Phosphate deposits can take 10–15 million years to accumulate in concentrations sufficient for commercial exploitation. It can therefore be considered a non-renewable resource. As global phosphate deposits become depleted, new sources will need to be found to maintain current agricultural production rates. It is estimated that, at current phosphorous usage rates, the world will run out of over the next several hundred years (Cordell et al., 2009).

Phosphates are normally found in waters less than 1,000m deep with continental shelf areas less than 200m being of greatest commercial interest. Offshore phosphate (technically phosphorite) mining has been proposed off the coasts of Mexico, Namibia and New Zealand. Currently, they have either failed to be permitted or have stalled due to environmental concerns. The companies involved are seeking to reverse those decisions.

Management of marine energy and mineral resources in deep-sea areas

The three main resource types under consideration for deep-seabed mining (DSM) are: seafloor massive sulphide deposits, usually created by hydrothermal vents; polymetallic nodules, found on the deep seafloor; and cobalt-rich ferromanganese crusts found on the flanks of some seamounts. Additionally, there has been some limited interest in deep-sea muds, found in national jurisdictions both on and off the continental shelf, which can contain rare earth elements and yttrium (Takaya et al., 2018).

The prospect of DSM, while attractive to some, is not without controversy. There is considerable concern over the environmental impacts, ranging from the assured destruction of benthic ecosystems directly mined, widespread smothering effects of sediment plumes on neighbouring areas, possible toxicity of crushed ores and sediments leaching into the water column, as well as noise and light pollution. Unlike on land, for most deep-sea ecosystems, environmental restoration of a mining site after operations have ceased would be very costly and impractical. Deep-sea ecosystems, including those associated with polymetallic nodules, are unaccustomed to broad-scale high-energy disturbances, and recovery is likely to be on timescales much longer than those on land – possibly centuries (or in the case of 're-growing' nodule habitat, millions of years). The ecological effects of small deep-sea disturbance experiments remain evident more than 25 years later (Jones et al., 2017). Indeed, the plough tracks created in 1988 at the 'DISCOL' site in the eastern Pacific are readily visible, as though made just a few weeks ago.

The International Seabed Authority (ISA), which is responsible for all DSM in the Area beyond national jurisdictions, has developed exploration regulations and is in the process of developing exploitation regulations. As of February 2019, the drafted regulations invoke the precautionary principle, but how it will be operationalized remains to be seen. Some scientists and policy-makers have expressed their concerns (e.g. Van Dover et al., 2017). In January 2018, European legislators voted in support of a non-binding resolution that calls for an international moratorium on commercial deep-sea mining exploitation licenses until such time as the effects of DSM on the marine environment, biodiversity and human activities have been researched and the risks are understood (Resolution P8 TA(2018)0004 (42)). Nevertheless, ISA preparations continue.

Seafloor massive sulphides

Seafloor massive sulphide (SMS) deposits are found where superheated subterranean waters have come into contact cold waters of the seafloor, resulting in the precipitation and deposition of minerals. Formed on the scale of decades to centuries, spread across a few hundred metres, SMS deposits are found anywhere with geothermal activity and include volcanic arcs, mid-oceanic ridges and their associated hydrothermal vents. First discovered in the late 1970s, hydrothermal vents have since been found in all ocean basins as a relatively common feature of mid-ocean ridges basin margins, typically at depths of 2,500–3,000m, both within and outside of national jurisdictions. However, very few of these have been found to contain volumes sufficient to support commercial mining operations (e.g. > one million tonnes of ore). Commercial exploration has to date focussed on relatively easy-to-find active vent sites and their environs, where the minerals are still being precipitated. Higher volumes are likely to exist in ancient sites, but these with their 'overburden' (covering) of several metres of seafloor sediments are much harder to find and survey. Although the mineral composition varies according to deposit, SMS can be rich in copper, lead, zinc, silver and gold, in concentrations several times higher than ores typically found on land.

The first commercial mining of SMS (or indeed any deep-sea mineral) was expected to have already commenced at depths of about 1,600m at two sites in Papua New Guinea's EEZ – Solwara 1 and Solwara 12 ('solwara' means 'salt water' in the local pidgin language). However, financial difficulties of the start-up company involved, Nautilus Minerals, has left the future of the project uncertain.

Polymetallic nodules

Potato-sized polymetallic nodules are the result of bio-chemical accretions that have formed over millions of years. They are normally found between 4,000–6,000m and, but for their depth, are readily accessible as they rest on top of the seabed, thus making them commercially attractive. Depending on their location, commercially relevant minerals include manganese, nickel, copper, cobalt and iron. They occur mostly in the Area beyond national jurisdictions but also within a few jurisdictions, such as that of the Cook Islands.

In 1967, when Ambassador Arvid Pardo of Malta addressed the UN General Assembly, it was from the chapter on polymetallic nodules from *The Mineral Resources of the Sea* (Mero, 1965) that he quoted and which galvanized the General Assembly into action. The historical (over-) estimates of the perceived riches of DSM were mainly about polymetallic nodules, and hence this resource has the longest history of commercial exploration and research. Unlike SMS and land-based mining, the resource is only on the surface. Therefore, the footprint of polymetallic nodule DSM is expected to be the largest of any kind of DSM (and larger than terrestrial mining) on the scale of several hundred square kilometres of seafloor each year per operation. Additionally, it is thought that the effects of sediment plumes arising from the collection of nodules and wastewater return could considerably amplify the footprint of affected biology beyond that of the immediate mined area. The extent and nature of sediment plume impacts, long suspected to be a major environmental stressor (Burns, 1980), has been a topic of recent and ongoing research projects.

Cobalt-rich ferromanganese crusts

Cobalt-rich ferromanganese crusts are similar in formation and composition to polymetallic nodules but formed when cold ocean bottom waters bio-accrete minerals onto hard-bottom substrates such as rocky outcrops or seamounts. Most crusts are only a couple centimetres thick but may be up to 25 centimetres. They are formed between depths of 400–4,000 metres and are most frequently found in the volcanic island arcs of the Pacific Ocean and Indian Ocean seamounts, both within and beyond national jurisdictions.

In addition to iron and manganese, crusts can contain cobalt, nickel, copper and some valuable exotic metals such as titanium, zirconium, cerium, platinum and tellurium, as well as some rare earth minerals, thus making this resource type potentially commercially important for the future (Hein et al., 2009). Currently, however, the technical difficulties in stripping the top few centimetres of hardened crust off the sides of deep-sea seamounts, and leaving the rest intact, has not been overcome at commercial scale. Seamounts have long been recognized as ecologically important 'oases' in the 'desert' of the open ocean. How the future mining of cobalt-rich ferromanganese crusts might affect local marine ecosystems would also need to be considered.

Discussion questions

- Will the Paris Accord climate targets drive new investments in deep-sea mining as nations transition from fossil-fuel based economies towards renewable energy sources that require rare earth metals?
- Explain what minerals are found on continental shelves versus deep-sea areas and why.
- Under what circumstances would deep-sea mining become profitable?
- Will offshore hydrocarbon development expand into the Arctic given climate change induced sea ice retreat?

Further reading

Anonymous (2018) 'Nautilus has many problems, including loss of an expensive ship', *Economist*, 06 December.

Antrim, C. (2005) 'What was old is new again: Economic potential of deep ocean minerals the second time around', *Oceans 2005*, vol 2, pp. 1311–1318.

Bahaj, A. S. (2011) 'Generating electricity from the oceans', *Renewable and Sustainable Energy Reviews*, vol 15, no 7, pp. 3399–3416.

Brown, C. G. (2018) 'Mining 2,500 fathoms under the sea: Thoughts on an emerging regulatory framework', *Ocean Science Journal*, vol 52, no 2, pp. 287–300.

BTM Consult (2013) *Offshore Report 2013*, emp.lbl.gov/sites/all/files/lbnl-6809e.pdf, accessed 23 February 2019.

Burns, R. E. (1980) 'Assessment of environmental effects of deep ocean mining of manganese nodules', *Helgoländer Meeresuntersuchungen*, vol 33, nos 1–4, pp. 433–442.

Collett, T., Bahkm, J.-J., Frye, M., Goldberg, D., Husebo, J., Koh, C., Malone, M., Shipp, C. and Torres, M. (2013) *Historical Methane Hydrate Project Review Report*, U.S. Department of Energy – National Energy Technology Laboratory, by the Consortium for Ocean Leadership, Washington, DC.

Cordell, D., Drangert, J.-O. and White, S. (2009) 'The story of phosphorus: Global food security and food for thought', *Global Environmental Change*, vol 19, pp. 292–305.

Delaney, M. L. (1998) 'Phosphorus accumulation in marine sediments and oceanic phosphorus cycle', *Biogeochemical Cycles*, vol 12, no 4, pp. 563–572.

Dunnett, D. and Wallace, J. S. (2009) 'Electricity generation from wave power in Canada', *Renewable Energy*, vol 31, no 1, pp. 179–195.

Hein, J. R., Conrad, T. A. and Dunham, R. E. (2009) 'Seamount characteristics and mine-site model applied to exploration- and mining-lease-block selection for cobalt-rich ferromanganese crusts', *Marine Georesources & Geotechnology*, vol 27, no 2, pp. 160–176.

Garnett, R. H. T. (2002) 'Recent developments in marine diamond mining', *Marine Georesources and Geotechnology*, vol 20, no 2, pp. 137–159.

International Tribunal for the Law of the Sea (ITLOS) (2011) *Responsibilities and Obligations of States Sponsoring Persons and Entities with Respect to Activities in the Area*, Advisory Opinion of the Seabed Disputes Chamber, www.itlos.org/fileadmin/itlos/documents/cases/case_no_17/adv_op_010211.pdf, accessed 26 February 2019.

Jones, D. O. B., Kaiser, S., Sweetman, A. K., Smith, C. R., Menot, L., Vink, A., Trueblood, D., Greinert, J., Billett, D. S. M., Arbizu, P. M., Radziejewska, T., Singh, R., Ingole, B., Stratmann, T., Simon-Lledó, E., Durden, J. M. and Clark, M. R. (2017) 'Biological responses to disturbance from simulated deep-sea polymetallic nodule mining', *PLoS One*, vol 12, no 2, p. e0171750.

Lozano-Minguez, E. (2011) 'Multi-criteria assessment of offshore wind turbine support structures', *Renewable Energy*, vol 36, no 11, pp. 2831–2837.

Mero, J. L. (1965) *The Mineral Resources of the Sea*, vol 1, Elsevier, New York.

Morgan, C. L. (2012) 'Deep-seabed mining of manganese nodules comes around again', *Sea Technology*, vol 53, no 4, p. 93.

Radzevičius, R., Velegrakis, A. F., Bonne, W., Kortekaas, S., Garel, E., Blažauskas N. and Asariotis, R. (2010) 'Marine aggregate extraction: Regulation and management in EU member states', *Journal of Coastal Research*, vol 51, pp. 151–164.

Sandrea, I. and Sandrea, R. (2007) 'Global offshore oil: Geological setting of producing provinces, E&P trends, URR, and medium term supply outlook', *Oil and Gas Journal*, vol 105, no 10, pp. 34–40.

Scott, S. D. (2007) 'The dawning of deep sea mining of metallic sulfides: The geologic perspective', Proceedings of the Seventh (2007) ISOPE Ocean Mining Symposium, Lisbon, Portugal, pp. 6–11.

Takaya, Y., Yasukawa, K., Kawasaki, T., Fujinaga, K., Ohta, J., Usui, Y., Nakamura, K., Kimura, J.-I., Chang, Q., Hamada, M., Dodbiba, G., Nozaki, T., Iijima, K., Morisawa, T., Kuwahara, T., Ishida, Y., Ichimura, T., Kitazume, M., Fujita, T. and Kato, Y. (2018) 'The tremendous potential of deep-sea mud as a source of rare-earth elements', *Scientific Reports*, vol 8, no 5763.

Teske, S., Florin, N., Dominish, E. and Giurco, D. (2016) *Renewable Energy and Deep Sea Mining: Supply, Demand and Scenarios*, Report prepared by ISF for J. M. Kaplan Fund, Oceans 5 and Synchronicity Earth, July 2016, 40 pp.

UN (1968) *United Nations General Assembly Resolution 2467 (XXIII)*, www.un.org/ga/search/view_doc.asp?symbol=A/RES/2467(XXIII)&Lang=E&Area=RESOLUTION, accessed 26 February 2019.

UN (1994) *Agreement Relating to the Implementation of Part XI of the United Nations Convention on the Law of the Sea of 10 December 1982*, www.un.org/depts/los/convention_agreements/convention_overview_part_xi.htm, accessed 26 February 2019.

Van Dover, C. L., Ardron, J. A., Escobar, E., Gjerde, K. M., Jaeckel, A., Jones, D., Levin, L. A., Niner, H., Pendleton, L., Smith, C. R., Thiel, T., Turner, P. J., Watling, L. and Weaver, P. P. E. (2017) 'Biodiversity loss from deep-seabed mining', *Nature Geoscience*, vol 10, pp. 464–465.

Chapter 12

Integrated approaches to ocean management

Putting policy into practice and managing across sectors

> **Key topics**
> - Integrated ocean management uses specific approaches and tools to assist with marine decision-making. These approaches may also be used to analyse existing and formulate new policies. They are characteristically multi-sector, forward-looking and inclusive and attempt to inform durable decisions.
> - Integrated approaches began with coastal zone management (CZM) in the 1970s, followed by ecosystem approaches to management (EAM), marine protected areas (MPAs), systematic conservation planning (SCP), marine spatial planning (MSP) and cumulative effects (CE).

Introduction

The application of marine policy has been an iterative process, predominantly evolving through trial and error and periodic refinements over time – mostly evolutionary rather than revolutionary. Borrowed initially from experiences in terrestrial environments, marine policy has grown into its own field and can now offer terrestrial planners new insights. Prior to the 1970s, marine issues (e.g. oil spills, overfishing) were addressed on a sector-by-sector basis as problems arose. There was little incentive or political appetite to address anticipated future problems or find solutions that solved more than one issue. However, there was increasing acknowledgement that for the global ocean, an integrated approach was required, as stated in the third paragraph to the 1982 UN Law of the Sea Convention (LOSC): 'Conscious that the problems of ocean space are closely interrelated and need to be considered as a whole'.

Early coastal management programs in the 1970s were modelled on land-use planning (also termed regional planning) that incorporated stakeholder engagement, scenario modelling and the identification of socio-economic and environment objectives to develop zoning plans to regulate future growth. These early plans attempted to link disparate sectors (e.g. housing, transport, employment, environmental quality) that were formerly managed in isolation. They were applied to large landscapes (e.g. maintaining ecological function in watersheds), political areas (e.g. growth plans for local and regional governments), environmental assessments (e.g. proposed mine developments) and protecting property and public safety (e.g. floodplain planning). With the advent of computer-assisted mapping (geographic information systems – GIS), they began to make use of geospatial data, involved multiple levels of government participation, attempted to maximize multiple objectives, engaged a range of user groups and the public and identified trade-offs (impacts to objectives). Additionally, these early terrestrial planning exercises were frequently designed to be updated (reviewed) as needed.

Integrated policy approaches versus integrated ocean management

Integrated policy approaches are distinct from integrated approaches to ocean management. Integrated policy approaches are attempts to realize efficiencies from aligning current sector-based policies (e.g. pollution prevention, fishing) towards more overarching societal goals (see Boxes 12.1 and 12.2). In contrast, integrated ocean management uses specific approaches and tools to assist with marine decision-making and the coordinated management of multiple coastal

---

Box 12.1   An example of an integrated policy approach: the European Commission Blue Growth Strategy

In 2007, the European Commission established the Integrated Maritime Policy to promote sustainable development and to protect European marine environments. In 2012, the Commission endorsed the development of a Blue Growth Strategy to increase jobs and growth in the European Union's marine sectors. A focus is to ensure sustainable growth and to address climate change. The Union defines its 'blue economy' as those activities that relate to shipping, fishing, aquaculture, energy development, coastal tourism and research and innovation. The blue economy represents 5.4 million full-time jobs with a gross added value of €500 million annually (European Commission, 2017). The Blue Economy Strategy includes the following key areas:

- Blue Energy: Focus on offshore wind development, tidal energy and thermal energy conversion. Currently, 150,000 jobs support offshore renewable energy in the European Union.
- Aquaculture: Focus on culturing new species, moving operations offshore and continuing to open up new markets for European Union aquaculture products.
- Maritime, coastal and cruise tourism: Ensuring environmental quality, infrastructure improvements, improved training and promoting off-season coastal tourism opportunities.
- Marine mineral resources: Promote European Union participation in seabed mineral exploration and development and extracting dissolved minerals from seawater.
- Blue biotechnology: Promoting industry development, commercialization of research and reducing barriers to attract investors.

Regional Blue Growth sea basin strategies are now underway for the Adriatic and Ionian Seas, Arctic Ocean, Atlantic Ocean, Baltic Sea, Black Sea, Mediterranean Sea and the North Sea. In addition, the European Commission continues to facilitate the Strategy through research, funding, regulatory reform, partnerships and identifying market opportunities.

Source: ec.europa.eu/maritimeaffairs/sites/maritimeaffairs/files/swd-2017–128_en.pdf

> **Box 12.2 An example of a troubled integrated policy approach: Australia's Ocean Policy**
>
> Australia's Ocean Policy (AOP) was announced in 1998 by the Commonwealth (federal) government with the goals of sustainably developing ocean resources, promoting industry competitiveness and protecting marine biological diversity. These goals were to be achieved through vertical integration between state, territory and Commonwealth government departments; horizontal integration across Commonwealth departments; and collaboration with all non-government ocean sectors (e.g. fishing, NGOs). Ecosystem-based planning and management was the mechanism, advanced through Regional Marine Plans, by which various actors would cooperate to achieve integrated oceans management across all 18.5 million square km of Australia's marine waters, including those under state jurisdiction.
>
> Globally lauded at its inception, after wide consultations, the AOP began very well but ultimately failed to achieve its goals for a number of reasons that, in retrospect, might have been avoided:
>
> - Lack of longer-term executive leadership and parliamentary support.
> - Lack of public visibility of ocean issues and therefore a lack of public support.
> - A decision to 'layer' the AOP on existing Commonwealth legislation rather than enable the AOP in a new statute, resulting in maintaining existing 'silos' within the Commonwealth government departments.
> - Lack of endorsement by state and territorial governments and, as such, limited investment in its success.
> - Coordination of the AOP was through a newly established National Oceans Office that, for political reasons, was located in remote Hobart, Tasmania. Isolated physically and politically, the NOO lacked sufficient funding and decision-making authority to advance the AOP. (It was closed down in 2005.)
> - Sectors lacked trust and incentives to participate in AOP implementation.
>
> At present, the National Marine Science Committee is all that remains of the original institutional ambitions of the AOP. Nevertheless, the AOP continues to inspire states to develop their own ocean policies, as well as providing a useful cautionary tale in the critical details of operationalization.
>
> Source: Vince (2018), www.environment.gov.au/archive/coasts/oceans-policy/publications/pubs/policyv1.pdf

and ocean activities. These approaches may also be used to analyse existing and formulate new policies. They are characteristically multi-sector, forward-looking, inclusive and transparent and attempt to inform durable decisions.

The integrated approaches discussed below are ordered chronologically to illustrate the linkages between approaches and how they have evolved. The following integrated approaches are therefore not mutually exclusive. Rather, they are complementary approaches that have been refined over time and adapted to different environments and spatial and temporal scales (see also Craig, 2012).

The application of integrated ocean management approaches should contextualize the problems under consideration. Technological and methodological improvements to (GIS) data collection, analysis and modelling, combined with continued improvements in the amount and quality of marine information, will continue to improve marine decision-making and further expand capabilities of the approaches discussed below.

Coastal zone management and integrated coastal management

The coastal zone is the part of the ocean with which we are most familiar. It is the most biologically and physiologically complex of all marine environments (see Chapter 2) and supports the greatest human use. Worldwide, coastal environments are characterized by high population growth rates, degraded fisheries, conflicts between users both within and between sectors and multiple levels of government with overlapping jurisdiction. Many inhabitants of the coastal zone in the developing world depend on these areas for their survival, whereas many developed nations often take coastal environments for granted. As such, coastal environments are a particular focus of marine policy. While this book is primarily concerned with the transnational and international aspects of ocean governance, coastal policy was the critical first step.

A variety of terms and definitions have been used to describe the areas of the coastal realm or coastal zone. Some of these terms are illustrated in Figure 12.1. While considerable disagreement exists on where the coastal zone begins and ends, a useful description is provided by the *Encyclopedia of the Earth*:

> The coastal ocean is the portion of the global ocean where physical, biological and biogeochemical processes are directly affected by land. It is either defined as the part of the global ocean covering the continental shelf or the continental margin. The coastal zone usually includes the coastal ocean as well as the portion of the land adjacent to the coast that influences coastal waters. It can readily be appreciated that none of these concepts has a clear operational definition.
>
> (editors.eol.org/eoearth/wiki/Main_Page)

Regardless of how the coastal zone is defined, approximately 40 per cent of the world's population and 60 per cent of the world's economic production are concentrated in a 100km swath along the world's coasts (Martínez et al., 2007; UN, 2017). Twenty-one of the world's 33 mega cities are coastal, and it is estimated that by 2020 up to 75 per cent of the world's 7.5 billion people may be living within 60km of the coastal zone. The confluence of such a significant proportion of the world's population on the narrow fringe separating the marine and terrestrial realms has significant ecological and socio-economic implications. Ultimately, this concentration of people inevitably culminates in a reduction of the ecological goods and services provided by coastal environments. Corresponding socio-economic ramifications (e.g. reductions in fish catches) and political conflicts between resource users are inescapable, resulting in destabilization of the social fabric, marginalization and unemployment.

Coastal zone management (CZM) emerged in the 1960s in response to the failure of sector-based management (e.g. fisheries management, pollution prevention) to address the environmental and socio-economic effects of multiple coastal and ocean uses. For example, the effects of overfishing on the health of fisheries was compounded by the degradation or loss of nearshore spawning and nursery habitats. Further, competition from introduced species and pollution from land- and ocean-based sources only exasperated the situation. While the notion of managing for cumulative effects (see below) was not yet fully embraced as a management principle, early CZM

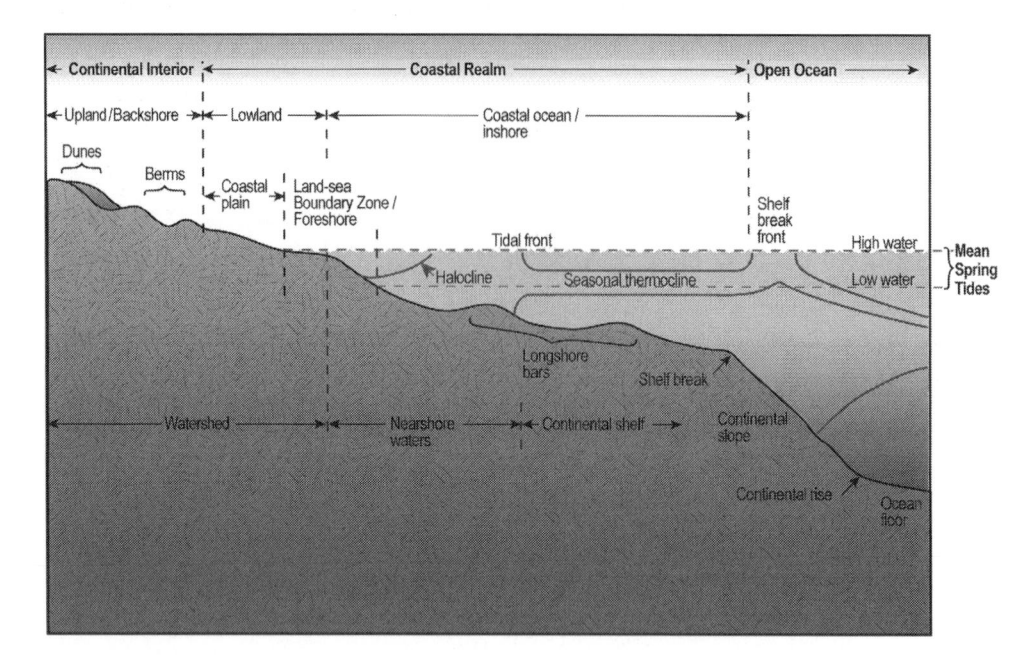

Figure 12.1 Various definitions and extents of the coastal realm and coastal zone

Source: Adapted from Ray and McCormick-Ray (2004) and University of Liverpool *Encyclopedia of Earth*

efforts fundamentally attempted to manage the many impacts of human activities on coastal and marine cultural, economic and environmental values. Given the fluid nature of the ocean, marine resource users, states and international organizations were quick to realize that an integrated approach was required to regulate human activities to maintain the vitality of ecological processes and socio-economic systems. Coastal management also became increasingly necessary as states began to exert their jurisdiction over their coastal resources through expansions of territorial waters and declarations of EEZs and extended continental shelves. This expansion resulted in the proliferation of laws at the local, provincial and national levels to manage activities in these areas moving further offshore. However, the resultant legal result was often a patchwork that left some competencies ambiguous or overlapping (e.g. aquaculture might fall under national fisheries legislation or provincial/local agricultural zoning, or both).

The need for an integrated approach to the management of coastal areas was recognized in discussions surrounding the early Law of the Sea Conventions. However, coastal management as a discipline was formally launched by the introduction of the US *Coastal Zone Management Act* in 1972 (NOAA, 1972). The *Act* defined the objectives, principles, concepts and guidelines for CZM in the US. Ten years later, the LOSC created a formal legal structure for regulating the transnational and international uses of the sea and mandated its jurisdictional zoning.

The LOSC (see Chapter 4) clarified several important jurisdictional issues that resulted in additional impetus for states to implement CZM. Most important, the LOSC extended the sovereignty of the territorial seas for coastal states to 12nmi and bestowed sovereignty and jurisdiction over state's EEZs. In return, the state must perform a number of obligations, including the conservation and management of living resources in the zone. This includes determining the allowable catch of the living resources in its EEZ and ensuring that the maintenance of the living resource within the EEZ is not endangered by over-exploitation (Chapters 6 and 8). Under the LOSC, the basic obligation of every nation is to 'protect and preserve the marine environment'. The agreement follows the basic principles of international environmental law and dictates that activities within a state's jurisdiction should be conducted in a manner that does not cause damage to other states. The LOSC also extends these principles to ABNJ (i.e. the high seas and deep seabed). The LOSC therefore evolved the former doctrine of unfettered freedom of the seas by applying several terms and conditions to these so-called freedoms. Lastly, the LOSC granted coastal states sovereignty over the mineral and petroleum resources of their continental shelves and gave them exclusive rights to the sedentary species of their seabed.

With their newly affirmed sovereignty under the LOSC, many nations developed CZM programs in the 1980s. Also contributing to the rise in CZM programs was the concept of 'sustainable development' expressed in the 1987 publication *Our Common Future* by the World Commission on Environment and Development. It outlined the following principles:

- Integration of conservation and development.
- Maintenance of ecological integrity.
- Economic efficiency.
- Satisfaction of basic human needs.
- Opportunities to fill other non-material human needs.
- Progress towards equity (present/future, cultural/economic) and social justice.
- Respect and support for cultural diversity.
- Social self-determination.

The final international agreement which set the stage for CZM was the UNCED (see Chapter 6). UNCED contained Agenda 21, which was a global commitment by coastal nations to apply integrated management and sustainable development principles to coastal and marine environments under their jurisdiction. Agenda 21 declared that all coastal nations would have an Integrated Coastal Management (ICM: what CZM is now generally referred as) program underway by the year 2000. Since UNCED, The Global Conference on Sustainable Development of Small Island Developing States (1994) and the International Coral Reef Initiative (1994) have also reinforced the need for coastal management (Cicin-Sain et al., 1995).

Since the 1980s, CZM has alternatively been referred to as integrated coastal zone management (ICZM), coastal area management (CAM), integrated management (IM), coastal resource management (CRM) and other terms based on geographic location and the state(s) sponsoring the programs. Fundamentally, however, all CZM-related programs have similar intents and purposes and are now most often referred to as integrated coastal management (ICM). Broadly stated, CZM/ICM is an attempt to integrate a number of often disparate ecological, political and socio-economic methods to move away from a sectoral management model (e.g. managing fisheries and foreshore development separately). ICM reflects the complex and integrated ecological nature of coastal environments, while also reducing overlap in traditional management methods. It creates a forum where coastal communities, and those that rely on the coastal environment, are provided access to the planning process, where the 'best available information' is assembled in order to make informed decisions on coastal use.

ICM is more akin to a philosophy on how coastal environments can be managed than a specific set of directives that are applied in all instances. ICM has been termed 'governance beyond government', meaning that governments are just one of the four types of participant in ICM. The other participants are civil society, the private sector and the scientific community (Cicin-Sain and Knecht, 1998; Bremer and Glavovic, 2013). The application of ICM depends on the type of problems to be addressed (e.g. pollution, overharvesting), the characteristics of the environment under consideration (e.g. tropical, temperate), the social and political will and resources available to address these problems. There are, however, some common goals of all ICM initiatives, which for coastal and marine areas are as follows:

- Achieve sustainable development.
- Balance and provide for the variety of (sometimes competing) human uses.
- Reduce vulnerability of habitats and their inhabitants to natural hazards.
- Maintain essential ecological processes, life support systems and biological diversity.

Regardless of the definition of ICM, there are a number of widely accepted principles of ICM that must be followed in order for it to be effective. ICM must:

- Be holistic, integrated and multi-sectoral in approach.
- Be consistent with, and integrated into, development plans.
- Be consistent with national environmental and fisheries policies.
- Build on, and integrate into, existing institutionalized programs.
- Be participatory.
- Be adaptive and rely on the 'learning by doing' model.
- Build on local/community capacity for sustained implementation.
- Build self-reliant financing mechanisms for sustained implementation.
- Address quality of life issues of local communities as well as conservation issues.
- Take a long-term perspective and consider the needs of present and future generations.

(Cicin-Sain and Knecht, 1998; Kay and Alder, 2005)

Success of integrated coastal management

Since the passage of the United States *Coastal Zone Management Act* in 1972, more than 95 nation-states have participated in over 150 ICM initiatives worldwide. In recent years, states have been cooperating with each other on regional ICM programs, such as through the UNEP Regional Seas Programme, which has over 140 participating states. The Mediterranean ICZM Protocol (adopted January 2008, entered into force March 2011) became the first international legal instrument mandating cooperative transnational ICM. The Protocol is unique in that it creates binding legal obligations over disciplines normally under national jurisdiction including administrative law, economic policies and urban planning (Rochette and Billé, 2012).

Overall ICM implementation, however, has been measured if not sluggish. Only moderate headway has been achieved towards the UNCED goal of all coastal states having integrated coastal management programs underway by the year 2000. A review of barriers to implementing ICM by Olsen (2003) concluded that the capacity of institutions (governance capacity) to implement ICM programs is a larger barrier than limitations on funding or scientific knowledge.

A 2006 study commissioned by the European Union determined that European ICM programs, while not implemented by all member states, raised awareness and preparedness of issues facing coastal areas. Programs could also improve the livelihood and employment in coastal areas, provide an opportunity to rethink approaches that attempt to reconcile socio-economic and environmental interests and provide a linkage between terrestrial and marine laws at the scale of 'regional seas'. Regional cooperation, stakeholder participation, science and information and monitoring ICM implementation were also singled out as areas that could be significantly improved (Rupprecht Consult, 2006).

A 2007 study of the success of the United States 1972 *Coastal Zone Management Act* concluded that the Act enabled voluntary partnerships between the federal and state government. Perhaps this is why 34 of 35 eligible states have participated in the program. Under the program, the federal government provides funding, oversight and a national vision for the program and yet leaves states free to design tailored programs that suit their needs (NOAA, 1972). The program has generally been viewed as successful with certain programs (e.g. water quality) demonstrating improvements in coastal ecosystem health (NOAA, 2013).

Thus, while ICZ has arguably not met all of its original expectations, it has nonetheless fostered greater jurisdictional cooperation, stakeholder engagement and appreciation that indeed, 'the problems of ocean space are closely interrelated and need to be considered as a whole'.

Ecosystem approaches to management

The notion that the disparate parts of an ocean area can be concurrently managed is over a century old (Baird, 1873; Link, 2005). However, it was not until the concept of the 'ecosystem' was championed by Tansley (1935) that scientists and decision-makers began to debate the precise definition of an ecosystem, whether and how ecosystems should be 'managed' and what aspects of traditional management paradigms need to be tweaked to make them 'ecosystem-based' (Leopold, 1949; Larkin, 1996).

The central premise of what are generally termed ecosystem approaches to management is that a global and holistic approach to marine management is required that simultaneously considers multiple ecological and socio-economic objectives in the management of either a geographic area (e.g. protected area) or ecosystem (however defined). These approaches strive to maintain the benefits humans derive from natural systems (ecological goods and services) while minimizing externalities associated with their use. There are over a dozen terms in use that fundamentally denote an approach to natural resource management that focus on sustaining ecosystems to meet both ecological and human needs for the future. Here, we will use the term 'ecosystem approaches to management' (EAM) since it is routinely applied in terrestrial and marine environments and is broader in scope than other terms in use (e.g. ecosystem-based management). See also Kidd et al. (2011) for further discussion of this issue. Given its fluid nature, coastal and marine ecosystems in particular captured the imagination of those promoting this broader, more holistic approach.

The term EAM, however, lacks consensus on its definition. As such, confusion arises over what exactly EAM is and what it entails for the marine environment. A myriad of definitions of the concept exist including: ecosystem-based management; integrated management; integrated oceans management; coastal zone management; integrated coastal zone management; and sustainable development. Broadly, EAM incorporates socio-economic, cultural and ecological inputs into management, conservation and decision-making processes and, unlike ICM discussed above, is intended to be applicable in coastal and non-coastal oceans. In this regard, COMPASS (2005), an American environmental non-government organization, offers a pragmatic definition of EAM:

> An integrated approach to management that considers the entire ecosystem, including humans with the goal to maintain an ecosystem in a healthy, productive, and resilient condition so that it can provide the services we want and need.

Many EAM principles have existed in international soft law for decades; however, it was not until the 1995 Food and Agriculture Organization (FAO) Code of Conduct for Responsible Fisheries and the UN Fish Stocks Agreement (also 1995) that the principles and procedures for what would become marine EAM were written down. Specifically, the UN Fish Stocks Agreement states that there is a 'conscious need to avoid adverse impacts on the marine environment, preserve biodiversity, and maintain the integrity of marine ecosystems' (FAO, 2005).

Marine EAM concepts have a growing presence in the scientific, peer-reviewed literature and a significant presence in the 'grey' literature of the applied conservation and management disciplines. Marine EAM academic- and application-related papers and reports are found in disciplines as broad as biology, economics, social science and Indigenous study. Some papers are descriptive, some are model-based, but few follow the traditional empirical development and testing of hypotheses. Ecosystem approaches are also a global phenomenon, with over 100 states either participating in EAM efforts or hosting EAM programs within their borders. Certain jurisdictions – most notably Australia, Canada and the United States – have EAM principles enshrined in certain marine statutes. Nonetheless, the absence of a single, authoritative (and pragmatic) guide to marine EAM continues to limit its application and use, particularly internationally.

In the face of climate change, EAM is an alternative to 'hard' engineering solutions (e.g. shoreline armouring). Termed 'ecosystem-based adaptation' (EbA), these approaches attempt to use natural coastal barriers, such as wetlands, mangroves and barrier beaches, to protect against storm surges and sea level rise.

Marine protected areas

Humans have long desired to protect natural areas from their own activities. Indeed, evidence suggests that the first protected areas were established in India 2000 years ago in order to preserve natural resources (Holdgate, 1999). In more recent times, European nobility excluded their 'commoners' from important hunting and fishing grounds. Based predominantly on resource value or ecological import, cultures worldwide have conferred sacred or 'taboo' status upon certain marine areas, reserving their use for special occasions only.

While a number of states introduced protected areas in the early 20th Century, the modern application of protected areas as a biodiversity management tool began in earnest with the inaugural World Conference on National Parks in 1962. This conference was probably the first time the need for protection of coastal and marine areas was internationally recognized. Prior to this, Beverton and Holt (1957) advocated for the use of closed areas in fisheries management, and a number of states had established smaller protected areas for protection of sites of cultural and historic significance. However, the need for a systematic and representative approach to establishing protected areas in marine environments was first clearly articulated at the International Conference on Marine Parks and Protected Areas, convened by the International Union for the Conservation of Nature (IUCN) in Tokyo in 1975 (Kenchington, 1996). The concept of marine protected areas (MPAs) gained traction in the 1980s with the publication of the IUCN/WWF/UNEP *World Conservation Strategy* (1980). Following a number of workshops on the establishment of MPAs, a 1994 IUCN report titled *Marine and Coastal Protected Areas: A Guide for Planners and Managers* was published (Salm and Clark, 1984).

MPAs are now internationally recognized as an essential and fundamental component of marine conservation. They are referred to within many international agreements, including the Convention on Biological Diversity (Article 8) and the accompanying Jakarta Mandate, the Global Program of Action, MARPOL 73/78 and the more recent IMO Guidelines and World Heritage Convention. The 2011 Aichi Biodiversity Targets under the Convention on Biodiversity (Chapter 6) and the UN Sustainable Development Goal 14 both commit states to protect 10 per cent of their marine areas. With these targets coming due in 2020, there are ongoing discussions about increasing the number. The UK and Seychelles, for example, have both suggested 30 per cent by 2050.

While the LOSC is silent on what conservation mechanisms (e.g. ecosystem approaches, MPAs) should be applied to meet the Convention's objectives to protect the marine environment, it is clear that states have specific duties. They are obliged to protect rare or fragile ecosystems, the habitats of depleted, threatened or endangered species and other forms of marine life (Articles 192 and 194), and to avoid interference with the ecological balance of the marine environment (Art. 145).

Today, there are over 15,000 MPAs worldwide in over 80 states that capture approximately 7 per cent (~26 million square km) of the ocean surface area with 1.9 million square km prohibiting all extractive activities (including fishing) (Figure 12.2). The amount of territorial waters designated as MPA ranges from nearly all (Ecuador, Slovenia) to nearly none (Cape Verde, Haiti) with six states with 50 per cent or more of their waters managed as MPAs (Boonzaier and Pauly, 2016; Fouqueray and Papyrakis, 2019). (Given the rapidly approaching target deadlines, these statistics are in a state of flux and should be seen as a snapshot only.)

There are, however, different types of MPAs with varying goals and objectives. While there continues to be some debate surrounding the definition of an MPA and, ergo, the purpose(s) of their existence (Toropova et al., 2010; Jones, 2014), two widely quoted definitions of an MPA are as follows:

> Any area of intertidal or subtidal terrain, together with its overlying water and associated flora, fauna, historical and cultural features which has been reserved by law or other effective means to protect part or all of the enclosed environment.
>
> (Kelleher and Kenchington, 1992)

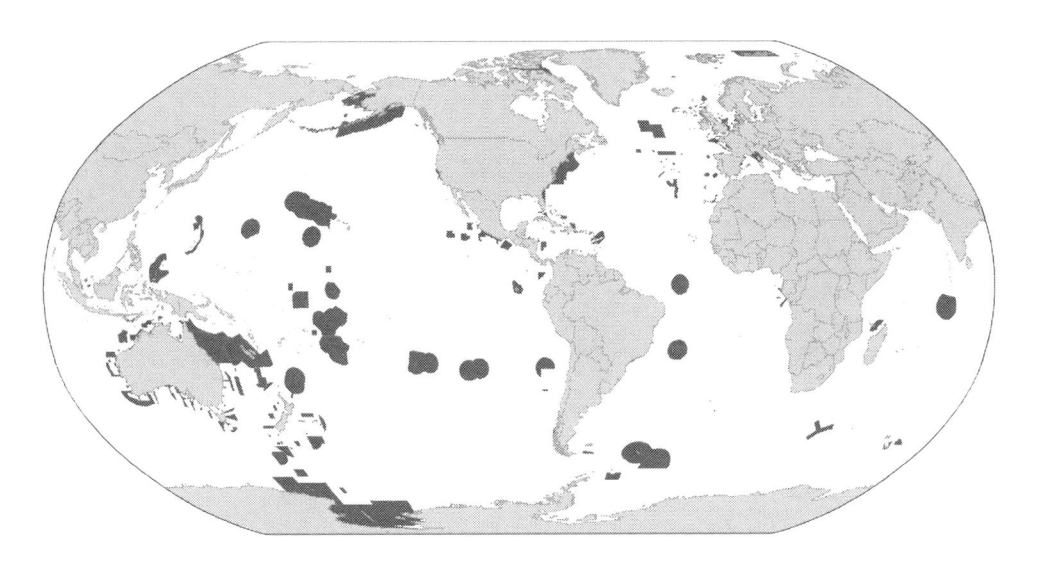

Figure 12.2 Global map of current marine protected areas

Source: UNEP-WCMC and IUCN (2019), Protected Planet: The World Database on Protected Areas (WDPA)/ The Global Database on Protected Areas Management Effectiveness (GD-PAME), www.protectedplanet.net. protectedplanet.net

> Any area of sea or ocean – where appropriate in combination with contiguous intertidal areas – together with associated natural and cultural features in the water column within, or on top of the seabed, for which measures have been taken for the purpose of protecting part or all of the enclosed environment.
>
> (Nijkamp and Peet, 1994)

While minor disagreement continues to exist on the exact definition of an MPA, for the purposes of this book MPAs are defined as areas set aside under legislation or international agreement to protect marine values. Two types of MPAs are generally recognized:

- Fully or strongly protected MPAs, also known as 'no-take' areas prohibit extraction or modification of biological (e.g. fish) and non-biological e.g. (oil and gas, minerals) resources. They may also restrict access or other human disturbance (e.g. limit tourism).
- A 'multiple-use' may permit the use and removal of resources, but where such use is controlled to ensure that long-term conservation goals are not compromised. Multiple-use MPAs generally have a spectrum of 'zones' within them, with some zones allowing greater use and removal of resources than other zones (e.g. no-take zones are commonly designated as one of the zones).

MPAs therefore range from small, highly protected 'no-take' reserves that sustain species and maintain natural resources to very large, multiple-use areas (see Box 12.3) in which the use and removal of resources is permitted yet controlled in order to ensure that conservation goals are achieved. For the very largest MPAs, the distinction between what is a protected area and what is simply good ocean management has become somewhat blurred (e.g. the Cook Islands' designation of its EEZ waters as the *Marae Moana* MPA). To reduce confusion, some authors suggest that these types of multiple-use MPAs are better labelled 'marine management areas' (MMAs), though the term has not yet received wide acceptance.

Box 12.3 Large Scale Marine Protected Areas

Large Scale Marine Protected Areas (LSMPAs) are those MPA greater than 100,000 square kilometres. The Great Barrier Reef (344,000 square km) was, for 23 years, the only LSMPA, but currently there are 26 designated LSMPAs with another 10 promised. Should promised LSMPAs be designated, 19.8 and 8.9 per cent of the world's EEZs and ocean area, respectively, would be MPAs. Australia (four), the United Kingdom (four) and the United States and France (two) currently have the most LSMPAs of any nation; however, Chile is expected to increase to five LSMPAs over the next several years. There is currently one high seas LSMPAs designated with another proposed.

Advantages of LSMPAs include the ability to connect coastal, pelagic and benthic environments to capture the life histories of most species, provide management control over a sufficient area to address known threats and instil resilience in the face of climate change. Critics of these areas note – often correctly – that LSMPAs without strong protections resemble status quo management, that they are established less for conservation purposes and more for meeting international commitments to protected area targets (Aichi Convention), that their placement is often driven by political rather than ecological needs and that conservation of these areas could be achieved just as easily through fisheries management regimes.

Source: O'Leary et al. (2018)

There have been many studies evaluating the performance of MPAs. Overall, MPAs have been demonstrated to be effective; however, this effectiveness ranges from near equivalency with the status quo to highly effective. Successful MPAs have many similarities: They were established with high levels of stakeholder participation, limited political interference and made legal through domestic laws. They have explicit objectives that are measurable, and, where monitoring programs identify a problem, they can be adaptively managed to improve outcomes. Adequate compliance and enforcement activities are undertaken and funded. In addition, 'no-take' MPAs that have been established for longer periods of time and are greater than 100 square km in area are generally found to be the most successful type of MPAs (Giakoumi et al., 2018).

Marine protected areas in ABNJ

As discussed in previous chapters, the high seas and deep seabed (ABNJ) comprise about 61 per cent of the world ocean by surface area and 90 per cent of marine habitat. They were, until the 20th Century, spared the impact of most human activities (save whaling) since they were mostly inaccessible and the waters of coasts and continental shelves met most human demands. Today, the global fisheries catch on the high seas is $US 16 billion, and there is growing interest in in deep-sea mining, oil and gas development and bioprospecting for pharmaceuticals. Furthermore, ocean acidification and ocean warming have placed several high seas fish stocks (e.g. tunas, whales) and habitats (e.g. deep-sea vents, feeding grounds) at risk (Drankier, 2012; Gjerde and Rulska-Domino, 2012). Currently, 0.5 per cent of the high seas is off limits to commercial exploitation (Heffernan, 2018).

Several international agreements provide a mandate for establishing MPAs in ABNJ. The LOSC does not specify MPAs as such but envisions that a combination of regional seas organizations and sectoral international organizations should be responsible for the protection of high seas marine environments. Under this model, RFMOs (also enabled by the UN Fish Stocks Agreement – see Chapter 6) may enact fisheries closures (spatial and temporal) and the International Seabed Authority has identified 'areas of particular environmental interest' (de facto MPAs, but without long-term protections) for the Clarion-Clipperton Zone in the equatorial North Pacific Ocean, with others planned for other regions in the coming years.

Article 8 of the CBD (see Chapter 6) presented contracting parties with several commitments. They are required to establish a system of protected areas to conserve biodiversity; develop guidelines for the selection, establishment and management of protected areas; and manage biological resources inside and outside of protected areas. The 1995 Jakarta Mandate on Marine and Coastal Biodiversity committed parties to institute a global network of MPAs by 2012. The 2008 CBD COP 9 (Decision IX/20) adopted seven scientific criteria for the identification of ecologically or biologically significant areas (EBSAs) for open-ocean and deep-sea habitats. Not MPAs, EBSAs were envisaged to be a precursor to enhanced management, including, but not limited to, MPAs. Through regional workshops, more than 300 EBSA have been identified worldwide, in national waters and areas beyond national jurisdictions (ABNJ). Their criteria are:

- Uniqueness of rarity.
- Special importance for life history of species.
- Importance for threatened, endangered or declining species and habitats.
- Vulnerability, fragility, sensitivity, slow recovery.
- Biological productivity.
- Biological diversity.
- Naturalness.

The IMO has a mandate to protect marine environments from shipping activities and may establish particularly sensitive sea areas (PSSAs) designated by the IMO Assembly. Special areas can also be introduced, which require an amendment to the MARPOL 73/87 Convention and respective Annex. While only two special areas and no PSSAs have been designated in ABNJ, the IMO is reviewing criteria for applying PSSAs and special areas in ABNJ.

In matters related to fishing, the FAO is responsible for the FAO Code of Conduct for Responsible Fishing (see Chapter 6). This, in turn, led to the development of the FAO Technical Guidelines on MPAs and fishers (the MPA Guidelines) (FAO, 2006). In response to UN General Assembly resolution 61/105, the FAO created guidelines for the identification of *vulnerable marine ecosystems* (VMEs) which are to be protected from the harmful effects of bottom fisheries in ABNJ. Because VME rules can change from year to year, most commentators do not view them as true MPAs (which should have some aspect of long-term protection); nevertheless, they are an *area-based management tool* (ABMT in UN parlance), another spatial tool in the toolbox.

Article V of the International Convention for the Regulation of Whaling (ICRW; see Chapter 8) permits the designation of sanctuaries. The only regulatory measures that can be taken involve prohibiting the harvest of all whale species at any time from a specified geographic area, irrespective of their conservation status. A sanctuary in the South Pacific sector of the Southern Ocean was established in 1949, although six years later it was deregulated on the advice of the IWC Scientific Committee. Since then, two additional sanctuaries have been adopted: the Indian Ocean Sanctuary (IOS) in 1979 (renewed in 2002) and the Southern Ocean Sanctuary (SOS) in 1994. Additional sanctuary proposals in the South Atlantic and the South Pacific Oceans have been tabled at recent IWC annual meetings.

While a number of international sectoral agreements permit the establishment of spatially defined ocean areas for conservation purposes, there is a paucity of integrated MPAs that protect marine areas from multiple sectors and their combined threats. To date, only the following four regional seas conventions (out of 13 or 18 depending on how 'regional seas' are defined) include provisions for the establishment of high seas MPAs:

- Convention for the Protection of the Marine Environment of the Northeast Atlantic of 1992 (OSPAR Convention) and its Annex V.
- Convention for the Protection of the Marine Environment and Coastal Area of the Southeast Pacific (Lima Convention), Protocol for the Conservation and Management of Protected Marine and Coastal Areas of the Southeast Pacific.
- Convention for the protection of the Mediterranean Sea against Pollution (Barcelona Convention) and its Protocol on Biodiversity and Specially Protected Areas (note: most countries have not yet claimed EEZs, which if they did, would leave no area as high seas).
- Antarctic Treaty System Protocol on Environmental Protection (note: some lawyers argue that the waters under this treaty that predates LOSC are not actually 'high seas' or ABNJ).

The 2002 World Summit on Sustainable Development committed states to initiate MPAs in ABNJ by 2012. This goal, unfortunately, has not been realized. Reasons for this failure are various:

- The lack of a global process and institution (implementation mechanism) for establishing MPAs in ABNJ.
- Intransigence by the current sector-by-sector players engaged in ABNJ management.
- The lack of agreed-upon environmental assessment procedures (and understanding of cumulative effects).

- The lack of regional governance bodies with a conservation mandate in ANBJ, and that most international conservation agreements and conventions have few, if any, powers of sanction (regulatory 'teeth').
- The lack of an internationally agreed-upon marine science advisory body for ABNJ.

One success, however, has been in the OSPAR (Northeast Atlantic) region where six MPAs were established in 2010, comprising 286,200 square km. The establishment was approved by 15 European states, and all of the MPA are located in ABNJ (Gjerde and Rulska-Domino, 2012). OSPAR, however, does not have the mandate to control fisheries. Nevertheless, through cooperation with the North East Atlantic Fisheries Commission (NEAFC), most of the areas now have overlapping fisheries closures.

The necessity of protecting biodiversity in ABNJ was recognized by the UN General Assembly in 2005; in 2006, an ad hoc working group was launched to study the issues. The Working Group held nine sessions between 2006 and 2015. It recommended, inter alia, that the UN General Assembly explore the development of an international legally binding instrument under LOSC. Negotiations for this new agreement are currently ongoing. It is widely viewed to be a 'package deal' that must address four themes together and as a whole: area-based management tools, including marine protected areas; marine genetic resources, including questions on the sharing of benefits; environmental impact assessments; and capacity building and the transfer of marine technology.

Marine planning

Systematic conservation planning

A recurrent theme throughout this book is the increasing conflict among the various communities and sectors (stakeholders) that depend on the oceans for their livelihood. As increased demands are placed on oceans to supply food, absorb wastes, provide energy and support transportation, the ongoing ability of certain ocean environments to provide ecosystem goods and services exceeds their capacity. Moreover, the concentration of multiple human activities, particularly in coastal environments, has the added risk of producing unexpected ecosystem responses (cumulative effects – see below). Marine space is becoming over-subscribed; planning therefore must be as spatially efficient as possible.

The increasing multiple stressors on marine environments, combined with most of the world ocean managed as a common property resource, is a twofold recipe for increased conflict between those that utilize the ocean and an overall decline in ocean health. While coastal management addresses balancing human activities with ocean health in nearshore areas, and while MPAs are a tool to protect certain geographic areas, marine management has, until recently, lacked a systematic approach to managing human activities to which everyone subscribes.

Historically (and often still today), protected areas were designated one at a time, usually to protect a particular species or habitat. However, in the 1990s, scientists and planners began to notice that when taken as a whole, these groups of protected areas did not adequately protect the whole of the ecosystems of a given region, hence overall long-term ecological health was not assured. A systematic approach to conservation planning was urged to transform collections of parks into viable ecological networks (Margules and Pressey, 2000). Since then, various steps have been suggested to facilitate systematic conservation planning (SCP). An indicative list is provided below (adapted from Sarkar and Illoldi-Rangel, 2010):

- Choose and delimit the planning region.
- Identify all stakeholders.
- Compile and assess data.
- Treat data and construct models as necessary.
- Identify and evaluate biodiversity surrogates.
- Set explicit biodiversity goals and targets.
- Review existing conservation areas for performance with respect to targets.
- Prioritize additional areas for conservation management.
- Assess biodiversity constituent and selected area vulnerabilities.
- Refine the network of selected areas.
- Carry out multi-criteria analysis.
- Implement conservation plan.
- Monitor network performance.

While different authors may endorse somewhat different steps from those above, widespread agreement remains on the need for conservation planning to be more systematic and spatially efficient. Usually this entails considering dozens (if not hundreds) of data layers and a wide variety of scenarios to be tested. Such an analysis is often beyond the capacities of basic GIS, or informed discussions. In the marine environment, the software tool Marxan (and Marxan with Zones; marxan.net/) has proven to be the most popular SCP tool, with a rich literature now available. Another tool worth considering is Zonation (www.syke.fi/en-US/Research__Development/Ecosystem_services/Specialist_work/Zonation_in_Finland/Zonation_software).

SCP can be thought of as two inter-linked processes operating at two different scales: the identification of individual ecologically important places and good overall network design. The CBD EBSA criteria, described earlier, can be thought of as fulfilling the first of these processes. The same CBD decision that adopted the EBSA criteria also outlines what a good MPA network should include:

- Ecologically and biologically significant areas (as described above).
- Representativity (including the range of biotic and habitat diversity of the region).
- Connectivity (in a connected network individual sites benefit one another).
- Replicated ecological features (more than one site containing examples of a given feature in the given biogeographic area).
- Adequate and viable sites (sites within a network should have size and protection sufficient to ensure the ecological integrity of the features for which they were selected).

Source: Abridged from CBD decision IX/20, Annex II.

Finally, there will need to be an SCP process that identifies both the ecologically important places and the network options, going through a series of iterative steps such as those outlined earlier in this chapter. While there are a variety of possible ways to approach this process, the CBD suggests four general stages (Box 12.4).

Marine spatial planning

Procedurally, marine spatial planning (MSP) can be very similar to SCP and can be viewed as a logical continuation of SCP, in that its remit is not just conservation but the area-based planning

> **Box 12.4 Four stages to be considered in the development of representative networks of MPAs**
>
> 1 *Scientific identification of an initial set of ecologically or biologically significant areas.* The EBSA (or similar) should be used, considering the best scientific information available and applying the precautionary approach. This identification should focus on developing an initial set of sites already recognized for their ecological values, with the understanding that other sites could be added as more information becomes available. (The vulnerability of a given site to human stressors is also usually a consideration in this stage, with more vulnerable sites given higher urgency.)
> 2 *Develop/choose a biogeographic, habitat and/or community classification system.* This system should reflect the scale of the application and address the key ecological features within the area. This step will entail a separation of at least two realms – pelagic and benthic. (As that ecosystem diversity is difficult to capture in a single system, more than one classification system for each could be used.)
> 3 *Drawing upon steps 1 and 2 above, iteratively use qualitative and/or quantitative techniques to identify sites to include in a network.* Their selection for consideration of enhanced management should reflect their recognized ecological importance or vulnerability and address the requirements of overall ecological coherence through representativity, connectivity and replication. (Here, the use of software optimization tools, mentioned above, can be very helpful.)
> 4 *Assess the adequacy and viability of the selected sites.* Consideration should be given to their size, shape, boundaries, buffering and appropriateness of the site-management regime. (Connectivity to other sites is theoretically also relevant here to ensure that they are supporting one another, but in practice this property has proven very difficult to assess. Therefore, physical site dimensions are usually all that is considered.)
>
> Source: Expanded from CBD decision IX/20, Annex III.

of all (or most) marine activities. It has been endorsed and adopted by a number of national and international organizations including the European Union, United Nations (UNESCO) and United States. MSP builds on the principles of SCP, coastal management and MPA identification in order to establish a framework for informational gathering and decision-making that can be applied to all marine areas (Douvere, 2008; Young et al., 2007). The most commonly used definition of MSP comes from the Intergovernmental Oceanographic Commission (IOC) of UNESCO:

> Marine spatial planning is a public process of analyzing and allocating the spatial and temporal distribution of human activities in marine areas to achieve ecological, economic, and social objectives that usually have been specified through a political process. Characteristics of marine spatial planning include ecosystem-based, area-based, integrated, adaptive, strategic and participatory.
>
> (Ehler and Douvere, 2009)

UNESCO goes on to note that MSP is not an end in itself but a practical way to create and establish a more rational use of marine space and the interactions among its uses, to balance demands for development with the need to protect the environment and to deliver social and economic outcomes in an open and planned way (msp.ioc-unesco.org/).

Unique to MSP is that information inputs and decision outputs are managed spatially, meaning the MSP process occurs within a geographic (mapped/geospatial) framework, based on the principles of ecosystem-based management and SCP (see above). MSP is therefore founded on the premise that human activities and their associated stresses along with marine habitats can be spatially represented (mapped). This permits the estimation of the location and severity of risks (both environmental and social). During the MSP process, different management alternatives (scenarios) are produced in order to assess their alignment with the overall vision for the marine region. MSP attempts to be inclusive of all marine stakeholders and governments, including Indigenous peoples, recreational users, environmental groups and traditional stakeholders – fishing and marine transportation interests being the foremost of these. Marine spatial plans are based on the best available science and information; they also exhort stakeholders to communicate and improve relationships with each other.

The tangible output from an MSP is an overall vision (map) for a particular marine region, a set of objectives and related performance measures, and recommendations on how stakeholders must conduct their activities to achieve the overall vision. Less tangible but an equally important benefit of MSP is that it creates a forum for communication between stakeholder groups and often allow groups to better understand each other's values and perspectives, leading to increased trust and a willingness to work out issues as they arise.

Key goals of MSP are as follows:

- Support sustainable, safe, secure, efficient and productive uses of the ocean and ocean users.
- Protect, maintain and restore coastal and ocean resources and maintain resilient ecosystems and improve and sustain the provision of ecosystem services.
- Provide for and maintain public access to the ocean.
- Promote compatibility among users and reduce user conflicts and environmental impacts.
- Improve the consistency, transparency, efficiency and durability of decision-making and regulatory processes.
- Increase certainty and predictability in coastal and ocean planning.
- Enhance interagency, intergovernmental and international communication and collaboration.
- Decrease user conflict, improve planning and regulatory efficiencies, decrease associated costs and delays, engage affected stakeholders and preserve critical ecosystem functions.

Source: Adapted from the US National Ocean Council (2013)

MSP is not a replacement for traditional marine resource management, which includes sector plans for fishing, transportation, energy development, conservation and other uses. Marine spatial plans ideally should inform the development or amendment of sector plans and more broadly identify the opportunities for each sector within a marine region such that the overall health and productivity of the area can be maintained.

Despite its explicit advantages, a number of barriers to implementing MSP exist, including:

- A lack of empirical evidence that MSP will successfully reduce user conflicts and improve ocean health (i.e. no 'guarantees' of success).
- Quality and quantity of spatial data are often poor.
- Complex yet unknown environmental interactions may occur even using the best available information.
- Fear of changes to the status quo by user groups and government departments.
- Difficult stakeholder relations, including distrust in the government process.
- The time and expense can be more than authorities can afford.

Table 12.1 Selected list of Marine Spatial Planning initiatives

Country	Agency	Project
Australia	Great Barrier Reef Marine Park Authority (GBRMPA)	GBRMPA zoning
	Department of the Environment, Water, Heritage and the Arts	Southeast Regional Marine Plan
		National Marine Bioregionalisation of Australia
Canada	Fisheries and Oceans Canada	Eastern Scotian Shelf Integrated Management (ESSIM) Plan
China	State Oceanic Administration	Territorial Sea zoning
Denmark, Germany and The Netherlands	Wadden Sea Secretariat	Trilateral Wadden Sea Cooperation Area
Germany	Federal Maritime and Hydrographic Agency	Spatial Plan for the North Sea
		Spatial Plan for the Baltic Sea
Norway	Ministry of the Environment	Integrated Management Plan of the Barents Sea
The Netherlands	Ministry of Transport, Public Works & Water Management – North Sea Directorate	Integrated Management Plan for the North Sea 2015 / National Waterplan

Source: Modified from www.unesco-ioc-marinesp.be/msp_references?PHPSESSID=5jhoa8anh8b2o1jbocn2riaf14

Overall, however, MSP continues to be adopted by an increasing number of states. In 2008, the European Commission adopted the *Roadmap for Maritime Spatial Planning: Achieving Common Principles in the EU*, which proposed a set of key principles for MSP (European Commission, 2008), and in 2014 passed the MSP Directive (Directive 2014/89/EU). In July 2010, the White House announced the Ocean Stewardship Executive Order 13547 that committed the United States to MSP (National Ocean Council, 2013), which was revoked by the Trump administration in 2018 (Executive Order 13840). MSP initiatives continue to operate in many other countries, however (Table 12.1).

Cumulative effects

The term 'cumulative effects' (or 'cumulative impacts') is used in natural resource management to connote situations where the seemingly rational regulation and management of individual sectors (e.g. fishing, transportation), when taken together, can result in unintended consequences to ecological systems and human well-being. Cumulative effects (CE) are changes to environmental, social or economic conditions (or values) caused by the combined effect of past, present and reasonably foreseeable actions or events. They may be localized (e.g. pollution discharge) or global (e.g. climate change) and may be acute (e.g. overharvesting a fishery) or enduring (e.g. disposed radionuclides).

The issue of CE in terrestrial natural resource management is well known. It arises from the difficulty in demonstrating that small, seemingly benign incremental land-use changes may combine over time to produce unexpected deleterious results (Theobald et al., 1997). Requiring an authority with an overarching mandate, CE does not naturally emerge from sector- and project-based approaches to decision-making used in most jurisdictions. Thus, individual decisions, such as fishery allocations – despite being made with the best available information – may still have unintended consequences on other aspects of marine systems and user groups.

CE analysis and management is still in its infancy. In a mathematical sense, stressors are often treated as additive (CE = stress a + stress b) where two or more stressors (e.g. fishing, pollution) are assumed to impact a resource (e.g. the fish stock) independently of one another but in the same way, such that they can be added up. In reality, however, stressors are seldomly purely additive. They can be antagonistic (CE < stress a + stress b) where two or more stresses interact to reduce the overall impacts on a particular value. Or they can be synergistic (CE > stress a + stress b) where two or more stresses interact to produce a negative response greater than the sum of their parts. A literature review by Crain et al. (2008) found that synergistic stresses were common in marine environments. Common examples of synergistic stresses included situations where the effects of toxins (pollution) on organisms were increased in the presence of changes in salinity, temperature or UV light exposure (as a result of climate change or ozone depletion). Furthermore, organisms usually have non-linear responses to stressors individually and especially when combined, such that when a threshold is reached, an organism may 'suddenly' respond in a fashion that was not evident at lower levels. For this reason, commonly used laboratory toxicity indicators (such as the LC50 – the lethal concentration that kills 50 % of the tested animals under controlled conditions) must be interpreted with caution in the field.

CE is a useful construct for considering how different activities interact to impact marine systems and the humans that depend on these systems, yet, in order for CE to inform management decisions, it must be nested within a broader consideration of scientific analysis, social choice and public policy development. **Cumulative effects assessment** (CAE) or **cumulative impacts analysis** are approaches that seek to quantify the overall impacts to a particular value or set of values from different activities where the activities are typically managed independent of each other. Many CAE methodologies describe the relationships between the source (e.g. dredging) of a pressure (e.g. habitat removal) and the pathway (e.g. mechanical disturbance), which is the mechanism by which a receptor (e.g. benthic community) is exposed to a pressure (Judd et al., 2015). Cumulative effects assessment therefore is part of the decision-making process and has been enshrined in domestic legislation in a number of jurisdictions (Canada, United States), particularly as a component of environmental impact assessment of major projects (Parkins, 2011).

Depending on the jurisdiction and problem(s)/areas(s) under consideration, cumulative effects assessment uses different nomenclature and descriptions of approaches (Weber et al., 2012). Assessment programs, however, have similar underlying methodologies that may include:

- **Identification of values**: Environmental and socio-economic (including cultural) values are identified to focus the cumulative effects assessment on preventing changes to important aspects of ecological and human systems.
- **Identification of indicators**: Indicators measure the condition or 'health' of the values under consideration.
- **Monitoring of values**: Values are monitored spatially and temporally to support decision-making and to better understand ecological and human systems.
- **Identification of ecological and social thresholds**: Finding the 'tipping points' where small changes in resource use produce large non-linear responses to ecosystem condition or human well-being.
- **Understanding adaptive capacity and resilience**: Determining the capacity of a system to withstand and recover from disturbance and perturbations.

- **Understanding patterns of cause and effect (or correlation):** Developing reliable generalizations about functional, measurable relationships between indicators and observed or forecast patterns of change.
- **Scenario analysis:** Used to evaluate options (often framed as trade-offs) and evaluate future outcomes.
- **Setting management direction:** Using the information gleaned in the above steps to inform management actions.

Historically, in marine environments, the primary impacts of human activities included overexploitation of key species, pollution and nearshore habitat loss; however, invasive species and climate change are becoming major concerns. Marine cumulative effects assessments have been piloted in areas such as the west coast of North America (Halpern et al., 2009; Ban et al., 2010; Wilsteed et al., 2018), but much more work still needs to be done.

Discussion questions

- How might the application of marine cumulative effects management differ from the application of terrestrial cumulative effects management?
- Explain why marine spatial planning approaches have not been widely implemented in less developed countries.
- Many authors contend that 30 per cent of the world ocean must be fully protected using marine protected areas. Do you think this will ever happen and why or why not?
- How might climate change considerations be incorporated into marine spatial planning approaches?

Further reading

Baird, S. F. (1873) 'Report on the condition of the sea fisheries of the south coast of New England in 1871 and 1872', in *Report of the United States Fish Commission*, vol 1, GPO, Washington, DC.

Ban, N. C., Alidina, H. M. and Ardron, J. A. (2010) 'Cumulative impact mapping: Advances, relevance and limitations to marine management and conservation, using Canada's Pacific waters as a case study', *Marine Policy*, vol 34, pp. 876–886.

Beverton, R. J. H. and Holt, S. J. (1957) 'On the dynamics of exploited fish populations', in *Fishery Investigations*, vol 19, Ministry of Agriculture, Fisheries, and Food, London.

Boonzaier, L. and Pauly, D. (2016) 'Marine protection targets: An updated assessment of global progress', *Oryx*, vol 50, no 1, pp. 27–35.

Bremer, S. and Glavovic, B. (2013) 'Mobilizing knowledge for coastal governance: Re-framing the science: Policy interface for integrated coastal management', *Coastal Management*, vol 41, no 1, pp. 39–56.

Cicin-Sain, B. and Knecht, R. W. (1998) *Integrated Coastal and Ocean Management: Concepts and Practices*, Island Press, Washington, DC.

Cicin-Sain, B., Knecht, R. W. and Fisk, G. W. (1995) 'Growth in capacity for integrated coastal management since UNCED: An international perspective', *Ocean and Coastal Management*, vol 29, nos 1–3, pp. 93–123.

Communication Partnership for Science and the Sea (COMPASS) (2005) 'EBM consensus statement', www.compassscicomm.org/ebm-consensus-statement-download, accessed 26 February 2019.

Craig, R. K. (2012) *Comparative Ocean Governance: Place-Based Protections in an Era of Climate Change*, Edward Elgar, Cheltenham.

Crain, C. M., Kroeker, K. and Halpern, B. S. (2008) 'Interactive and cumulative effects of multiple human stressors in marine systems', *Ecology Letters*, vol 11, pp. 1304–1313.

Douvere, F. (2008) 'The importance of marine spatial planning in advancing ecosystem-based sea use management', *Marine Policy*, vol 32, pp. 762–771.

Drankier, P. (2012) 'Marine protected areas beyond national jurisdiction', *The International Journal of Marine and Coastal Law*, vol 27, pp. 291–350.

Ehler, C. and Douvere, F. (2009) 'Marine spatial planning: A step-by-step approach toward ecosystem-based management', UNESCO-IOC, unesdoc.unesco.org/ark:/48223/pf0000186559, accessed 23 February 2019.

European Commission (2008) *Communication from the Commission: Roadmap for Maritime Spatial Planning: Achieving Common Principles in the EU*, COM (2008) 791 Final, http://eur-lex.europa.eu/LexUriServ/LexUriServ.do?uri=COM:2008:0791:FIN:EN:PDF, accessed 14 February 2019.

European Commission (2017) 'Report on the Blue Growth Strategy towards more sustainable growth and jobs in the blue economy', SWD (2017) 128, ec.europa.eu/maritimeaffairs/sites/maritimeaffairs/files/swd-2017-128_en.pdf, accessed 10 January 2019.

FAO (2005) *Progress in the Implementation of the Code of Conduct for Responsible Fisheries and Related Plans of Action*, FAO, Rome.

FAO (2006) *Report and Documentation of the Expert Workshop on Marine Protected Areas and Fisheries Management: Review of Issues and Considerations*, FAO Fisheries Report No. 825, FAO, Rome.

Fouqueray, M. and Papyrakis, E. (2019) 'An empirical analysis of the cross-national determinants of marine protected areas', *Marine Policy*, vol 99, pp. 87–93.

Giakoumi, S., McGowan, J., Mills, M., Beger, M., Bustamante, R. H., Charles, A., Christie, P., Fox, M., Garcia-Borboroglu, P., Gelcich, S., Guidetti, P., Mackelworth, P., Maina, J. M., McCook, L., Micheli, F., Morgan, L. E., Mumby, P. J., Reyes, L. M., White, A., Grorud-Colvert, K. and Possingham, H. (2018) 'Revisiting "success" and "failure" of marine protected areas: A conservation scientists perspective', *Frontiers in Marine Science*, vol 5, no 223.

Gjerde, K. M. and Rulska-Domino, A. (2012) 'Marine protected areas beyond national jurisdiction: Some practical perspectives and moving ahead', *The International Journal of Marine and Coastal Law*, vol 27, pp. 351–373.

Halpern, B. S., Kappel, C. V., Sekloe, K. A., Micheli, F., Ebert, C. M., Kontgis, C., Crain, C., M., Martone, R. G., Shearer, R. and Teck, S. J. (2009) 'Mapping cumulative human impacts to California Current marine ecosystems', *Conservation Letters*, vol 2, no 3, pp. 138–150.

Heffernan, O. (2018) 'How to save the high seas', *Nature*, vol 557, pp. 153–155.

Holdgate, M. (1999) *The Green Web: A Union for World Conservation*, Earthscan, London.

Jones, P. (2014) *Governing Marine Protected Areas*, Routledge, London.

Judd, A. D., Backhaus, T. B. and Goodsir, F. (2015) 'An effective set of principles for practical implementation of marine cumulative effects assessment', *Environmental Science and Policy*, vol 54, pp. 254–262.

Kay, R. and Alder, J. (2005) *Coastal Planning and Management*, Taylor & Francis, London.

Kelleher, G. and Kenchington, R. A. (1992) *Guidelines for Establishing Marine Protected Areas*, IUCN, Gland, Switzerland.

Kenchington, R. A. (1996) 'A global representative system of marine protected areas', in R. Thackway (ed.), *Developing Australia's Representative System of Marine Protected Areas*, Department of the Environment, Sport and Territories, Canberra, Australia.

Kidd, S., Plater, A. and Frid, C. (eds.) (2011) *The Ecosystem Approach to Marine Planning and Management*, Routledge, London.

Larkin, P. A. (1996) 'Concepts and issues in marine ecosystem management', *Reviews in Fish Biology and Fisheries*, vol 6, pp. 139–164.

Leopold, A. (1949) *A Sand County Almanac and Sketches Here and There*, Oxford University Press, Oxford.

Link, J. S. (2005) 'Translating ecosystem indicators into decision criteria', *ICES Journal of Marine Science*, vol 62, no 3, pp. 569–576.

Margules, C. R. and Pressey, R. L. (2000) 'Systematic conservation planning', *Nature*, vol 405, pp. 243–253.

Martínez, M. L., Intralawan, A., Vázquez, G., Pérez-Maqueo, O., Sutton, P. and Landgrave, R. (2007) 'The coasts of our world: Ecological, economic and social importance', *Ecological Economics*, vol 63, pp. 254–272.

National Oceanic and Atmospheric Administration (1972) *Coastal Zone Management Act of 1972*, coast.noaa.gov/czm/media/CZMA_10_11_06.pdf, accessed 24 February 2019.

National Oceanic and Atmospheric Administration (2013) *CZMA Section 312 Evaluation Summary Report-2013*, coast.noaa.gov/czm/media/evaluation-process.pdf, accessed 24 January 2019.

National Ocean Council (2013) *CMSP National Goals and Principles*, obamawhitehouse.archives.gov/sites/default/files/microsites/ceq/sap_2_cmsp_full_content_outline_06-02-11_clean.pdf, accessed 23 February 2019.

Nijkamp, H. and Peet, G. (1994) *Marine Protected Areas in Europe*, Commission of European Communities, Amsterdam.

O'Leary, B. C., Ban, N. C., Fernandez, M., Friedlander, A. M., García-Borboroglu, P., Golbuu, Y., Guidetti, P., Harris, J. M., Hawkins, J. P., Langlois, T., McCauley, D. J., Pikitch, E. K., Richmond, R. H. and Roberts, C. M. (2018) 'Addressing criticisms of large-scale marine protected areas', *BioScience*, vol 68, no 1, pp. 359–370.

Olsen, S. B. (2003) 'Frameworks and indicators for assessing progress in integrated coastal management initiatives', *Ocean & Coastal Management*, vol 46, pp. 347–361.

Parkins, J. R. (2011) 'Deliberative democracy, institution building, and the pragmatics of cumulative effects assessment', *Ecology and Society*, vol 16, no 3, pp. 20–31.

Ray, G. C. and McCormick-Ray, J. (2004) *Coastal Marine Conservation: Science and Policy*, Blackwell, Malden, MA.

Rochette, J. and Billé, R. (2012) 'ICZM protocols to regional seas conventions: What? Why? How?', *Marine Policy*, vol 36, pp. 977–984.

Rupprecht Consult (2006) 'Evaluation of Integrated Coastal Zone (ICZM) management in Europe: Final report', Rupprecht Consult, Cologne, Germany, ec.europa.eu/environment/iczm/pdf/evaluation_iczm_report.pdf, accessed 24 February 2019.

Salm, R. V. and Clark, J. R. (1984) *Marine and Coastal Protected Areas: A Guide for Planners and Managers*, IUCN, Gland, Switzerland.

Sarkar, S. and Illoldi-Rangel, P. (2010) 'Systematic conservation planning: An updated protocol', *Natureza & Conservação*, vol 8, no 1, pp. 19–26.

Tansley, A. G. (1935) 'The use and abuse of vegetational terms and concepts', *Ecology*, vol 16, pp. 284–307.

Theobald, D. M., Miller, J. R. and Hobbs, N. T. (1997) 'Estimating the cumulative effects of development on wildlife habitats', *Landscape and Urban Planning*, vol 39, pp. 25–36.

Toropova, C., Meliane, I., Laffoley, D., Matthews, E. and Spalding, M. (2010) *Global Ocean Protection: Present Status and Future Possibilities*, IUCN, Gland, Switzerland, www.iucn.org/content/global-ocean-protection-present-status-and-future-possibilities, accessed 26 February 2019.

UN (2017) 'Ocean Fact Sheet', www.un.org/sustainabledevelopment/wp-content/uploads/2017/05/Ocean-fact-sheet-package.pdf, accessed 04 January 2019.

UNEP (2008) *Protocol on Integrated Coastal Zone Management in the Mediterranean*, www.pap-thecoastcentre.org/razno/PROTOCOL%20ENG%20IN%20FINAL%20FORMAT.pdf, accessed 23 February 2019.

Vince, J. (2018) 'The twenty year anniversary of Australia's Oceans Policy: Achievements, challenges and lessons for the future', *Australian Journal of Maritime & Ocean Affairs*, vol 10, no 3, pp. 182–194.

Weber, M., Krogman, N. and Antoniuk, T. (2012) 'Cumulative effects assessment: Linking social, ecological and governance dimensions', *Ecology and Society*, vol 17, no 2, article 22.

Wilsteed, E. A., Birchenough, S. N. R., Gill, A. B. and Jude, S. (2018) 'Structuring cumulative effects assessment to support regional and local marine management and planning obligations', *Marine Policy*, vol 98, pp. 23–32.

Young, O., Oshrenko, G., Ekstron, J., Crowder, L., Ogden, J., Wilson, J., Day, J., Douvere, F., Ehler, C., McLeod, K., Halpern, B. and Peach, R. (2007) 'Solving the crisis in ocean governance: Place-based management of marine ecosystems', *Environment*, vol 49, pp. 21–30.

Conclusions
A report card on ocean management

Albert Einstein, with a conjecture analogous to Bertrand Russell's contention that 'wisdom' is a prerequisite to peacefully harness human innovation, famously surmised that, 'The significant problems we have cannot be solved at the same level of thinking with which we created them'. In terms of this book, the first 'significant problem' addressed by states was the overharvesting of fish and marine mammals in the early 20th Century. Although there was substantial denial and foot-dragging concerning the problem, the agreements that were eventually negotiated prevented the near certain extirpation of many important fish and marine mammal populations.

In 1951, Rachel Carson brought attention to the vulnerability of the world ocean to pursuits extending beyond fishing (Carson, 1951). The recognition that the management of coastal ocean should be more coordinated spawned coastal zone management as a discipline, which has since been adopted worldwide. This led to the development of other integrated management tools such as marine protected areas and marine spatial planning. The mid-20th Century also gave rise to the Antarctic Treaty and the need to establish the Antarctic as a place of scientific research and international concord. It has been over 25 years since the publication of seminal 'calls to action' (e.g. Thorne-Miller and Catena, 1991; Norse, 1993) that alerted us to the unceasing degradation of the ocean environment and the increasing prominence of globally 'wicked problems' (e.g. climate change) that demanded our immediate, cooperative, undivided attentions.

Extensive progress has since been made towards rectifying many of these thorny problems; establishing marine protected areas, reducing greenhouse gasses from shipping and steadfastly regulating pollution are merely three examples of headway made in managing the oceans benevolently and efficaciously. The Law of the Sea Convention, Antarctic Treaty and major pollution conventions (London Dumping Convention, MARPOL) have improved both the health of the oceans and the plight of people who depend on them. Maritime transportation is governed by a widely accepted set of rules that have fostered exponential growth in international trade. Nearly all international conventions recognize the needs of less developed, geographically disadvantaged and landlocked states. Finally, most states have fully embraced the need to cooperate on a regional basis to address shared ocean issues.

Significant challenges persist. IUU fishing, piracy, poverty, worker safety, flags of convenience, invasive alien species and an incomplete comprehension of the biological and physical functions of the ocean are but some. The effects of climate change on the ocean – ocean acidification, warming and sea level rise – are poorly understood but likely to be severe. Likewise are the environmental and social consequences of rapid human population growth along the world's coastlines. Certain regions, most notably the South China Sea, continue to be at the nexus of jurisdictional disagreements, leading to political and economic friction, as well as potential conflict. Yet pessimism need not prevail; states have demonstrated that cooperation with each other is not only possible but can lead to mutually improved economic, environmental and political outcomes.

Indeed, since the first edition of this text was published in 2014, there have been changes, many positive and some negative, in how humanity perceives and behaves with respect to the health of the ocean. A non-exhaustive list of the improvements over the past decade includes:

- Much greater awareness of ocean issues across the globe.
- The 2015 United Nations Sustainable Development Goal (SDG) 14 – thereby recognizing ocean issues alongside other established international priorities.
- The 2018 Commonwealth Blue Charter signed by all 53 Commonwealth countries to actively cooperate to solve ocean-related problems and meet commitments for sustainable ocean development.
- The various plastic pollution declarations at the national and international level.
- Increased funding towards oceans issues (e.g. from the Global Environmental Fund, World Bank and Norway).
- New and much more ambitious marine protected area (MPA) targets are being taken seriously, corresponding with a global increase in MPAs.
- The establishment of the UN inter-governmental conference for the negotiation of a new instrument to protect biodiversity in areas beyond national jurisdictions (BBNJ).
- The International Maritime Organization's ambitious target to reduce CO_2 emissions by 50 per cent (of 2008 levels) by 2050.
- The 2016 UNFCCC Paris Agreement and an increasing recognition of the relevance of ocean issues at the UNFCCC.
- Substantial progress being made by states in meeting their 2020 SDG-14 and Aichi targets.

Yet, there are aspects of ocean governance that are troubling as compared to 2014. A non-exhaustive list includes:

- Back peddling by some parties to the UNFCCC.
- An understanding that a 2.0°C increased in global temperatures will be more damaging than previously thought.
- Antarctica and the Arctic are melting faster than anticipated.
- Marine plastics are more extensive and more damaging that predicted.
- Loss of sovereignty due to sea level rise is a real legal issue (previously many states assumed that the LOSC protected them).
- Weak management of MPAs continues pretty much unabated.

So is the fragile world of ocean policy half empty, half full, cracked, leaking or broken? Or does it really matter? As long as the ocean sustains most of this planet's life and life support, surely the least we can do is manage to manage ourselves properly when in its presence.

Further reading

Carson, R. (1951) *The Sea around Us*, Oxford University Press, Oxford.
Norse, E. A. (ed.) (1993) *Global Marine Biodiversity: A Strategy for Building Conservation into Decision Making*, Island Press, Washington, DC.
Thorne-Miller, B. and Catena, J. (1991) *The Living Ocean: Understanding and Protecting Marine Biodiversity*, Island Press, Washington, DC.

Index

Note: Page numbers in italics indicate figures and page numbers in bold indicate tables on the corresponding pages.

Abidjan Convention 160–161
ABNJ *see* areas beyond national jurisdiction (ABNJ)
Achille Lauro 240
Action Plan for the Conservation of the Marine Environment and Coastal Areas in the Red Sea and Gulf of Aden 162
Action Plan for the Protection, Management and Development of the Marine and Coastal Environment of the Eastern African Region 162
Action Plan for the Protection and Development of the Marine Environment and Coastal Areas of the West and Central African Region 161
Action Plan for the Protection of the Marine Environment and Coastal Areas of the Southeast Pacific 161
admiralty law 63
Agenda 21 207
aggregate sand and gravel deposits 280–281
Agreement on Sand and Gravel Extraction 281
Agreement on the Conservation of Albatrosses and Petrels (ACAP) 259
Agreement on the Conservation of Polar Bears 267
Agreement to Promote Compliance with International Conservation and Management Measures by Fishing Vessels on the High Seas 202, 208
Aichi Biodiversity Targets 296
albedo 169, 273
Albert, R. J. 127
Aleut International Association (AIA) 269
ambient standards 126
Amundsen, Roald 257
analytic hierarchy process (AHP) 113–114
anoxic areas 24

Antarctica: Chinese interest in 273; collaborative-conservationist scenario for 263; collaborative-exploitative scenario for 263; exploration of 257; individualistic-conservationist scenario for 263; individualistic-exploitative scenario for 263; international agreements 262–265; jurisdiction in 68, 84; ownership regimes 38, 46; Regional Seas Programme in 158
Antarctic and Southern Ocean Coalition (ASOC) 259
Antarctic Convergence 257, 258
Antarctic Ocean *see* Southern Ocean
Antarctic Region 256, 258, 263
Antarctic Seals agreement 264
Antarctic Treaty Consultative Meeting (ATCM) 259
Antarctic Treaty System: administration of 259; articles of **260**; boundaries of 257; challenges of 262–263; Chinese views on 273; decision-making in 262; governance and 258–259; marine protected areas in 300; mineral resources and 85; organizations advising 262; parties to 259, **260**–**261**
archipelagic states 76
Arctic Athabaskan Council (AAC) 269
Arctic Contaminants Action Program (ACAP) 270
Arctic Council 268–270, **270**, 271, **271**
Arctic Environmental Protection Strategy (AEPS) 268–269
Arctic Monitoring and Assessment Programme (AMAP) 270
Arctic Ocean: challenges of 272; characteristics of 266; climate change impacts on 257, 268; food fisheries in 179; as global commons 272; high seas fishing in 269; human impact on 267; national interests in **268**; pollution in

267; sea ice decreases in 273; sovereignty and 267; warming in 168–169, 250, 257, 273
Arctic Region: characteristics of 256; climate change impacts on 272–274; as global commons 257; human impact on 266; Indigenous peoples and 266–267, 271; industrial expansion challenges in 271–272; joint management of 268–270; jurisdiction of 68, 257, 267–270, 272; map of 266; sovereignty and 267, 272
Area 74, 280
areas beyond national jurisdiction (ABNJ) 120, 123, 127, 208, 278, 299–301
Athens Convention relating to the Carriage of Passengers and their Luggage by Sea (PAL) 233–234
Atlantic Ocean 5, 13, 19, 24, 27, 46
Australia 45, 124, 290, 296
Australia's Ocean Policy (AOP) 290

bag limits 181
baleen whales 203
ballast water 127, 153–155, 244
Ban Amendment 156
Barcelona Convention 159, 164, 281
Basel Convention on the Control of Transboundary Movements of Hazardous Wastes and their Disposal 155–156, 252
bathymetry 20–21
Bay of Fundy 29
Belt and Road Initiative 273
Benguela current 28
benthic realm 15–16, *17*, 31, **31**, 33–35
Bering Sea Arbitration (1893) 38
Beverton, R. J. H. 296
biodiversity audits 138
biodiversity conservation 142, 150
biological pollution 6–7
Black Sea 164
Blue Growth Strategy 289
boundary differences 19
broadcasting 72
Brundtland Commission 147–148
Brussels Convention 243
Bucharest Convention 164
bulk cargo services 225
bulk carriers 223
by-catch 5, 123–124, 137, 178–179
Byers, M. 257

cables 71–73
Canada 45, 124, 250, 267–268, **268**, 280, 296
cargo ships 223, 225
Caribbean Action Plan 162
Caribbean Environment Program (CEP) 162
Carson, Rachel 3

Cartagena Convention 162
Cartagena Protocol on Biosafety 151
catch limits 180–182, 213–215
catch per unit effort (CPUE) 183
catch share programs 213–215
Central Africa 160–161
chemical impacts 170
China 222, 252, 272–273, 278
Churchill, R. 81
circulation: climate change impacts on 168–171; convergences/divergences 28; currents and 29; deep ocean 29; differences in 17–18; surface ocean 26–27, *27*, 28
CITES *see* Convention on International Trade in Endangered Species of Wild Fauna and Flora (CITES)
citizen actors 97
civil law 39, 47
Civil Liability Convention 247
classical food web 33
CLCS *see* Commission on the Limits of the Continental Shelf (CLCS)
clean development mechanisms (CDM) 174
climate change: Arctic Region 257, 268, 272–274; chemical impacts of 170; circulation and 168–171; coral reef bleaching and 1, 10, 151; extinction and 10, 168; high-energy events and 30; human impact on 10, 168; hypoxic areas and 170–171; international law and 171–175; marine environment impacts 168–171; marine organisms and 170–171; sea ice decreases and 273; shipping emissions and 249–251; shipping industry impact on 11; socio-economic impacts of 171
CO_2 168, 170, 250
Coalition of Legal Toothfish Operators (COLTO) 259
coastal area management (CAM) 293
coastal biodiversity 150
coastal management 288, 291
coastal resource management (CRM) 293
coastal states: continental shelf rights 71–72, 277, 279–280; Exclusive Economic Zone (EEZ) 70–71, 279–280, 292–293; geographically disadvantaged states' rights and 78; high seas and 72–73; landlocked states' rights and 78; least developed countries' rights and 80; marine jurisdiction and 63, 65, 68–69; marine scientific research and 80; rights of passage through straits and 75; territorial waters and 46–47, 65–66, 68, 70
coastal upwelling 28
coastal zone 20–21, 291, 292
coastal zone management (CZM) 291–295
Coastal Zone Management Act 292, 294–295
cobalt-rich ferromanganese crusts 284–285

314 Index

Code of Conduct for Responsible Fisheries (FAO) 193, 202, 208–210, 295
Code of Practice for the Investigation of Crimes of Piracy and Armed Robbery Against Ships 242
coercive regulation 124
collaborative-conservationist scenario 263
collaborative-exploitative scenario 263
colonization 221
Comité Maritime International (CMI) 230
command and control regulation 124
commercial shipping 227
Commission on the Limits of the Continental Shelf (CLCS) 67, 71, 263
Committee for Environmental Protection (CEP) 262–264
Committee on Fisheries (COFI) 196
common law 39
common ownership 38
common property characteristics 119
community development quotas (CDQs) 213
COMPASS 295
compensation point 21
Conference on the Protection of the Area from Pollution by Oil 244
Conflict of Maritime Laws 63
Conservation of Antarctic Fauna and Flora 265
Conservation of Arctic Flora and Fauna (CAFF) 270
constituent policies 95
Container Security Initiative (CSI) 243
container ships 223, 243
contaminants of emerging concern (CECs) 6–7
continental shelves: aggregate sand and gravel deposits on 280–281; coastal state rights on 71–72, 277, 279–280; Gardiner/Hedberg lines 71; international agreements for 280–281; jurisdiction of 277; LOSC and 71, 277–278; methane hydrates 282; mineral resources and 279–280; offshore gas extraction on 281–282; offshore oil drilling on 277–278, 281–282; phosphate deposits in 283; renewable energy and 278, 282–283
contracts of carriage 227–228
Convention for the Conservation of Antarctic Marine Living Resources (CCAMLR) 193
Convention for the Conservation of Antarctic Seals (CCAS) 259, 264–265
Convention for the Pacific Settlement of International Disputes 81
Convention for the Prevention of Marine Pollution by Dumping of Wastes and Other Matter (1993) 179
Convention for the Protection, Management and Development of the Marine and Coastal Environment of the Eastern African Region 162
Convention for the Protection and Development of the Marine Environment of the Wider Caribbean Region 162
Convention for the Protection of Natural Resources and Environment of the South Pacific Region 163, **163**
Convention for the Protection of the Marine Environment and Coastal Area of the Southeast Pacific 161, 300
Convention for the Protection of the Marine Environment of the Northeast Atlantic 281, 300
Convention for the Protection of The Mediterranean Sea Against Pollution (the Barcelona Convention) 159, 300
Convention for the Regulation of Meshes of Fishing Nets and the Size Limits of Fish 195
Convention for the Suppression of Unlawful Acts against the Safety of Maritime Navigation (SUA) 240–242
Convention on a Code of Conduct for Liner Conferences 235–236
Convention on Biological Diversity (CBD) 4, 142, 150–151, 202, 207, 296, 302
Convention on Facilitation of International Maritime Traffic 243
Convention on Fishing and Conservation of the Living Resources of the High Seas 66, 204–205
Convention on International Trade in Endangered Species of Wild Fauna and Flora (CITES) 205
Convention on the Conservation of Antarctic Marine Living Resources (CCAMLR) 257, 259, 264–265
Convention on the Continental Shelf 66
Convention on the International Regulations for Preventing Collisions at Sea (COLREG) 232
Convention on the Prevention of Marine Pollution by Dumping of Wastes and Other Matter *see* London Dumping Convention (1972)
Convention on the Protection of Migratory Species of Wild Animals (CMS) 259
Convention on the Protection of the Black Sea Against Pollution 164
Convention on the Safe and Environmentally Sound Recycling of Ships 252
Convention on the Territorial Sea and Contiguous Zone 66
cooperative instruments 120
coral reefs 1, 10, 127, 151, 169
co-regulation 124
Corfu Channel case (1946) 47
Coria, J. 131

Coriolis parameter 28
corporate environmental responsibility (CER) 137
Costanza, R. 112
cost-benefit analysis 101, 111–113
cost-effectiveness analysis 114–115
Council of European Ministers 281
Council of Managers of National Antarctic Programs (COMNAP) 259, 262
Council of Regional Organisations (CROP) of the Pacific 163
counterfactual analysis 115, *116*
Cousteau, Jacques 3
criteria ranking 111
cruise fleet 223–225
cumulative effects (CE) 305–306
cumulative effects assessment (CAE) 306–307
cumulative impacts analysis 306
currents 22, 24–29, 32, 135, 283
customary law 40

dangerous maritime cargoes (DMCs) 239
d'Arge, R. 112
deep ocean circulation 29
deep-seabed mining (DSM) 9, 14, 85, 278–279, 284–285
deep-sea communities 31
deep sea fisheries 199
deep-sea muds 284
Deepwater Horizon 125–126, 160
de Groot, R. 112
Denmark 268, **268**, 277, 282
depth 21–22, 33
Der Blaue Engel environmental labelling 251
desalinization 160
design standards 126
dispute resolution 44–45, 63, 67
dissolved gases 24
distributive policies 95
Djibouti Code of Conduct 242
domestic law 39–40, 45
dominance rule 111
driftnet fishing 67

Eastern Africa 162–163
Ecolabel Index 139
eco-labels 138–139, 251
ecological fiscal reform (EFR) 131
ecological goods and services 119
ecological responses 18–19
ecological tax reform (ETR) 131
economic-environmental indicators 117–118
economic instruments: marine policy and 120; market-based 128–129; market-based advantages/disadvantages **129**–**130**; market-enhancement 128, 133–134; price-based 128, 131; property rights 134–135; quantity-based 128, 131–132; subsidies and tax concessions 134; tradable permits 132–133
ecosystem approaches to management (EAM) 295–296
ecosystem approach to fisheries (EAF) 189, 191–192, 195
ecosystem approach to fisheries management (EAFM) 189, *190*, 191–193, 195
ecosystem-based adaptation (EbA) 296
ecosystem-based fisheries management (EBFM) 189
effectiveness evaluations 103–106
efficiency evaluations 103, 106–107
efficiency rule 106
ELECRE (Elimination and Choice Expressing Reality) 114
El Niño Southern Oscillation (ENSO) cycle 169
Emergency Prevention, Preparedness and Response (EPPR) 270
emission standards 106, 126
emissions trading 174, 251
Emmerson, C. 257
employment subsidies 134
enclosed seas 73
Energy Efficiency Design Index (EEDI) 250
enterprise allocations (EAs) 213
environmental education 136
environmental perturbations 19
environmental policy 118; *see also* marine environmental protection
environmental treaties 37
epontic community 32
estuarine environments 16, 21
euphotic zone 21, 35
European Agricultural Guidance and Guarantee Fund (EAGGF) 95
European Code of Conduct for Coastal Zones 281
European Commission 282, 289, 305
European Environment Agency 106
European Parliament 269
European Regional Development Funds (ERDF) 95
European Social Fund (ESF) 95
European Union: emissions trading for shipping in 251; fisheries management and 199, 211; integrated coastal management in 294; international law and 45; offshore activities and 282; redistributive policies in 95; shipping and 222; Waste Shipment Regulation and 156
euryhaline species 23
Exclusive Economic Zone (EEZ): coastal states and 70–71, 279–280, 292–293; continental

shelves and 21, 135, 280; extent of 70, 70; fisheries management and 195, 206, 212, 215; regulation in 123; right of passage in 75
export subsidies 134
extinctions 10, 168
Exxon 126
Exxon Valdez 125

fairness evaluations 103, 107
FAO *see* Food and Agriculture Organization (FAO)
Farber, S. 112
ferry fleet 223
financial caps 63
fisheries: by-catch and 5, 123–124, 137, 178–179; certification programs 139–140; deep sea 199; depletion of 1, 4–5, 119, 178–179, 213; gear types for 180; global food security and 4, 177–178; individual transferrable quotas in 38; international law and 195–196, 202; marine capture 4–5, 38, 178; market failure and 213; recruitment overfishing 183–184; regulation in 127; spawning potential ratio (SPR) 184; spawning stock biomass (SSB) 184; spawning stock biomass per recruit (SSBR) 184; tradable permits in 132; types of 179
Fisheries Case (1951) 47
fisheries management: approaches to **194**; catch limits 180–182, 213–215; catch per unit effort (CPUE) 183; catch share programs 213–215; challenges of 186–189; components of 180–181; defining 184; development of 179; ecosystem approach to 189, 190, 191–193, 195; FAO fishing areas 197; fishery-dependent/independent data 183; food fisheries 179; gear types in 180; industrial fisheries 179; input controls 181–182; international agreements for 202–211; LOSC and 195, 198–199, 204–208; maximum sustainable yield (MSY) 185, 187–188; multispecies 186; North Atlantic Ocean 178; optimum sustained yield (OSY) 185–186; organizations overseeing 196, 198–199, **200–201**; output controls 182; policy and 195–196, 198; recreational fisheries 179; regional bodies 198, 198, 199; regulatory instruments 179; rights-based 212–213; shellfish fisheries 179; single-species approach to 184–189; stock assessment 182–184, 186–187
fishing: driftnet 67; high seas and 73, 195–196, 206; illegal, unreported and unregulated (IUU) 196, 210–211; Southern Ocean 258; subsidies in 134
Fish Stocks Agreement 202, 208–209
F_{MSY} 185–187, 191

Food and Agriculture Organization (FAO) 4, 178, 193, 196, 197, 208–209, 295, 300
food fisheries 179
formative evaluations 101
Forum Fisheries Agency (FFA) 163
free-riders 136
Freestone, D. 80
Frid, C. 295
fringing communities **31**, 34
functionalist states 66
Future We Want, The (WSSD) 151

Gardiner line 71
general cargo ships 223, 225
geographically disadvantaged states (GDS) 78
Germany 222, 251
Gilbert, N. 263
global commons 135, 257, 272
Global Conference on Sustainable Development of Small Island Developing States 293
global food security 4, 177–178
Global Programme of Action (GPA) for the protection of the Marine Environment from Land-based Activities 156–158, 296
Gorbachev, Mikhail 268
Gordon, H. S. 213
Grasso, M. 112
Great Barrier Reef 244, 298
Great Britain 38, 81, 147, 195
Greece 222
green GDP 117
greenhouse gas emission 168, 172–173, 249–251
Greenpeace 3, 204
green shipping practices (GSPs) 251
gross domestic product (GDP) 117
Grotius, Hugo 46–47, 68, 146
Guerry, A. D. 189
Guidelines for the Management of Marine Sediment Extraction (ICES) 281
Gulf of Aden 162
Gulf of Guinea 242
Gulf of Mexico 24, 46, 277, 281
Gulf Stream 27–28
Gwich'in Council International (GGI) 269

habitat loss/degradation 8–9
Hague Rules 230, 234
halibut fishing 214
Hamburg Rules 234
Hannesson, R. 215
Hannon, B. 112
Hardin, G. 119
hard law 40–41
harmful algal blooms (HABs) 7
hazardous wastes 156
Hedberg formula 71

high seas 67, 72–73, 195–196, 206
H.M.S. *Challenger* 74
holoplankton 32
Holt, S. J. 296
Hong Kong Convention 252
hot pursuit 73
Humbolt current 28
hurricanes 30
Hydrocarbons Licensing Directive 282
hydrocarbon spills *see* oil spills
hypoxic areas 170–171

ice cover 22, 169, 267
Iceland 204, 268
ice scour 22
ichthyoplankton 32
illegal, unreported and unregulated (IUU) fishing 196, 210–211
Ilulisaat Declaration 269
impact analysis 115
impact effectiveness 105
impact evaluations 101
India 252
Indian Ocean 28–29, 46, 85, 242, 257, 285
Indian Ocean Sanctuary (IOS) 300
Indigenous peoples: Arctic Council and 269; Arctic Region and 266–267, **268**, 271; biodiversity conservation and 150; contaminant levels 5; marine property rights and 135; marine spatial planning (MSP) and 304; market-based instruments (MBIs) and 128; subsistence whaling by 179
individual fishing quotas (IFQs) 181
individualistic-conservationist scenario 263
individualistic-exploitative scenario 263
individual transferrable quotas (ITQs) 38, 213
individual vessel quotas (IVQs) 181, 213–214
industrial fisheries 179
informational instruments 120, 135–140
information supply 135–136
in-kind subsidies 134
institutional effectiveness 104
integrated coastal management (ICM) 291–295
integrated management (IM) 293
integrated marine policy 288–291
integrated ocean management 289–291
Intergovernmental Panel on Climate Change (IPCC) 168
internal waters 68
International Arctic Science Committee 268
International Association of Antarctica Tour Operators (IAATO) 259
International Commission for the Conservation of Atlantic Tunas (ICCAT) 187
International Commission for the Northwest Atlantic Fisheries (1949) 179

International Conference for Safety of Life at Sea (SOLAS) 231–232
International Conference on Marine Parks and Protected Areas 296
International Conference on Responsible Fishing 209
International Convention for the Control and Management of Ships' Ballast Water and Sediment 153–155
International Convention for the Northwest Atlantic Fisheries 195
International Convention for the Prevention of Pollution from Ships (MARPOL) 3, 126, 179, 244–245
International Convention for the Prevention of Pollution of the Sea by Oil 244
International Convention for the Regulation of Whaling (ICRW) 179, 202–204, 300
International Convention for the Safety of Life at Sea 229
International Convention on Arrest of Ships 233
International Convention on Civil Liability for Oil Pollution Damage 247
International Convention on Load Lines 232–233
International Convention on Maritime Liens and Mortgages 230
International Convention on Oil Pollution Preparedness, Response, and Co-operation (OPRC) 249
International Convention on Standards of Training, Certification and Watchkeeping for Seafarers (STCW) 238
International Convention on the Establishment of an International Fund for Compensation for Oil Pollution Damage 248–249
International Convention relating to Intervention on the High Seas in Cases of Oil Pollution Casualties 246–247
International Convention Relating to Stowaways 243
International Convention Relating to the Arrest of Sea-Going Ships 230
International Coral Reef Initiative 293
International Council for the Exploration of the Sea (ICES) 5, 195
International Court of Justice (ICJ) 42, 44–45, 47, 82–84, 147, 204
International Labour Organization 229–230
international law: areas beyond national jurisdiction (ABNJ) and 120, 123, 127; climate change and 171–175; codification of 47; compliance with 45–46; defining 93; domestic law and 45; fisheries and 195–196, 202; international personality and 41;

Index

marine environment 147–148; objectives of 40; operation of 43–45; public/private 40; regulation in 123–124; resolutions and 42; rights of archipelagic states 76; role of the state in 37, 41; seafarer and 236–238; shipping companies and 234–236; ships and 230–234; soft/hard 40–41; sources of 42; terminology of 43
International Law Commission (ILC) 47
international marine law *see* marine law
International Maritime Organization (IMO) 11, 152, 153, 154, 229
International Oil Pollution Compensation Fund (IOPC Funds) 249
International Organization for Standardization (ISO) 138–139
International Pacific Halibut Commission (IPHC) 198, 214
international personality 41
International Private Maritime Law 63
International Safety Management Code (ISMC) 234
International Seabed Authority (ISA) 67, 71, 74, 85, 278, 280, 284
International Ship and Port Facility Security Code (ISPS) 239–240
International Smart Gear Competition 137
international treaties: development of 44; dispute resolution and 44–45; international law and 37, 42, 44–45; in marine law 46, **48–62**, 147–148
International Tribunal for the Law of the Sea (ITLOS) 67, 85–91
International Union for the Conservation of Nature (IUCN) 296
International Whaling Commission 187, 202–204
International Whaling Convention (IWC) 202–204, 258, 300
intertidal environments 16
Inuit Circumpolar Council (ICC) 269
IPCC *see* Intergovernmental Panel on Climate Change (IPCC)
Islamic law 39–40
islands 73
ISO 14024 standard 139

Jakarta Mandate on Marine and Coastal Biological Diversity 150, 207, 296, 299
Japan 204, 222, 282
Jeddah Amendment 242
Jeddah Convention 162
Jewish law 39
Johannesburg Declaration on Sustainable Development 151
Joyner. C. C. 257
jurisdiction 63, 68–73, 75

Khian Sea 156
Kidd, S. 295
Kofinis, C. 101
Kuwait Regional Convention for Co-operation on the Protection of the Marine Environment from Pollution 160
Kyoto Protocol 173–174

land-based sources of pollution 156–157
landlocked states 77, *77*, 78
Large Marine Ecosystems (LMEs) 26
Large Scale Marine Protected Areas (LSMPAs) 298
Lasswell, H. 97
law: administration of 38–39; civil 39, 47; common 39; customary 40; defining 93; domestic 39–40; international 40–41; Islamic (Sharia) 40; legal systems 39, **39**, 40; public policy and 96, 120; *see also* international law; marine law
Law of the Sea Convention (LOSC): archipelagic states' rights 76; Arctic Region and 267; Area in 280; coastal zone management and 292–293; continental shelves and 71, 277–278; development of 3, 47, 65–66; dispute resolution and 67; fisheries management and 195, 198–199, 204–208; geographically disadvantaged states' rights 78; institutions of 67; integrated approach of 288; jurisdiction of 68–69, 69, 70–74; landlocked states' rights 77, *77*, 78; least developed countries' rights 77, *77*, 78–79; limitations of 80–81; marine environmental protection and 148–150; marine scientific research and 80; marine transportation and 228–229; negotiations for 179; organizations overseeing law in 85–91; Permanent Court of Arbitration (PCA) and 81–82; piracy and 67, 241; principles of 67; prohibition of reservations in 44; rights of passage through straits 74–75; signing of xii; small developing states' rights 79–80; Southern Ocean 259; states in breach of 81
League of Nations 47
least developed countries (LDCs) 77, 78–79
legal systems 39, **39**, 40, 42, 47, 63
liability 125–126
Liggett, D. 263
light 16, *17*
Lima Convention 161, 300
Limburg, K. 112
liner cargo services 225
Lishman, J. M. 127
littoral zone 21
London Dumping Convention (1972) 3, 152–153, 179

Index

London International Overfishing Conference 195
London Protocol 152, 174, 179
longitudinal diversity gradients 19, *19*, 20, *20*
LOSC *see* Law of the Sea Convention (LOSC)

Madrid Protocol 263
Malanczuk, P. 46
Mare Clausum (Closed Sea) (Selden) 46
Mare Liberum (Free Sea) (Grotius) 46, 68
Marine and Coastal Protected Areas (IUCN) 296
marine capture fisheries 4–5, 38, 178
marine conservation awards 137
marine debris 7–8, 152
marine eco-labelling 139
marine ecosystem approaches to management 295–296
marine environment: benthic realm 15–16, *17*; biological 30–31, **31**, 32–35; circulation differences 17–18; climate change impacts on 10–11, 168–171; communities in **31**; compared with terrestrial 14–19; contributions of 2–3; cumulative effects assessment (CAE) 305–307; ecological responses to 18–19; environmental perturbations 19; estuarine 16, 21; habitat loss/degradation 8–9; human impact on xii, 1, 3–11; impacts of shipping on 243–251; importance of 2–3; intertidal 16; introduced species in 10; light in 16, *17*; longitudinal diversity gradients 19, *19*, 20, *20*; market-based instruments (MBIs) and **129–130**; oceanographic characteristics of 22–25; oceanographic processes 26–30; open access 38, 211–213; pelagic realm 15, *17*; phosphorus in 283; physical properties of 15; physiographic characteristics of 20–22; pollution and 5–8; primary production in 18; public benefits of 119; taxonomic differences in 18; temperature differences in 16; temperature increases in 168–170; *see also* oceans
marine environmental protection: ballast water and 153–155; binding conventions for 152–156; biodiversity and 150; conventions pertaining to 148–152; hazardous wastes and 156; land-based sources of pollution and 156–157; non-binding agreements and 156–157; regional agreements for 157–164; treaties for 147–148; waste dumping and 152–153; *see also* pollution
marine fauna 20, 169
marine flora 169–171
marine law: arbitration in 81; development of 46–47; differences with other legal systems 47, 63; dispute resolution and 63; environmental protection 147–148; financial caps in 63; international agreements 37, 46–47, **48–62**; jurisdiction in 63; limitation of liability in 247; organizations overseeing 81–91; ownership regimes 38; self-regulation and 63; types of 63; *see also* Law of the Sea Convention (LOSC)
marine management areas (MMAs) 298
marine oil spills *see* oil spills
marine placer deposits 281
marine planning 301–305
marine policy: characteristics of 120; common property characteristics and 119; cooperative instruments for 120; ecological goods and services 119; economic instruments for 120, 128–129, 131–135; goals of xiii; informational instruments for 120, 135–140; instruments for **121**, **122**; integrated approach of 288–291; liability and 125–126; need for 119; negative externalities and 119–120; performance criteria for **105**; property rights 134–135; regulatory instruments for 120, 122–127; regulatory standards 126–127; voluntary instruments for 141–142
marine property rights 134–135, 212–213
marine protected areas (MPAs): Antarctic Region 263; areas beyond national jurisdiction (ABNJ) and 299–300; conservation mechanisms for 297, 301; defining 297–298; development stages of 303; evaluation of 299; global map of 297; international agreements 296, 299–301; marine planning 301–305; recognition of 296
marine regional agreements 157–158
marine renewable energy 278, 282–283
marine scientific research 80
marine spatial planning (MSP) 302–305, **305**
Marine Stewardship Council (MSC) 139
marine transportation law 226–230
Maritime Labour Convention (MLC) 236–238
maritime law *see* marine law
maritime security 238–243
maritime states 65
market actors 97
market-based instruments (MBIs) 104, 107, 128–129, **129–130**
market-enhancement instruments 128, 133–134
market price support 134
MARPOL *see* International Convention for the Prevention of Pollution from Ships (MARPOL)
maximum sustainable yield (MSY) 81, 185, 187–188
May, J. V. 97
Mediterranean ICZM Protocol 294
Mediterranean Sea 29, 158–160

Mediterranean Strategy for Sustainable Development 160
Mero, J. L. 285
meroplankton 32
metals 7
methane hydrates 9, 282
Mexico 283
microbial food web 33
Migratory Bird Treaty Act (MBTA) 125
Minas Basin 29
Mineral Resources of the Sea, The (Mero) 285
minerals 74, 85, 280–281
mining 9, 278; *see also* deep-seabed mining (DSM)
Montreal Declaration on the Protection of the Marine Environment from Land-Based Activities 157
Morgan, F. 263
multi-attribute utility theory (MAUT) 113
multi-criteria decision analysis (MCDA) 113–114
multispecies fishery management 186
municipal wastewater 157
Muslim law 39–40
MV *Sambo Dream* 241

Naeem, S. 112
Nagoya Protocol on Access and Benefit Sharing 151
Nairobi Convention 162–163
Nairobi International Convention on the Removal of Wrecks 230
Namibia 281, 283
nannoplankton 32
nationally determined contributions (NDCs) 4
National Programmes of Action (NPAs) 157
natural resource management 305
Nauru 90
Nauru Ocean Resources Inc. 90
Nautilus Minerals 278
nektonic community 32
netplankton 32
neuston 32
New Zealand 193, 215, 283
North American Comb Jelly (*Mnemiopsis leidyi*) 10, 154
North Atlantic Coast Fisheries dispute (1910) 81
North Atlantic Fisheries Organization (NAFO) 5
North Atlantic Ocean 27, 29, 178
North East Atlantic Fisheries Commission (NEAFC) 179, 195, 301
North East Atlantic Fisheries Convention (1959) 179, 195
Northern Sea Route 250, 267, 272, 272, 273

North Pacific Fur Seal Convention 195
North Pacific Ocean 27, 29
North Sea 178, 195
North Sea Fisheries Conference 195
Northwest Atlantic Fisheries Organization (NAFO) 178
Northwest Passage 250, 267, *272*
Norway 204, 268, **268**
Noumea Convention 163, **163**
nutrients 7, 23–24

ocean fertilization 174
ocean management 289–290
oceanographic processes: coastal upwelling 28; currents 29; deep ocean circulation 29; high-energy events in 30; oceanic convergences/divergences 28; stratification 30; surface ocean circulation 26–27, *27*, *28*; tidal amplitude 29; water motion 16–17, 25
Ocean Pathway Partnership 175
oceans: acidification of 170–171; bathymetry 20–21; circulation differences 17–18; climate regulation and 168; depth 21–22; dissolved gases 24; health of 3; horizontal division in 20–22; importance of 2; salinity 23; size of 15; water motion 16–17; *see also* marine environment
Oceans and Coastal Areas Programme Activity Centre (OCA/PAC) 157
Ocean Stewardship Executive Order 13547 305
offshore gas extraction 281–282
offshore mineral exploitation 278
offshore oil drilling 9, 277–278, 281–282
offshore wind production 282–283
Oil Pollution Act of 1990 (OPA 90) 125
Oil Pollution Convention (OPC) 244–245
oil spills: compensation levels 248; conventions pertaining to 244–249; largest marine **246**; liability and 125–126, 247–249; pollution from 8, 125, 160, 244–246, **246**, 247–249; regulation of 245–248
O'Neill, R. V. 112
open access 38, 211–213
optimum sustained yield (OSY) 185–186
Organization for Economic Cooperation and Development (OECD) 228
oscillation periods 29
Osherenko, G. 257
OSPAR (Northeast Atlantic) 301
Ottawa Declaration 269
Our Common Future (WCED) 147–148, 293
outcome evaluations 101
outranking 113–114
Overfishing Convention 195
overharvesting 4–5
oxygen 24

Pacific halibut fishery 214
Pacific Islands Development Forum 163
Pacific Islands Development Program (PIDP) 163
Pacific Ocean 13, 19, 24, 27, 46, 285
Pacific Regional Environment Programme 163
Pakistan 252
Papua New Guinea (PNG) 278
Pardo, Arvid xiii, 279, 285
Paris Agreement (2015) 4, 174–175
Paruelo, J. 112
passenger shipping services 223–225
Patagonian toothfish 86–87, 210–211, 215, 258
Patton, C. V. 94
pelagic realm 15, *17*, 25, 30–31, **31**, 32–33
performance-based regulation 124
performance criteria 105, **105**, 140
Permanent Court of Arbitration (PCA) 81–82
Persian Gulf 160
persistent organic pollutants (POPs) 6
Peruvian anchovy fishery 214
pesticides 6
petroleum extraction *see* offshore oil drilling
pharmaceuticals 6–7
phosphate deposits 283
phytoplankton 32
picoplankton 32
Pigouvian taxes 131
pipelines 8, 71–73
piracy: defining 241–242; high seas and 72; historical 241; international law and 44, 47; LOSC and 67, 241; maritime security and 238–239, 241–242
planktonic community 32–33
Plan of Implementation of the WSSD 151
Plater, A. 295
Plato 94
polar bears 274
Polar Oceans 1, 9, 22, 273
Polar Regions 256
policy: defining 94; evaluation of 101–102; formulation of 99–100; implementation of 100–101
policy analysis: aims of 96; cost-benefit analysis 111–113; cost-effectiveness 114–115; counterfactual analysis 115, *116*; criteria ranking 111; decision-making models 108, 110; dominance rule 111; economic-environmental indicators in 117–118; multi-criteria decision analysis 113–114; prospective 97; quantitative *vs.* qualitative 115, 117; retrospective 97; screening 114
policy cycle: actors and institutions in 99; challenges with 102; decision-making in 100; policy development and 97–98; policy evaluation in 101–102; policy implementation in 100–101; problem definition in 98–99; proposal of solution in 99–100; stages of **98**
policy instruments: cooperative 120; economic 104, 106–107, 120, 128–129, 131–135; government mandates and 108; informational 120, 135–140; marine policy 120–122, **122**; market-based 104, 107, 128; market-enhancement 133–134; political acceptability and 108; quantity-based 131–133; regulatory 104, 120–127; selection criteria for 103, **104**, 109; taxonomy of **121**; types of **121**; voluntary 141–142
political acceptability evaluations 103, 107–108
pollution: Arctic Ocean 267; biological 6–7; hydrocarbon spills 8; impact of 5–8; land-based sources of 156–157; Law of the Sea Convention (LOSC) and 148–150; marine debris 7–8; marine resources and 119–120; metals 7; nutrients 7; oil spills 8, 125, 160, 244–246, **246**, 247–249; pharmaceuticals and 6–7; regional agreements for 158; shipping and 244–245; synthetic compounds 6–7; *see also* marine environmental protection
polymetallic nodules 35, 85, 284–285
Port State Measures Agreement 202, 207, 210–211
practical salinity units (psu) 23
pressure 21
price-based instruments 128, 131
primary production 18
private law 40
private ownership 38
problem definition 98–99
process-based regulation 124
process evaluations 101
procurement subsidies 134
product certification 138–140
production subsidies 134
Programme for the Environment of the Red Sea and Gulf of Aden (PERSGA) 162
Proliferation Security Initiative (PSI) 242
PROMETHEE (Preference Ranking Organizational Method for Enrichment Evaluation) 114
property rights 134–135, 212
prospective policy analysis 97
Protection of the Arctic Marine Environment (PAME) 270
Protocol for the Protection of the Mediterranean Sea against Pollution Resulting from Exploration and Exploitation of the Continental Shelf and the Seabed and its Subsoil 281
Protocol for the Suppression of Unlawful Acts against the Safety of Fixed Platforms Located on the Continental Shelf 240–242

Protocol on Environmental Protection to the Antarctic Treaty 259, 263
Protocol on Preparedness, Response and Co-operation to Pollution Incidents by Hazardous and Noxious Substances (OPRC-HNS Protocol) 249
Protocol relating to Intervention on the High Seas in Cases of Marine Pollution by Substances other than Oil 247
Public International Maritime Law 63
public law 40
public outreach 136
public ownership 38
public policy: assessment criteria 103–108; considerations in selecting 102–104, **104**, 105, 108, **109**, 110; constituent 95; cooperative 120; decision-making models **110**; defining 94–95; development of 97–98, **98**, 99–102, 108; distributive 95; economic 120; effectiveness of 103–106; efficiency of 103, 106–107; fairness of 103, 107; impact effectiveness 105; informational 120; institutional effectiveness 104; instruments for 120–121, **121**; law and 96, 120; performance criteria for 105, **105**; political acceptability of 103, 107–108; rational model of 108, 110; redistributive 95; regulatory 95, 120; role of science in 118; stakeholders in 97, 99; symbolic 95; see also policy analysis

quantitative vs. qualitative indicators 115, 117
quantity-based instruments 128, 131–133

rare earths 279
Raskin, R. G. 112
recreational fisheries 179–180
recruitment overfishing 183–184
redistributive policies 95
Red Sea 162
Refuse Act 125
Regional Convention for the Conservation of the Red Sea and Gulf of Aden Environment 162
regional fisheries management organizations (RFMOs) 198–199
Regional Organization for the Protection of the Marine Environment (ROPME) 160
Regional Seas Conventions and Action Plans (RSCAPs) 158
Regional Seas Programme (UNEP) 157–158, 159, 294
Regional Seas Programme: Black Sea 164
Regional Seas Programme: Eastern Africa 162–163
Regional Seas Programme: Kuwait/ROPME 160
Regional Seas Programme: Mediterranean 158–160
Regional Seas Programme: Red Sea and Gulf of Aden 162
Regional Seas Programme: Southeast Pacific 161
Regional Seas Programme: South Pacific 163
Regional Seas Programme: West and Central Africa 160–161
Regional Seas Programme: wider Caribbean 162
regulatory instruments: command and control 124; co-regulation 124; enforcement of 125–126; fisheries discards and 123–124; international law and 123–124; marine policy and 120, 122–124; performance-based 124; process-based 124; success of 127
regulatory policies 95
regulatory standards 126–127
religious law 39
renewable energy 278, 282–283
research 136
retrospective policy analysis 97
Reykjavik Conference on Responsible Fisheries in the Marine Ecosystem 211
Reykjavik Declaration on Responsible Fisheries 211
RFMO see regional fisheries management organizations (RFMOs)
rights-based fishing 212–213
rights of passage 74–75
Rio + 10 151
Rio + 20 151
Rio 'Earth Summit' 150
Roadmap for Maritime Spatial Planning (European Commission) 305
ROPME see Regional Organization for the Protection of the Marine Environment (ROPME)
Rotterdam Rules 234
Royal Charter 236
Russell, B. xii
Russia 250, 267–268, **268**
Russian Arctic Indigenous Peoples of the North (RAIPON) 269

Saami Council (SC) 269
salinity 23
Sargasso Sea 27
Sargassum algae 27
Sawicki, D. S. 94
Saxena, J. 127
Schulz, E. 111
Schulz, W. 111
Scientific Committee on Antarctic Research (SCAR) 259, 262
scientific knowledge 118
screening policy 114
Seabed Disputes Chamber 89–91
sea cucumber fisheries 127

seafarer law 236–238
seafloor massive sulphide (SMS) deposits 284–285
sea ice 273
seals 264
seawater composition 23–24
Secretariat of the Pacific Community (SPC) 163
Selden, John 46
self-regulation 63
semi-enclosed seas 73
shellfish fisheries 179
ship breaking 251–252
ship classifications **224**
shipping: bulk cargo services 225; climate change and 11; commercial 227; contracts of carriage 227; control of 222; energy efficiency targets for 250–251; environmental impact of 243–251; fleet size **223**; global trade and 222; greenhouse gas emission 249–251; green practices 251; historical 221; international law and 230–234; liner cargo services 225; marine law and 72; maritime security 238–243; oil spills and 245–246, **246**, 247–249; organizations overseeing 228–230; passenger services 223–225; piracy and 241–242; regulation of 226; routeing 250; seafarer law 236–238; ship size 222–223, **224**; stowaways 243; terminology of 228; tramp services 225–226; types of contracts 228; types of ships 223–226
shipping companies 234–236
ship recycling 251–252
Singapore 222
single-species approach to fisheries 184–189
small developing states (SIDs) 79, 79, 80
soft law 40–41
SOLAS *see* International Conference for Safety of Life at Sea (SOLAS)
Somalia 241–242
South Atlantic Ocean 27, 46, 300
South China Sea 241
Southeast Pacific 161
Southern Ocean: challenges of 262–263; collaborative-conservationist scenario for 263; collaborative-exploitative scenario for 263; environmental protection in 245; fishing in 258; human activities in 257–258; individualistic-conservationist scenario for 263; individualistic-exploitative scenario for 263; international agreements 258–259, 260–261, 262–265; jurisdiction of 259; map of 258; marine protected areas (MPAs) in 263; mineral resources in 85; overfishing in 262; regulation of 257; whaling in 83, 204, 262
Southern Ocean Sanctuary (SOS) 300

South Pacific Ocean 27, 29, 300
South Pacific Region 163
South Pacific Regional Environment Programme (SPREP) 163
spawning potential ratio (SPR) 184
spawning stock biomass (SSB) 184
spawning stock biomass per recruit (SSBR) 184
specialist ships 223
species introduction 10
state actors 97
states: Arctic 266; coastal 65, 68–71; domestic law and 45; functionalist 66; international law and 41–45; maritime 65; multilateral treaties and 44; recognition of 41; unilateralist 66
States Parties to the Convention (SPLOS) 67
Stavins, R. N. 132
stenohaline species 23
Sterner, T. 131
Stokke, O. S. 257
storm surges 30
stowaways 243
straits 74–75, 75
Strategic Action Plan for the Conservation of Marine and Coastal Biodiversity in the Mediterranean (SAP BIO) 160
Strategic Action Plan for the Environmental Protection and Rehabilitation of the Black Sea 164
stratification 30
subpolar gyre 27
subsidies 134
subtropical gyre 27
sulphur dioxide (SO_2) 132
summative evaluations 101
surface ocean circulation 26–27, 27, 28
Sustainable Development Working Group (SDWG) 270
Sutton, P. 112
Sweden **268**
symbolic policies 95
synthetic compounds 6–7
systematic conservation planning (SCP) 301–302, 304

tankers 223
Tansley, A. G. 295
tax concessions 134
tax credits 134
tax deferrals 134
tax exemptions 134
temperatures: climate change impacts on 168–169; depth and 21; marine environment and 16, 22–23, 168–169; terrestrial environment and 16
terrestrial environment 14–19
territorial user rights in fisheries (TURFs) 213

territorial waters 68–70
terrorism 239, 241–242
Tetley, W. 63
Theodoulou, S. Z. 101
tidal amplitude 29
tidal energy production 283
tidal mixing 30
Titanic 231
Tonga 90
Tonga Offshore Mining Ltd. 90
toothed whales 203
Torrey Canyon 246
total allowable catch (TAC) 180–181, 213–215
toxic waste dumping 156
tradable permits 131–133
trade routes 221
tramp ships 225–226
transparency mechanisms 137–138
Treaty of Tordesillas 46
tributyltin (TBT) 127
trip limits 181
Truman Proclamation 47, 68
tsunamis 30
Turkey 252

UNCED *see* UN Conference on Environment and Development (UNCED)
UN Charter 44, 82
UN Commission for International Trade Law (UNCITRAL) 234
UN Conference on Environment and Development (UNCED) 150, 207, 293
UN Conference on the Human Environment (UNCHE) 147–148
UN Conference on the Law of the Sea (UNCLOS I) 66
UN Conference on the Law of the Sea (UNCLOS II) 66
UN Conference on the Law of the Sea (UNCLOS III) *see* Law of the Sea Convention (LOSC)
UN Convention on Contracts for the International Carriage of Goods Wholly or Partly by Sea (Rotterdam Rules) 234–235
UN Declaration of Human Rights 243
UN Environment Program (UNEP) 147, 157–159, *159*, 160–164
UNEP *see* UN Environment Program (UNEP)
UNESCO 303
UNESCO Convention on Underwater Cultural Heritage 80
UN Fish Stocks Agreement (UNFSA) 202, 208–209, 295
UN Framework Convention on Climate Change (UNFCCC) 4, 171–175
UN General Assembly (UNGA) 41–42, 47, 67, 171, 279, 285, 300
unilateralist states 66
United States: Arctic Region and 266, **268**, 269; cap and trade systems 132; climate change agreements and 173–174; coastal zone management in 294–295; deep-seabed mining (DSM) and 279; ecosystem approaches to management (EAM) in 296; fisheries management in 5, 178, 181, 188, 213–214; international law and 45, 195; introduced species in 154; jurisdiction and 47, 65, 68; marine aggregate mining by 280; marine arbitration 81; marine oil spill liability in 125–126; marine spatial planning in 303, 305; maritime security and 242–243; process-based regulation in 124; regulatory instruments 124; sustainability reporting in 138
UN Liner Code 235–236
Unpopular Essays (Russell) xii
UN Secretary-General 37, 67
UN Security Council (UNSC) 45, 67
UN Sustainable Development Goal 14 xiv, 296, 311

van den Belt, M. 112
Vidas, D. 257
Vienna Convention on the Law of Treaties 43
voluntary instruments 141–142
vulnerable marine ecosystems (VMEs) 300

warships 68, 72–73
waste dumping 152–153, 156
Waste Shipment Regulation 156
water density 24–25
water masses 24–25
water mixing 30
water motion 16–17, 25
water spouts 30
wave energy 283
weapons of mass destruction (WMD) 242
Weddell Sea 29
Weitzman, Martin 106, 111
Weitzman Theory 106
West Africa 160–161, 241, 243
whaling 4, 83, 179, 202–204, 262
wicked problems 94
Wildavsky, A. 97
windfarms 9, 134, 277
wind mixing 30
wind turbines 282–283
World Commission on Environment and Development (WCED) 147–148, 293
World Conference on National Parks 296
World Conservation Strategy (IUCN/WWF/UNEP) 296

World Heritage Convention (1972) 150, 296
World Summit on Sustainable Development
  (WSSD) 151–152, 300
World Trade Organization (WTO) 228
World Wildlife Fund (WWF) 137, 269

Young, O. R. 257

Zebra Mussel (*Dreissena polymorpha*) 154
zoobenthos communities 34–35
zooplankton 32–33

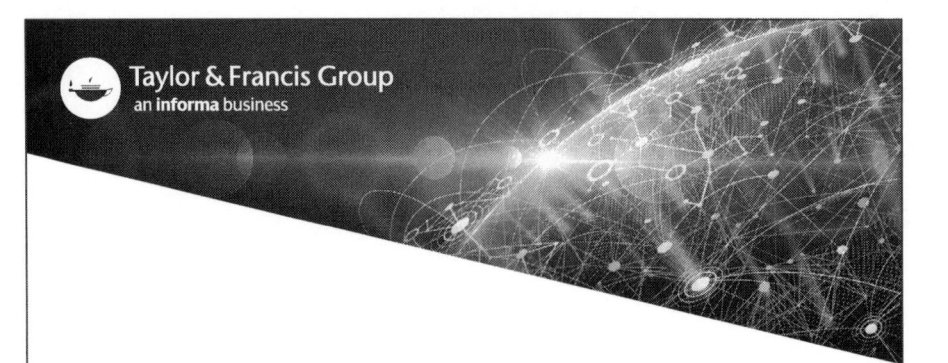

Printed in Great Britain
by Amazon